Chromatographic Theory and Basic Principles

CHROMATOGRAPHIC SCIENCE

A Series of Monographs

Editor: JACK CAZES
Silver Spring, Maryland

Volume 1: Dynamics of Chromatography
J. Calvin Giddings

Volume 2: Gas Chromatographic Analysis of Drugs and Pesticides
Benjamin J. Gudzinowicz

Volume 3: Principles of Adsorption Chromatography: The Separation of Nonionic Organic Compounds (out of print)
Lloyd R. Snyder

Volume 4: Multicomponent Chromatography: Theory of Interference (out of print)
Friedrich Helfferich and Gerhard Klein

Volume 5: Quantitative Analysis by Gas Chromatography
Joseph Novák

Volume 6: High-Speed Liquid Chromatography
Peter M. Rajcsanyi and Elisabeth Rajcsanyi

Volume 7: Fundamentals of Integrated GC-MS (in three parts)
Benjamin J. Gudzinowicz, Michael J. Gudzinowicz, and Horace F. Martin

Volume 8: Liquid Chromatography of Polymers and Related Materials
Jack Cazes

Volume 9: GLC and HPLC Determination of Therapeutic Agents (in three parts)
Part 1 edited by Kiyoshi Tsuji and Walter Morozowich
Part 2 and 3 edited by Kiyoshi Tsuji

Volume 10: Biological/Biomedical Applications of Liquid Chromatography
Edited by Gerald L. Hawk

Volume 11: Chromatography in Petroleum Analysis
Edited by Klaus H. Altgelt and T. H. Gouw

Volume 12: Biological/Biomedical Applications of Liquid Chromatography II
Edited by Gerald L. Hawk

Volume 13: Liquid Chromatography of Polymers and Related Materials II
Edited by Jack Cazes and Xavier Delamare

Volume 14: Introduction to Analytical Gas Chromatography: History, Principles, and Practice
John A. Perry

Volume 15: Applications of Glass Capillary Gas Chromatography
Edited by Walter G. Jennings

Volume 16: Steroid Analysis by HPLC: Recent Applications
Edited by Marie P. Kautsky

Volume 17: Thin-Layer Chromatography: Techniques and Applications
Bernard Fried and Joseph Sherma

Volume 18: Biological/Biomedical Applications of Liquid Chromatography III
Edited by Gerald L. Hawk

Volume 19: Liquid Chromatography of Polymers and Related Materials III
Edited by Jack Cazes

Volume 20: Biological/Biomedical Applications of Liquid Chromatography IV
Edited by Gerald L. Hawk

Volume 21: Chromatographic Separation and Extraction with Foamed Plastics and Rubbers
G. J. Moody and J. D. R. Thomas

Volume 22: Analytical Pyrolysis: A Comprehensive Guide
William J. Irwin

Volume 23: Liquid Chromatography Detectors
Edited by Thomas M. Vickrey

Volume 24: High-Performance Liquid Chromatography in Forensic Chemistry
Edited by Ira S. Lurie and John D. Wittwer, Jr.

Volume 25: Steric Exclusion Liquid Chromatography of Polymers
Edited by Josef Janča

Volume 26: HPLC Analysis of Biological Compounds: A Laboratory Guide
William S. Hancock and James T. Sparrow

Volume 27: Affinity Chromatography: Template Chromatography of Nucleic Acids and Proteins
Herbert Schott

Volume 28: HPLC in Nucleic Acid Research: Methods and Applications
Edited by Phyllis R. Brown

Volume 29: Pyrolysis and GC in Polymer Analysis
Edited by S. A. Liebman and E. J. Levy

Volume 30: Modern Chromatographic Analysis of the Vitamins
Edited by Andre P. De Leenheer, Willy E. Lambert, and Marcel G. M. De Ruyter

Volume 31: Ion-Pair Chromatography
Edited by Milton T. W. Hearn

Volume 32: Therapeutic Drug Monitoring and Toxicology by Liquid Chromatography
Edited by Steven H. Y. Wong

Volume 33: Affinity Chromotography: Practical and Theoretical Aspects
Peter Mohr and Klaus Pommerening

Volume 34: Reaction Detection in Liquid Chromatography
Edited by Ira S. Krull

Volume 35: Thin-Layer Chromatography: Techniques and Applications, Second Edition, Revised and Expanded
Bernard Fried and Joseph Sherma

Volume 36: Quantitative Thin-Layer Chromatography and Its Industrial Applications
Edited by Laszlo R. Treiber

Volume 37: Ion Chromatography
Edited by James G. Tarter

Volume 38: Chromatographic Theory and Basic Principles
Edited by Jan Åke Jönsson

Volume 39: Field-Flow Fractionation: Analysis of Macromolecules and Particles
Josef Janča

Volume 40: Chromatographic Chiral Separations
Edited by Morris Zief and Laura Crane

Other Volumes in Preparation

Quantitative Analysis by Gas Chromatography, Second Edition, Revised and Expanded
Josef Novák

Chromatographic Theory and Basic Principles

Edited by

Jan Åke Jönsson
University of Lund
Lund, Sweden

MARCEL DEKKER, INC. New York and Basel

Library of Congress Cataloging-in-Publication Data

Chromatographic theory and basic principles.

Includes bibliographies and index.
1. Chromatographic analysis. I. Jönsson, Jan Åke,

QD79.C4C483 1987 543'.089 87-6864
ISBN 0-8247-7673-9

MARCEL DEKKER, INC.
270 Madison Avenue, New York, New York 10016

Current printing (last digit):
10 9 8 7 6 5 4 3 2 1

PRINTED IN THE UNITED STATES OF AMERICA

Foreword

Chromatographic theory is the basis for the development of chromatographic instrumentation and, ultimately, for the application of chromatography to all types of separation problems. In the 1960s the use of gas chromatography (GC) for the separation of volatile organic molecules became widespread. In the 1970s high-performance liquid chromatography (HPLC) took its rightful place in the analytical laboratory for the separation of the nonvolatile, thermally labile, ionic, and high-molecular-weight compounds which could not readily be separated by GC. Now HPLC and GC systems are necessary in all types of tasks. Unfortunately, most "users" of chromatographic systems do not understand the basic principles underlying why solutes in a mixture separate or the processes that counteract the separations and hinder resolution.

All chromatographic separations are based on a common concept: the distribution of components in a mixture between two immiscible phases, one a stationary phase and one a moving phase. However, for each chromatographic mode the nature of the physicochemical processes can be different.

In this book, we have an excellent presentation of the fundamental theory of chromatography common to all chromatographic modes as well as the basic features that distinguish one mode from another. The material is presented in readable form understandable to users of chromatography; thus, based on theory, they can make intelligent decisions as to the type of chromatography best suited to their particular separation problem. Only when users of chromatographic techniques comprehend the basic principles that underlie the chromatographic separations will they be able to take full advantage of this powerful technique in order to obtain the highest resolution of the solutes of interest in the shortest possible time. In addition to being an excellent reference, this book can be used in analytical chemistry courses on chromatography.

Phyllis R. Brown
University of Rhode Island
Kingston, Rhode Island

Preface

Chromatography constitutes a family of closely related methods for separation and analysis of a wide variety of chemical sample types. Gas chromatography, ion exchange chromatography, and thin-layer chromatography (to name a few of these techniques) are all, in spite of their different technical realizations, based on the same simple principle, namely the distribution of the molecules under study between two media, one stationary and the other moving. Essentially the same principles govern the movements and shape of the chromatographic peaks (or bands, or spots), even if the nature of the underlying physicochemical processes are often widely different.

Thus, the theory of chromatography should not be considered separately from that of gas chromatography, of steric exclusion chromatography, and so on. Instead, the common features must be stressed and only the distinguishing details should be treated separately.

This book is an attempt to deal with the theory of chromatography in such a way. Its aim is to answer, or at least illustrate, the two more or less independent main questions of chromatographic theory:

1. "Which parameters determine the traveling velocity of the sample in the chromatographic system?" Equivalent formulations include: "Which parameters determine the retention time (or the R_f value)?" This question deals with the processes that create separation and it is addressed by the retention theory.

2. "Which parameters determine the width of the peaks (of the bands, the spots, etc.)?" Equivalent: "Which parameters determine the theoretical plate number (the HETP, the efficiency)?" Here the processes that counteract separation are in focus.

The first chapter provides a basic overview of terminology as well as some common concepts that help tie together the various aspects of chromatographic theory presented in the rest of the book. The second chapter, which is the most widely applicable one, covers different peak broadening and peak shaping mechanisms that, to various extents, operate in all chromatographic techniques. The remainder of the book is devoted to thorough descriptions of retention theories applicable to common versions of chromatography. These chapters have been written by a number of authors, each having approached the subject from his own viewpoint, with effort made by the editor to impose unification and consistent terminology.

Throughout, the use of chromatography for chemical analysis is emphasized, but preparative separations and physicochemical measurements are also discussed. One-phase techniques such as electrophoresis and field flow fractionation, although closely related to chromatography, are not considered.

It is a pleasure to sincerely thank all the authors who have devoted much time and effort in preparing contributions to this book. In spite of their heavy workloads and numerous other obligations, the collaboration has been remarkably agreeable and efficient.

The book contains one of the last contributions to science by Dr. Josef Novák. All of us in the chromatographic community with whom Josef generously shared his knowledge and experience will remember him for his unusual warmth and friendliness.

Several colleagues, especially Dr. Lennart Mathiasson and Dr. Per Lövkvist, deserve warm thanks for advice, help, and criticism during the course of this project. Ms. Annica Bergkvist and Ms. Nancy Simonsson are also thanked for skillful and patient typing. Finally, I want to thank my family for all their patience and support.

Jan Åke Jönsson

Contents

Contributors

Lave Fischer Market Manager, AB Sangtec Medical, Bromma, Stockholm, Sweden

Dennis R. Jenke Research Scientist, Sterile Fluids Research and Development Department, Travenol Laboratories, Inc., Morton Grove, Illinois

Jan Åke Jönsson Docent, Department of Analytical Chemistry, Chemical Center, University of Lund, Lund, Sweden

Lennart Mathiasson Docent, Department of Analytical Chemistry, Chemical Center, University of Lund, Lund, Sweden

Josef Novák† Institute of Analytical Chemistry, Czechoslovak Academy of Sciences, Brno, Czechoslovakia

Gordon K. Pagenkopf Professor, Department of Chemistry, Montana State University, Bozeman, Montana

Władysław Rudziński Professor, Department of Theoretical Chemistry, Institute of Chemistry, Maria Curie-Skłodowska University, Lublin Poland

---

†Deceased

Chromatographic Theory and Basic Principles

1
Common Concepts of Chromatography

Jan Åke Jönsson
University of Lund
Lund, Sweden

I. HISTORICAL PERSPECTIVE

The aim of this book is to present in some detail the contemporary state of the theory and basic principles of chromatography. In many cases the concepts and results (but usually not the terminology) date back several decades. Therefore, a brief review of the main events of the history of chromatography, with some emphasis on theory, should be useful. More comprehensive reviews of the field can be found in the literature [1].

Chromatography was originally invented by the Russian scientist Tswett in 1903 [2]. He used the new tool in his studies of chlorophyll and other plant pigments. His technique can be described as column liquid-solid adsorption chromatography.

After Tswett's work, the technique was used only occasionally until the early 1940s, when the famous paper by Martin and Synge appeared [3]. They presented the invention of liquid-liquid partition chromatography, both in columns and in planar form (paper chromatography). They also provided a theoretical framework for the basic chromatographic process. For their work, which had a large influence, especially in biochemistry, Martin and Synge received the 1952 Nobel Prize in chemistry.

Another main step in the progress of chromatography was the presentation of gas-liquid chromatography by James and Martin [4]. In their papers on the subject, a fundamental difference from liquid chromatography, namely, the compressibility of the mobile phase, was treated in a definite way. The introduction of gas chromatography had an unprecendented impact on the analytical chemistry of organic compounds. Some further milestones were as follows:

The theoretical and experimental development of open tubular gas chromatographic columns in 1956 by Golay [5], dramatically increasing efficiency and separation power
The presentation of steric exclusion chromatography, allowing easy separation of macromolecules (Porath and Flodin in 1949 [6])
The publication in 1961 of the extensive treatment of chromatographic theory in Giddings' *Dynamics of Chromatography* [7]
The introduction of modern ("high-performance") liquid chromatography in the early 1970s, permitting a vastly extended range of components to be efficiently separated (several fundamental works, notably by Knox, Huber, and Scott, see Chap. 2, Sec. V.D)

Fundamental theory largely evolved in parallel with technical development: in the 1940s, the emphasis was on fundamental processes, in the 1950s and 1960s, the theory of gas chromatography was developed, and during the 1970s mainly liquid chromatography was studied. In recent years the area seems to have become somewhat less active.

In addition to those mentioned above, a large number of persons have made significant contributions to the theory of chromatography, among whom are E. Glueckauf, J. J. van Deemter, F. Helfferich, G. Guiochon, and H. Purnell.

The historical development of chromatography is summarized in a book [8] that describes many of the individuals who deserve thanks for developing this technique.

II. BASIC TERMINOLOGY

The theory of chromatography has, as described above, evolved over a relatively long time. However, it has also evolved along several different lines: liquid chromatography, gas chromatography, and other techniques have been treated more or less separately. Therefore, there are still confusion and inconsistency in everyday chromatographic terminology. More careful attention to these matters, especially by experimentalists, would be desirable.

The terminology of chromatography has been subject to several reviews [9] and recommendations. In this book, the recommendations of the ASTM [10,11] are adopted, even if the rules of IUPAC [12] are equally consistent and useful. The differences apply to details only.

A. Classification of Techniques

The Chromatographic System

A chromatographic system consists of two immiscible phases: a mobile fluid phase that streams over a stationary phase. The mobile phase is a liquid, a gas, or a supercritical fluid. The stationary phase may be a liquid supported by a porous, inert material or by the inner wall of a tubular column, and may also consist of molecules chemically bonded to such a material or to the wall of the column. The stationary phase may be an adsorptive or inert solid, usually porous, or it may be an ion-exchange resin or a gel. The stationary phase can be arranged in a column, a tube through which the mobile phase flows. It may also constitute a planar layer or a paper.

Substances to be separated should have different relative affinities for the stationary and the mobile phases. Thus, a substance with relatively higher affinity for the stationary phase moves with a lower velocity through the chromatographic system than does a substance with lower affinity. This difference in migration velocity ultimately leads to physical separation of the components in a sample.

Chromatographic Techniques

Among chromatographic methods, there are a great number of combinations of phases and of technical realizations. This gives rise to a multitude of techniques with individual names and acronyms. A number of these names, as approved by ASTM or IUPAC, are listed and briefly explained in Table 1. Note that the common term "HPLC" is missing, since it is ambiguous and is for various other reasons not recommended. (For example, does the "P" stand for performance, pressure, or, perhaps, price?) Another commonly used term that is not recommended in the official nomenclature is "capillary column." The word *capillary* implies that the diameter of such columns is small, although their essential feature is that they are open (i.e., not filled with particles).

Experimental Methods

An alternative classification refers to the general method of chromatographic development. Here the distinction is among elution, frontal, and displacement chromatography.

Elution Chromatography. In column chromatography, the elution method is characterized by the introduction of a small volume of the sample to be analyzed into the flowing mobile phase (*eluent*) and the observation of the various components of the sample, after their passage through the chromatographic column, in the form of concentration bands or peaks separated in time. In planar chromatography the sample is applied in a small area, followed by the passage of mo-

Table 1. Names and Acronyms for Chromatographic Techniques as Approved by ASTM or IUPAC

<u>Names describing phases</u>		
Gas chromatography	GC	Mobile phase is gas
Gas-liquid chromatography	GLC	GC with liquid stationary phase
Gas-solid chromatography	GSC	GC with solid stationary phase
Liquid chromatography	LC	Mobile phase is liquid
Liquid-liquid chromatography	LLC	LC with liquid stationary phase
Liquid-solid chromatography	LSC	LC with solid stationary phase
Reversed-phase chromatography		LC where the mobile phase is more polar than the stationary phase
Supercritical fluid chromatography	SFC	Mobile phase is a supercritical fluid
<u>Names describing mechanisms of retention</u>		
Adsorption chromatography		Separation is based on adsorption to a solid (GSC or LSC)
Partition chromatography		Separation is based on fluid-phase partition equilibria (GLC or LLC)
Ion-exchange chromatography	IEC	Version of LSC in which separation is based on ion exchange
Steric-exclusion chromatography	SEC	Version of LC in which separation is based on exclusion effects in a porous solid or gel
Gel-permeation chromatography	GPC	Same as SEC, but applies to gel only
<u>Names describing experimental techniques</u>		
Column chromatography	CC	The chromatographic system is arranged in a cylindrical tube
Open-tube chromatography		CC, in which the column has an "unobstructed central channel" [12]

Table 1. (Continued)

Planar chromatography		The chromatographic system is arranged as a plane
Paper chromatography		The stationary phase is, or is supported by, a paper
Thin-layer chromatography	TLC	The stationary phase is a layer of absorbent material spread on an inactive sheet or plate

Source: From Refs. 10-12.

bile phase and by observation of the components as physically separated spots. Elution is the totally predominant method, and its use is usually implicit in this book and in other discussions of this method.

Frontal Chromatography. This technique is characterized by feeding the sample continuously into the chromatographic bed (i.e., column or layer). The result is observed as a series of concentration steps (frouts), each corresponding to a separate component of the sample. Here only the most quickly moving component is (partly) physically separated from the other substances. This technique is used very little for its own sake, but sampling or cleanup processes are often described in terms of frontal chromatography.

Displacement Chromatography. Displacement chromatography is a variant of the elution method in which the mobile phase contains components that are more strongly retained by the stationary phase than the sample components under study. The latter are then forced out, or displaced, from the stationary phase and subsequently eluted. This technique is used mainly for preparative purposes.

B. The Concepts of Linearity and Ideality

The basis for most chromatographic techniques is a phase equilibrium. The analyte is distributed between the mobile and the stationary phase. We may write its concentration in the two phases as C_M and C_S, respectively. If, in a chromatographic system, the quotient C_S/C_M is independent of concentration—that is, if an isotherm plot of C_S versus C_M is linear (and passes through the origin)—the chromatographic system is classified as *linear*. In the other cases, as when the quotient varies with concentration and the isotherm is curved, the system is *nonlinear*.

Linearity is more or less a prerequisite for efficient chromatography and is usually tacitly assumed. The peaks are symmetrical, with easily defined retention volumes and plate numbers, and the mathematics is manageable. Theoretical predictions are accurate and consistent with experimental results. In contrast, nonlinearity is a disturbing feature, making peaks asymmetric with varying retention times, the mathematics is intractable or even impossible to handle, and theoretical models are very inaccurate. In the chapters to follow (especially Chaps. 2 and 5), these concepts are discussed in detail.

The term "ideal" implies that there is no peak-broadening mechanism operating. In reality, only *nonideal* chromatographic systems exist.

These terms are often combined: nonideal, linear chromatography is the common and preferred case, and ideal, nonlinear chromatography is a hypothetical construction, necessary for obtaining crude descriptions of nonlinear systems.

III. CHARACTERISTICS OF CHROMATOGRAPHIC SYSTEMS

A. Fundamental Parameters

Retention Parameters

In column elution chromatography, the dominating method, the results of a chromatographic run are obtained in the form of a chromatogram, which is a plot of the response of a detector device and is usually proportional to the flux of material through the detector as a function of time or volume. The time between the instant of sample introduction into the column and the peak resulting from a component of the sample is called the retention time t_R. The corresponding volume of gas or liquid is the retention volume V_R. If the flow rate F_c is constant, it holds that

$$V_R = t_R F_c \tag{1}$$

There is some ambiguity about which point on the peak should be used to define the retention time. Often the maximum is used, but the mean is generally considered to be more exact from a thermodynamic point of view. It turns out (see Chap. 2, page 54) that in those cases in which it is possible to make a clear statement, the median is preferable to both the maximum and mean. With narrow peaks, the differences between these points are usually negligible.

If the sample component under study is not retained by the stationary phase but moves entirely within the eluent, its retention time is the *mobile phase holdup time* t_M. The corresponding volume V_M is the volume of the mobile phase in the column (neglecting extracolumn volumes).

By subtraction we obtain the *adjusted retention time* and *volume*:

$$t'_R = t_R - t_M \tag{2a}$$

$$V'_R = V_R - V_M \tag{2b}$$

We also define a *capacity factor* k as

$$k = \frac{t'_R}{t_M} = \frac{V'_R}{V_M} \tag{3}$$

Thus,

$$t_R = t_M(1 + k) \tag{4a}$$

$$V_R = V_M(1 + k) \tag{4b}$$

With porous column packings and if the molecular size of the substance is comparable to the diameter of the pores, the determination of t_M is not straightforward, since molecules of various sizes experience different holdup values. This is discussed in Chap. 2, pages 34-36. The effect leads to a separation of molecules of different sizes as in the technique of size-exclusion chromatography (see Chap. 8).

In gas chromatography, expansion of the mobile phase (the carrier gas) must be considered. This introduces the *pressure correction factor* j, which depends on the decrease in pressure over the length of the column. In this chapter, we will assume that the gas volumes being considered are already corrected in this way whenever applicable. We therefore have the "real" volume, corresponding to the physical dimensions of the column. For example, consider the mobile phase volume V_M. At outlet pressure, the portion of gas contained in this volume is expanded, with the result that any type of volume measurement made after the column exit gives a larger result than the proper V_M. This is corrected by multiplying the measured volume by j.

The capacity factor k, although basically a normalized retention parameter, can, in the usually assumed case of linear chromatography, be related to an equilibrium constant. In partition chromatography, the *distribution coefficient* K for the two-phase equilibrium that exists (either gas-liquid or liquid-liquid) is given by

$$K = \frac{C_S}{C_M} \tag{5}$$

where C_S and C_M designate the concentration of solute in the two phases.

The basic equation of retention in linear chromatography is

$$V_R = V_M + V_S K \tag{6}$$

where V_S is the volume of the stationary phase. Combining Eqs. (4) and (6), we arrive at

$$k = \frac{V_S}{V_M} K \tag{7}$$

This equation is not, as is often believed, an alternative definition of the capacity factor but follows from the assumption of linear chromatography.

An equilibrium constant for an adsorption equilibrium may be stated as in Eq. (5), with C_S a surface concentration. Then Eq. (7) will read

$$k = \frac{A_S}{V_M} K \tag{8}$$

where A_S is the surface area of the adsorbent.

In linear chromatography, the velocity u_{eff} of the peak center through the column is a constant if the linear mean flow rate $\bar{u}$ is constant:

$$u_{eff} = \frac{\bar{u}}{1 + k} = \bar{u}R \tag{9}$$

In fact, this is only another version of Eq. (4a), as $t_R = L/u_{eff}$ and $t_M = L/\bar{u}$. R is the *retention ratio*, which is sometimes used instead of the capacity factor as a normalized measure of retention. It is equal to $1/(1 + k)$.

In Chaps. 3-8, the theories of retention for the most common chromatographic techniques are examined in detail in order to provide an understanding of the various specific factors governing the value of k.

Dispersion Parameters

A chromatographic peak is characterized not only by its position (i.e., retention time or volume) but also by its width. This is most rigorously expressed in terms of the variance σ^2 of the peak, the second statistical central moment (Chap. 2, page 31). As a suitable dimensionless dispersion parameter, the *number of theoretical plates* n is defined as

$$n = \frac{t_R^2}{\sigma^2} \tag{10}$$

With this definition, σ^2 should be expressed in units of time (squared).

The *height of a theoretical plate* (HETP) h is defined as

$$h = \frac{L}{n} \tag{11}$$

where L is the length of the column.

Additionally, the *number of effective plates* N is sometimes defined as:

$$N = \frac{(t_R')^2}{\sigma^2} = \left(\frac{k}{1 + k}\right)^2 n \tag{12}$$

The term "theoretical plate" is taken from plate theory, which relates the theory of chromatography to the theory of distillation.

In nomenclatural recommendations [10-12], the plate number n is defined as

$$n = 5.54\left(\frac{t_R}{w_h}\right)^2 \tag{13}$$

where w_h is the peak width at half-height. The numerical factor should be 8 ln 2, which is closer to 5.55, but the number 5.54 is stated in all texts. This has no practical significance. Equation (13) shows the usual way by which n is calculated from an experimental chromatogram. However, the use of this expression in the official nomenclature is unfortunate, since it implies that the chromatographic peak is gaussian, an assumption that at best is only a fair approximation.

The plate height can also be expressed as

$$h = \frac{\partial \sigma_z^2}{\partial z} \tag{14}$$

where z is the distance of the plate from the column inlet and σ_z^2 is correspondingly expressed in units of length squared. Accordingly, h in this definition is the rate of increase in peak variance along the column.

The plate height and plate number depend on many factors. These are treated in detail in Chap. 2, where complete equations for h and n are found for a number of chromatographic techniques. Of interest here is the dependence of h and n on the capacity factor. Examples of such dependence are seen in Fig. 1. Note that the effective plate number N also depends on k; the definition in Eq. (12) does not remove that dependence, as is sometimes believed. However, all

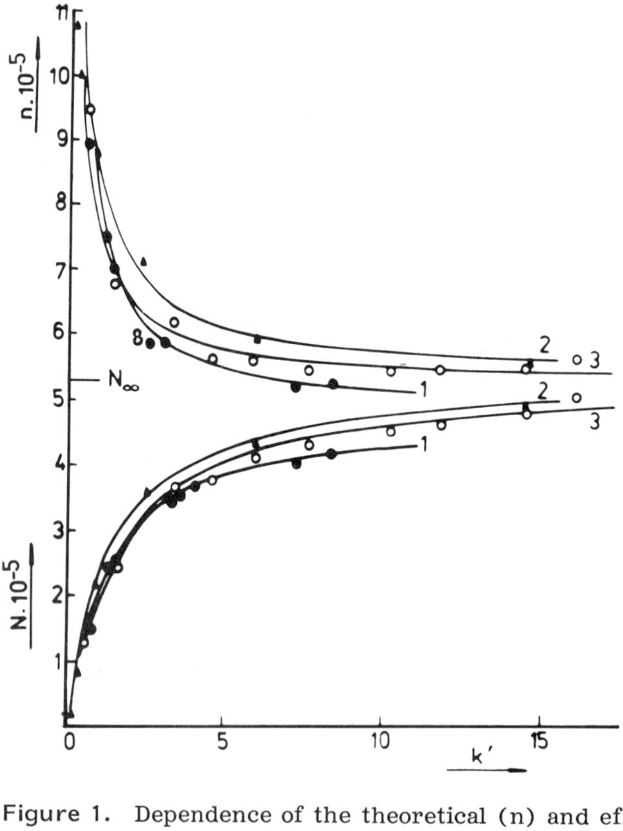

Figure 1. Dependence of the theoretical (n) and effective (N) plate numbers on the capacity ratio k for branched alkanes (1), normal alkanes (2), and aromatics (3) on a 191-m, squalane-coated open tubular column at 58°C. (After Krupčík et al. [13], with permission.)

curves, for both n and N, tend to a common, limiting value which is largely independent of experimental factors.

Resolution

From Eq. (9), it follows that if two substances with differing capacity factors are simultaneously introduced into a column, the distance between the centers of the two peaks formed will increase linearly with time:

$$\Delta z = z_1 - z_2 = t(u_{eff,1} - u_{eff,2}) = t\bar{u}\left(\frac{1}{1 + k_1} - \frac{1}{1 + k_2}\right) \tag{15}$$

The subscripts 1 and 2 refer to the two peaks, with $k_1 < k_2$.

At the same time, the width and length units for each peak increase, according to Eq. (14), only with the square root of t:

$$\sigma_{z,i} = \sqrt{t\bar{u}h_i} \qquad i = 1.2 \tag{16}$$

From Eqs. (15) and (16) we see that the resolution, in terms of the distance between the peak divided by the peak width, increases with the square root of time. Thus, the separation between two peaks constantly increases as the peaks travel along the column and is principally limited only by the available time and column length.

In column elution chromatography, the predominant chromatographic method, the chromatogram is recorded as a function of time instead of length. Thus, to define resolution in a rational way, with this method, the retention differences and peak widths should be expressed in time units. This leads to the standard definition of chromatographic *resolution* R_s, which is given by

$$R_s = \frac{t_{R,2} - t_{R,1}}{2(\sigma_1 + \sigma_2)} \tag{17}$$

where the subscripts 1 and 2 again refer to two peaks with $k_1 < k_2$. The denominator is the mean of the peak widths, each of which is taken as 4σ. (For a gaussian peak, about 95% of the peak area is contained between the limits defined by the mean $\pm 2\sigma$.)

The term R_s from Eq. (17) may be rewritten in various ways. By noting from Eq. (10) that $\sigma = t_R/\sqrt{n}$, expressing the retention time as in Eq. (4a) and defining $\alpha = k_2/k_1$, we arrive at the resolution equation

$$R_s = \frac{k_1(\alpha - 1)}{[2 + k_1(1 + \alpha)]} \frac{\sqrt{n}}{2} \tag{18}$$

We may define a mean capacity factor $\bar{k} = (k_1 + k_2)/2$. With this, we obtain

$$R_s = \frac{\alpha - 1}{\alpha + 1} \frac{\bar{k}}{1 + \bar{k}} \frac{\sqrt{n}}{2} \tag{19}$$

We may also assume that k_1 and k_2 are very close, which is reasonable, since we are usually interested in the resolution of relatively close peaks only. With this assumption, the resolution may be written:

$$R_s \approx \frac{\alpha - 1}{\alpha} \frac{k_2}{(1 + k_2)} \frac{\sqrt{n}}{4} \tag{20}$$

Note that Eq. (19) is algebraically equivalent to the definition of R_s in Eq. (17), but Eq. (20) is an approximation. The resolution may also be expressed in a simpler form by using the concept of the effective plate number N [see Eq. (12)]:

$$R_s = \frac{\alpha - 1}{\alpha + 1} \frac{\sqrt{N}}{2} \tag{21}$$

This expression is the rationale for the definition of N.

Equations (18) through (20) are different versions of the most widely used separation criterion. The term R_s is the product of three factors, each of them emphasizing a different aspect of the separation process:

Selectivity, or difference in migration velocities of the two components, reflected as different values of α
Retention, reflected as different values of $\bar{k}$
Column efficiency, measured as the plate number n

The quantity R_s depends strongly on the value of α, especially when α is close to unity. The retention factor $\bar{k}/(1 + \bar{k})$ approaches unity as $\bar{k}$ is increased. When $\bar{k}$ exceeds 5, for example, not much is gained by a further increase. R_s increases only with the square root of the plate number, which means that in order to double the resolution, a column of approximately four times greater length must be used. Consequently, with a reasonably optimized chromatographic system (with respect to $\bar{k}$ and n), the potentially most rewarding approach to increasing the resolution of the system is to increase the selectivity.

Solving Eq. (19) for n, we obtain the *required number of theoretical plates* n_{req} necessary to effect a chosen resolution:

$$n_{req} = R_s^2 4 \left(\frac{\alpha - 1}{\alpha + 1}\right)^2 \left(\frac{\bar{k} + 1}{\bar{k}}\right)^2 \tag{22}$$

Similarly, we obtain from Eq. (21):

$$N_{req} = R_s^2 4 \left(\frac{\alpha - 1}{\alpha + 1}\right)^2 \tag{23}$$

giving the required number of effective plates. Often, these terms refer to a separation with $R_s = 1$, which corresponds to a nearly complete separation, in which the distance between the peak centers is equal to 4σ (i.e., the peak width at the base).

In the official nomenclatural recommendations, the definitions of R_s, n_{req}, and N_{req} are again based on unnecessary approximations.

Extent of Separation

The definition of resolution according to Eq. (17) rigorously applies to gaussian peaks only. When peaks are asymmetric, the use of R_s may give unexpected and inconsistent results. Additionally, R_s is not directly related to the purity of the separated fractions.

A different criterion for separation was defined by Rony [14]. Realizing that the object of separation is the physical division of one sample into two (or more) fractions, he divided a chromatogram with two peaks into two regions. The cut point between the regions may principally be placed anywhere, but its natural position is somewhere between the two peaks. In elution chromatography, the regions separated by the cut point correspond to time periods. If the column effluent is collected in two test tubes, for example, the two regions will be physically separated. Rony [14] defined the *extent of separation* ξ as

$$\xi = |X_{11} - X_{21}| = |X_{22} - X_{12}| \tag{24}$$

Here, X_{ij} is the fraction of component i in region j. The values of ξ range from 0 to 1: if the separation is complete, all of component 1 and none of component 2 is in region 1. Thus, $X_{11} = X_{22} = 1$, and $X_{21} = X_{12} = 0$, so that $\xi = 1$. With no separation at all, there is an equal amount of both components in both regions, so that all $X_{ij} = 0.5$, leading to $\xi = 0$.

This concept is very general and may be used, for example, to find the optimum cut point between two not necessarily gaussian peaks or to generally optimize chromatographic systems. For the latter application, the *rate of separation* r_s can be defined as

$$r_s = \frac{\partial \xi_{opt}}{\partial t} \tag{25}$$

where ξ_{opt} is the optimum value of ξ, which is obtained with the cut point that gives maximal extent of separation.

It turns out [14] that procedures for optimizing chromatographic systems on the basis of the extent of separation lead to results that are very similar or even identical to those obtained with the conventional concept of resolution. The value of the extent of separation concept lies mainly in its wider applicability to all classes of separation techniques.

B. Measures of Column Performance

When preparing chromatographic columns intended to have a large separation capacity, an important question is how the performance of a certain column should be specified. The specification that is sought may be general, meant to express the overall quality of the column either absolutely or in relation to what can theoretically be obtained, or more specialized, expressing the suitability of the column for a certain type of analysis. Many column performance measures of various kinds have been defined, but agreement about their applicability and relevance is limited. This is due to a disagreement about the meaning of the term "performance," which in some cases refers simply to a high number of theoretical plates but may also take several additional parameters into consideration, such as time and pressure. A survey of this complex matter follows.

Plate Numbers

The *theoretical plate number* n and the *effective plate number* N were defined in Eqs. (10) and (12), respectively. Both depend on the properties of the compound studied and on the chromatographic conditions that exist. In Chap. 2, various contributions to the plate number are discussed in detail for various techniques. Accordingly, a plate number gives only a relatively rough idea of the potential performance of a column, and the same is true for the corresponding plate height h, defined in Eq. (11). Columns offered for sale by most suppliers are, despite these shortcomings, characterized by a single plate number only.

The Limiting Plate Number

If the width of each of a number of peaks produced by a homologous series of chemical compounds is plotted against the capacity factor of each peak, a straight line is often obtained. This linear relation, which was first observed a long time ago [15], has no exact theoretical foundation but is approximately valid in practice, provided the chromatographic system is working properly. On the basis of such a linear relation, a procedure for evaluating column quality, known as the "abt concept," was suggested by Kaiser [16]. The introduction of this concept caused considerable discussion, sometimes of a slightly unworthy and personal nature. Some of this discussion is contained in Refs. 17-19.

One of the generally accepted ideas within the abt concept is related to the so-called real plate number. The term "real" is illogical—a better value [17] would be the "limiting" plate number n_{lim}. If the following linear equation is fitted to data obtained in the manner described earlier:

$$w_h = b_0 + ak \tag{26}$$

and t_M is the holdup time, then n_{lim} is defined as

$$n_{lim} = 5.54\left(\frac{t_M}{a}\right)^2 \tag{27}$$

where the numerical factor arises from the use of w_h instead of σ in Eq. (26) [see Eq. (13)].

It can be shown [19] that the theoretical plate number n approaches n_{lim} when k approaches infinity. The same is true for the effective plate number:

$$\lim_{k \to \infty} n = \lim_{k \to \infty} N = n_{lim} \tag{28}$$

Thus, n_{lim} is an expression for the column efficiency that is independent of the capacity factor.

Krupčík and co-workers [13] calculated n_{lim} from the series

$$n = n_{lim} + \frac{b}{k} + \frac{c}{k^2} + \cdots \tag{29}$$

which is equivalent to the value derived from Eqs. (26) and (27). See Fig. 1, where n_{lim} is marked as N_∞.

After the apparent settling of the controversy over the abt concept, the entire idea of the limiting plate number was recently claimed as a "new measure of column efficiency" [20]. The publication that carried this claim, obviously unrelated to Kaiser's work, neatly presents once again, with relevant experimental verification, the complete theory related to n_{lim}.

The Separation Number

The usual definition of chromatographic resolution R_s was given in Eq. (17). That quantity can be used for the characterization of columns if the resolution of some specified test substances is given. Kaiser suggested the so-called *separation number* or, in German, *Trennzahl*, for this purpose [21]. This quantity, described as the number of peaks that may be placed between the peaks of two consecutive members of a homologous series, is defined by

$$TZ = \frac{t_{R(z+1)} - t_{R(z)}}{w_{h(z+1)} + w_{h(z)}} - 1 \tag{30}$$

where w_h is the peak width at half-height and the indices z and z + 1 refer to the carbon numbers of the test substances. These are usually normal alkanes, but this is not necessary; other homologous series can be used with similar results [22].

Figure 2 illustrates the concept of separation number: if peaks 1 and 12 arise from two consecutive normal alkanes, the situation in Fig. 2 corresponds to TZ = 10.

Ettre [22] showed that Eq. (30) is equivalent to

$$TZ = \frac{R_{s(z+1)/z}}{1.177} - 1 \tag{31}$$

where $R_{s(z+1)/z}$ is the resolution between two successive alkanes.

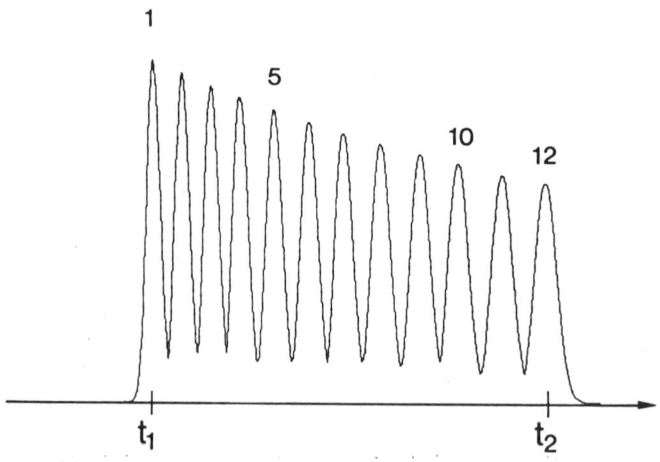

Figure 2. Simulated hypothetical chromatogram of 12 peaks having equal areas and a resolution of approximatively unity. This illustrates the concepts of separation number TZ and peak capacity (PC) (see text).

Thus the resolution of the peaks located between the two homologs is 1.177, which means a separation of 4.7σ or a little better than the resolution of unity (4σ) that is usually connected with the definition of n_{req} [Eqs. (22) and (23)]. This inconsistency is unfortunate but not very serious.

A quantity very similar to the separation number is the *effective peak number* (EPN) [23]. It differs only from the separation number TZ in that it assumes a resolution of unity. Thus,

$$EPN = R_{s(z+1)/z} - 1 \tag{32}$$

and

$$EPN = 1.177 + 0.177 \tag{33}$$

If TZ were not already in widespread use, clearly EPN would be preferred instead.

There has been some debate about the usefulness of the separation number for column characterization. The reader is referred to Refs. 15 and 16 and other works cited therein. We will not go into the details of this matter, but will note only that the obvious variation of TZ with temperature and capacity factors was considered a definite disadvantage in one study [24], motivating the epithet "rubber ruler." Other authors [25] claimed that exactly this property of the TZ concept could provide a distinct advantage. This illustrates that the question of describing the performance of a column with a single numerical value cannot be addressed quite objectively.

Peak Capacity

The concept of separation number, as described in the preceding section, may be generalized to refer to the possible number of peaks that could occur (with a given resolution) between any two given points of time. This number is the *peak capacity* (PC). Clearly, the separation number is a special case of the peak capacity and is obtained when the two given points of time correspond to the retention times of two adjacent homologs.

To evaluate an expression for the peak capacity, we observe from Eq. (10) that the peak width parameter σ at any time in a chromotogram is expressed by

$$\sigma = \frac{t}{\sqrt{n}} \tag{34}$$

where n is generally not a constant but varies with k and, consequently, with time. If we further assume a resolution of unity between two peaks, the width of each peak is 4σ. Thus, during the infinitesimal time interval dt, the number of eluted peaks dPC is given by

$$dPC = \frac{\sqrt{n}}{4t}\, dt \tag{35}$$

Integrating between the actual time limits, we obtain

$$PC = 1 + \frac{1}{4} \int_{t_1}^{t_2} \frac{\sqrt{n}}{t}\, dt \tag{36}$$

The origin of the digit 1 is the following: the first peak is centered around t_1, so that only half of it is included in the integral. The same applies to the last peak, centered around t_2. These two half-peaks must consequently be added to the integral. This is at variance with the definition of the separation number, where a similar 1 is subtracted in order to *not* include the end peaks, which are the homolog standards. This can be seen in Fig. 2, where PC = 12 if t_1 and t_2 are the retention times of peaks 1 and 12, respectively.

To evaluate PC, we may proceed in either of two ways. The most simple is to neglect the variation of n with time [26]. This clearly gives a rough approximation in many cases, especially when dealing with small retention times (see Fig. 1). The peak capacity is then given by

$$PC = 1 + \frac{\sqrt{\bar{n}}}{4} \ln \frac{t_2}{t_1} \tag{37}$$

Here, $\bar{n}$ is some "mean" plate number, to correct for the approximation than n is constant. Equation (37) may equally well be formulated with a quotient of retention volumes instead of times.

If we consider the available "separation space" as the time interval from t_M to $8t_M$ (corresponding to a variation of k between 0 and 7), we obtain the following rough relation between plate number and maximum peak capacity:

$$PC_{max} \approx \frac{\sqrt{\bar{n}}}{2} \tag{38}$$

In the previously cited study, Krupčík and co-workers [13] evaluated Eq. (36) using a reasonable expression for the dependence of n on the capacity factor k and obtained a complex expression for PC suitable only for numerical evaluation. They compared that expression with Eq. (37) and with the separation number TZ and concluded that, for k values below 2, both Eq. (37) and TZ provide erroneous results.

In most chromatographic techniques, the peak capacity is principally unlimited; with a large enough time interval, PC may reach any value. In size-exclusion chromatography, however, all peaks must elute within a limited volume interval (see Chap. 8). Then the peak capacity is limited and related to the ratio of the total void volume (the upper limit of the retention volume) to the interstitial volume (lower limit). This ratio is usually about 2.3, leading to

$$PC_{SEC} = 1 + 0.2\sqrt{n} \tag{39}$$

under the same assumptions as for Eq. (37).

It should be noted that the concept of peak capacity (and also of separation number) is highly idealized (a point that is immediately obvious in Fig. 2). It gives the number of peaks that may be separated within some time limits, provided that each of the peaks emerges after exactly the proper retention time needed to fulfill the conditions of unity resolution. Such mixtures do not exist in reality.

Evaluation of the number of components with random retention times that could be resolved on a certain column is a statistical question that was addressed by Davis and Giddings [27]. They found that, with random spacing between peaks, the obtainable number of resolved peaks is at most 37% of the peak capacity. However, more than half of these peaks may contain two or more unresolved sample components. This applies to favorable cases; in general, the situation is still less satisfying. The peak capacity is apparently not a valid measure of the true number of components that may be separated on a given column.

A related problem is the estimation of the total number of components actually present in a sample by counting the (usually smaller) number of peaks in a chromatogram. Theoretical studies [28] show that this issue is indeed relevant and that the column efficiency needed for the resolution of complex mixtures is easily grossly underestimated.

Utilization of Theoretical Efficiency

All the measures of column performance that were discussed above express the efficiency of a column in an absolute way. Thus, they principally tell the user, in various ways, the extent to which a desired separation may be carried out. There is another type of measure, primarily concerning the producer or developer of columns, that expresses the column efficiency in relative terms or in comparison with the principally obtainable performance. Such a measure was defined by Ettre [29,30], who suggested the term *utilization of theoretical efficiency* (UTE) to define it:

$$\text{UTE} = \frac{h_{\text{theor,min}}}{h_{\text{exp,min}}} \tag{40}$$

where the theoretically obtainable minimum plate height $h_{\text{theor,min}}$ should be calculated from the relevant dispersion equation, and $h_{\text{exp,min}}$ is the experimentally obtained plate-height minimum.

For open-tube gas chromatographic columns, the UTE concept is in widespread use, usually under the slightly improper term "coating efficiency." For these columns, the Golay equation [Eq. (90), Chap. 2] is used to calculate $h_{\text{theor,min}}$ [5]. In this calculation, it is normally assumed that the effects of pressure gradients and the contribution of diffusion in the stationary phase may be neglected. Then the UTE is calculated as

$$\text{UTE} = \frac{R\sqrt{\dfrac{11k^2 + 6k + 1}{3(1 + k)^2}}}{h_{\text{exp,min}}} \tag{41}$$

where R is the column radius and k is the capacity factor. This expression can be cirticized [31], since the quantity calculated from it conforms poorly with several experimental observations, probably because of the crude assumptions inherent in Eq. (41).

It should be observed that these assumptions remove the influence of the stationary phase, the quality of which is said to be described in this way (hence the term "coating efficiency"). Clearly this is illogical, and more detailed expressions, based directly on the Golay equation, should be used. The use of the complete expression [Eq. (90) in Chap. 2] leads to a relatively cumbersome expression. Several successive simplifications for deriving UTE were discussed by Cramers and co-workers [32]. A reasonable compromise, useful if the ratio of the inlet and outlet pressures does not exceed 1.5, leads to

$$\text{UTE} = \sqrt{\frac{\dfrac{(11k^2 + 6k + 1)R^2j^2}{3(1 + k)^2} + \dfrac{16k}{3(1 + k)^2}\dfrac{D_{M,0}}{D_S}d_f^2jj_1}{h_{\text{exp,min}}}} \tag{42}$$

Here $D_{M,0}$ and D_S are the diffusion coefficients of the solute in the gas phase (at ambient pressure) and in the stationary phase, respectively, d_f is the thickness of the stationary phase film, and j and j_1 are pressure correction factors, defined in Eqs. (92) and (93) in Chap. 2.

The application of Eq. (42) is not completely straightforward since data on D_S are often not available. Nevertheless, only by the corrected expression for UTE, as given by Eq. (42), can this quantity be used as a theoretically sound means for expressing column performance, especially with respect to the properties of the stationary liquid phase.

When referring to other types of chromatographic columns, Eq. (42) must be modified accordingly. Sufficient information about the dispersion equations relevant for different types of chromatography are given in Chap. 2.

Dynamic Measures of Column Efficiency

All the quantities described above for expressing the efficiency of chromatographic columns are of a static nature: they do not take into account the time needed to obtain the stated efficiency (not even the peak capacity). However, since time certainly is a very important experimental parameter, a few expressions have been devised to state the column efficiency in a dynamic way.

The simplest quantity of this kind is the *rate of production of theoretical plates* ν, which is defined as

$$\nu = \frac{n}{t_R} \tag{43}$$

A similar rate of production of effective plates can also be used. It is easily seen that

$$\nu = \frac{u}{h(1 + k)} \tag{44}$$

and that this quantity is therefore directly related to the basic chromatographic parameters discussed earlier. It has been used mainly for the optimization of high-speed chromatography [33,34].

Golay [5] defined a "performance index," additionally considering the pressure drop over the column. High performance was defined as the simultaneous achievement of a high plate number in a short time with a small pressure drop. A simplified expression for a *performance index* π was given by Bristow and Knox [35], who suggested simply the product of the number of plates generated per unit of time multiplied by the number of plates generated per unit of pressure drop. Clearly, high performance requires that both these factors be large.

$$\pi = \left(\frac{n}{t_R}\right)\left(\frac{n}{\Delta P}\right) = \frac{n^2}{t_R\ \Delta P} \tag{45}$$

where ΔP is the pressure drop over the column. This quantity is not dimensionless and continues to depend on experimental conditions. This led to the definition of the dimensionless *separation impedance* E [35], which is defined as

$$E = \frac{1}{\pi\eta(1 + k)} = \frac{t_M \Delta P}{n^2 \eta} \tag{46}$$

where η is the viscosity of the mobile phase. This number expresses the "difficulty" of achieving n theoretical plates and is thus low for high performance. A "good" value for a packed column is E $\approx$ 2000; for an open tubular column (where ΔP is much smaller), other conditions being the same, E $\approx$ 10 in favorable cases [36].

It was shown that

$$E = \frac{h^2 \Phi}{d_p^2} \tag{47}$$

where Φ is a flow resistance parameter:

$$\Phi = \left(\frac{d_p}{L}\right)^2 \Delta P \frac{t_M}{\eta} \tag{48}$$

Here d_p is the particle diameter (or the column diameter in the case of open tubular columns) and L is the column length. An extensive discussion of these quantities was presented by Knox [36].

The separation impedance is a very general measure, summarizing much of the behavior of a chromatographic column. This might be a disadvantage [35]: by itself, a high value of E provides little information about what is wrong, but only shows that something is wrong. Other less sophisticated measures of column performance, which are related to only one aspect of performance, might have a better diagnostic value.

REFERENCES

1. E. Heftmann, History of Chromatography and Electrophoresis, *Chromatography, Part A: Fundamentals and techniques* (E. Heftmann, ed.) Elsevier, Amsterdam, p. A19, (1983).
2. K. Sakodynskii, The life and scientific works of Michael Tswett, *J. Chromatogr.*, 73:303 (1972).
3. A. J. P. Martin and B. L. M. Synge: A new form of chromatogram employing two liquid phases, *Biochem. J.*, 35:1358 (1941).

4. A. T. James and A. J. P. Martin: Gas-liquid partition chromatography: The separation and micro-estimation of volatile fatty acids from formic acid to dodecanoic acid, *Biochem. J.*, 50:679 (1952).
5. M. J. E. Golay, Theory of chromatography in open and coated tubular columns with round and rectangular cross-sections, *Gas Chromatography 1958*, (D. H. Desty, ed.), Butterworths, London, pp. 36-55 (1958).
6. J. Porath and P. Flodin, Gel filtration: A method for desalting and group separation, *Nature*, 183:1657 (1959).
7. J. C. Giddings, *Dynamics of Chromatography. Part I*, Marcel Dekker, New York, 1965.
8. L. S. Ettre and A. Zlatkis (eds.), *75 Years of Chromatography —A Historical Dialogue*, Elsevier, Amsterdam, 1979.
9. L. S. Ettre, The nomenclature of chromatography. I. Gas Chromatography, *J. Chromatogr.*, 165:235 (1979). II. Liquid Chromatography, *ibid.*, 220:29 (1981). III. General Rules for Furture Revisions, *ibid.*, 220:65 (1981).
10. *Gas Chromatography Terms and Relationships ASTM E355*, American Society for Testing and Materials, Philadelphia. (Originally published in 1968; latest revision 1977.)
11. *Liquid Chromatography Terms and Relationships ASTM E682*, American Society for Testing and Materials, Philadelphia. (Originally published in 1979.)
12. IUPAC Recommendations on nomenclature for chromatography, *Pure Appl. Chem.*, 37:447 (1974).
13. J. Krupčík, J. Garaj, P. Čellár and G. Guiochon, Calculation of the peak capacity in capillary gas chromatography, *J. Chromatogr.*, 312:1 (1984).
14. P. R. Rony, The extent of separation: Time normalization and minimum time analysis in elution chromatography, *J. Chromatogr. Sci.*, 9:350 (1971).
15. B. D. Blaustein and G. M. Feldmann, Peak width vs retention time in gas liquid chromatography on packed columns, *Anal. Chem.*, 36:65 (1964).
16. R. E. Kaiser, Zur richtigen Messung und Bewertung von Gütetrennzahlen in der Chromatographie: Die real Trennstufenzahl, die Trennzahl, die Dosiergütekennzahl. *Chromatographia*, 9:337 (1976). In English: *Chromatographia*, 10:323 (1977).
17. G. Guiochon, On the fallacies of the "ABT-concept," *Chromatographia*, 11:249 (1978).
18. R. E. Kaiser, Why only 50% of the concept? *Chromatographia*, 11:257 (1978).
19. T. W. Smuts, T. S. Buys, K. de Clerk, and T. G. duToit, Interpretation of the real plate number concept. A fundamental analysis, *J. HRC CC, High Resolut. Chromatogr. Commun.* 1:41 (1978).

20. J. Ceulemans, The number of theoretical plates at infinite capacity: A better measure of the efficiency of gas chromatographic columns, *J. Chromatogr. Sci.*, 22:296 (1984).
21. R. Kaiser, Neuere Ergebnisse zur Anwendung der Gas-Chromatographie. *Z. Anal. Chem.*, 189:1 (1962).
22. L. S. Ettre, Separation values and their utilization in column characterization, Part I, *Chromatographia*, 8:291 (1975); Part II, *Chromatographia*, 8:355 (1975).
23. R. A. Hurrell and S. G. Perry: Resolution in gas chromatography, *Nature*, 196:571 (1962).
24. J. Krupčík, J. Garaj, G. Guiochon, and J. M. Schmitter, On the use of the separation number as a criterion for the evaluation of gas chromatographic columns in isothermal conditions. *Chromatographia*, 14:501 (1981).
25. K. Grob, Jr., and K. Grob, Evaluation of capillary columns by separation number or plate number, *J. Chromatogr.*, 27:291 (1981).
26. E. Grushka, Chromatographic peak capacity and the factors influencing it, *Anal. Chem.*, 42:1142 (1970).
27. J. M. Davis and J. C. Giddings, Statistical theory of component overlap in multicomponent chromatograms, *Anal. Chem.*, 55:418 (1983).
28. D. P. Herman, M.-F. Gonnord, and G. Guiochon, Statistical approach for estimating the total number of components in complex mixtures for nontotally resolved chromatograms, *Anal. Chem.*, 56:995 (1984).
29. L. S. Ettre, Support-coated open tubular columns, *Gas Chromatography 1966* (I. B. Littlewood, ed.), Institute of Petroleum, London, pp. 115-118 (1967).
30. L. S. Ettre, The evolution of open tubular columns, *Applications of Glass Capillary Gas Chromatography* (W. G. Jennings, ed.), Marcel Dekker, New York, pp. 1-47 (1981).
31. C. A. Cramers, F. A. Wijnheijmer, and J. A. Rijks, An analysis of coating efficiency as a measure for capillary column performance, *Chromatographia*, 12:643 (1979).
32. C. A. Cramers, F. A. Wijnheijmer, and J. A. Rijks, Optimum gas chromatographic conditions in wall-coated capillary columns, *J. HRC CC, High Resolut. Chromatogr. Commun.*, 2:329 (1979).
33. J. L. DiCesare, M. W. Dong, and L. S. Ettre, Very-high-speed liquid column chromatography. The system and selected applications, *Chromatographia*, 14:257 (1981).
34. R. J. Jonker, H. Poppe, and J. F. K. Huber, Improvement of speed of separation in packed column gas chromatography, *Anal. Chem.*, 54:2447 (1982).

35. P. A. Bristow and J. H. Knox, Standardization of test conditions for high performance liquid chromatographic columns, *Chromatographia*, 10:279 (1977).
36. J. H. Knox, Kinetic factors influencing column design and operation, *Techniques in Liquid Chromatography* (C. F. Simpson, ed.), Wiley Heyden, Chichester, pp. 31-56 (1982).

2

Dispersion and Peak Shapes in Chromatography

Jan Åke Jönsson
University of Lund
Lund, Sweden

I. INTRODUCTION

When a sample zone travels down a chromatographic column or layer, its width is continuously increased owing to a multitude of different mechanisms collectively known as dispersion processes. These include diffusion of solute along the column or layer, resistance to mass transfer between and within phases, and the influence of various flow inequalities and disturbances. The separation resulting from the different velocities of sample components is counteracted by dispersion, which decreases the resolution of these compounds. The matter of dispersion in chromatography was extensively treated during the development of the technique. The most thorough treatment was made by Giddings [1]. Some more recent reviews are those of Horváth and Melander [2], Perry [3], and Knox [4]. More or less detailed discussions on the matter are found in most textbooks on instrumental analytical chemistry.

The success of a chromatographic analysis, however, depends not only on the mere width of peaks but also to a great extent on their shape. Peaks that are not narrow and symmetrical, but wide or in various ways deformed ("tailing," "leading," and so on), limit the separation and prevent accurate quantification. This also makes sample identification by comparison of retention parameters (time, volume, index, and others) more difficult and arbitrary. The problem of peak distortion has attracted considerably less attention than that of dispersion, although an instructive review without mathematical detail was recently published [5].

It is possible to analyze peak shapes in order to obtain various physicochemical data, notably sorption isotherms and kinetic data. In this way, peak shapes can indicate the detailed types of retention mechanisms involved in the separation. This can give important clues toward the successful solution of difficult separation problems. More detailed discussions of this topic can be found in Chap. 5 and in an authoritative book on the subject [6].

Much work has been done on the problem of fitting mathematical peak models to experimental data in order to analyze unresolved chromatograms. Here it is of great importance that the great shape equations that are used reliably describe the physical reality.

In this chapter some of the existing theories about dispersion and peak shape will be reviewed, with emphasis on analytical gas and liquid column chromatography. Thin-layer chromatography is not explicitly treated. An article by Guiochon and Siouffi [7] treats this in an excellent way.

The main approach throughout this chapter is within the framework of the "rate theory," i.e., based mainly on partial differential equations, to describe the various processes in the chromatographic column.

An alternative way of formulating the theory of chromatography is the stochastic method, which employs discontinuous random walk models and statistical techniques. This was the main approach taken by Giddings, whose famous book, *Dynamics of Chromatography* [1], contains much of today's chromatographic theory, derived using stochastic methods.

II. TOOLS FOR THE DESCRIPTION OF PEAK SHAPE AND DISPERSION

A. Statistical Moments

The obvious way in which to characterize the shape of a chromatographic peak is with an explicit function by which the concentration of sample as sensed by the detector is expressed as a function of time or volume. Unfortunately, in many cases of interest, the peak shape equation is obtained in an indirect way, for example in the Laplace domain, and its transformation to an explicit form may be impossible. In such cases an alternative way to characterize the peak is to state statistical moments for the sample distribution. The moments define, in principle, the shape of the distribution and provide easily understood qualities, such as width, skew, and excess.

The concept of statistical moments is often used in this context, largely because of its mathematical advantages. The accurate calculation of moments from experimental data is, however, not easy, since it is very sensitive to experimental shortcomings, such as electrical noise and baseline drift. In spite of this, a considerable number of articles have utilized statistical moments to compare theoretical and practical concepts [8-10].

A chromatographic peak can be considered a probability density distribution of molecules over time. Then c(t) is the probability that a molecule in the sample is eluted in the time interval [t, t + dt], where t is the time from the introduction of the sample into the column.

From the definition of c(t) as a density distribution, we have

$$m_0 = \int_0^\infty c(t)\,dt = 1 \qquad (1)$$

Throughout most of this chapter there is no need to consider the amount of sample (i.e., the actual value of the peak area). For simplicity we assume that all peak distributions are normalized according to Eq. (1). In nonlinear chromatography (Sec. VI), peak profiles are concentration dependent, and normalization in the sense of Eq. (1) is therefore not possible.

The definition of the nth *zero-point moment* m'_n is

$$m'_n = \int_0^\infty c(t)t^n\,dt \qquad n \geqslant 1 \qquad (2)$$

where m_0 is the area of the peak, by definition [Eq. (1)] equal to 1. The term m'_1 is the time coordinate of the center of gravity of the peak, one of the definitions of retention time (see p. 50).

The definition of the nth *central moment* m_n is

$$m_n = \int_0^\infty c(t)(t - m'_1)^n\,dt \qquad n \geqslant 2 \qquad (3)$$

Central moments can be calculated from zero-point moments by the formula

$$m_n = \sum_{i=0}^{n} \frac{n!}{i!(n-i)!}(-m'_1)^i m'_{n-1} \qquad (4)$$

The second central moment m_2 is the variance σ_t^2 of the distribution c(t), a measure of the width of the peak and, consequently, of dispersion. It is discussed in more detail in the next section.

For a gaussaian distribution, $m_3 = 0$. The *skew* *S* is defined as

$$S = \frac{m_3}{m_2^{3/2}} \tag{5}$$

The quantity m_4 expresses the degree of peak flattening. The *excess* E is defined as

$$E = \frac{m_4 - 3m_2^2}{m_2^2} \tag{6}$$

For a gaussian distribution, E = 0. A negative value of E signifies a peak that is more flat than a gaussian.

Figure 1 shows how the peak shape is influenced by different values of the skew and excess.

Higher central moments than m_4 are sometimes referred to. Their intuitive meaning is less clear than that of the moments described above. Also, their calculation from experimental data is very inaccurate.

In the foregoing definitions and discussions, the independent variable is time. In chromatography, volume and distance are also pertinent variables. This introduces only trivial modifications in the equations defining the moments.

Moments can be calculated from experimental peaks with the use of a computer by digitizing the chromatogram with equidistant intervals in the independent variable. Calculation of the statistical moments from digitized data is most often done with a straightforward numerical

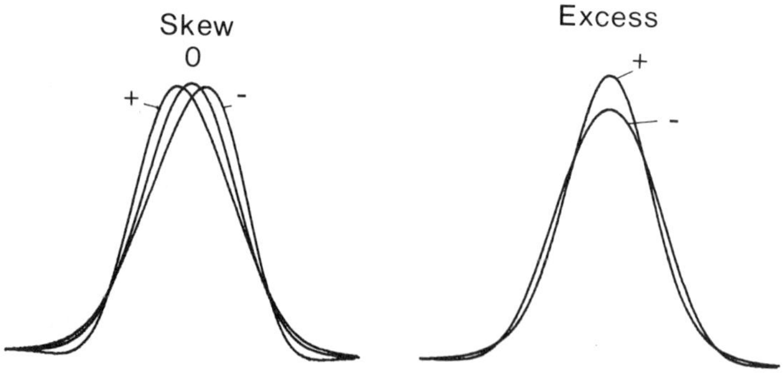

Figure 1. Peaks with different skew and excess values.

integration, corresponding to Eqs. (2) and (3) [9]. The crucial problem in the process is to define the baseline. A slight error in the baseline level will greatly influence the ordinate values on the peak flanks. This will create large errors, especially in the higher moments. This problem is beyond the scope of this chapter but has been discussed in detail by others [11].

Moments can also be calculated by curve fitting [12] and other algorithms [13], assuming some special peak shape. This approach may give better precision than the integration methods, but the selection of a predetermined shape will introduce important and fundamental systematic errors when applied to peaks that do not conform to this shape. Such methods cannot be recommended for more careful studies of peak shape by the use of statistical moments.

As will be seen below, the fitting of a general equation, such as a Gram-Charlier series (pages 37–38), to an experimental peak may be a useful method for the calculation of statistical moments.

B. Measures of Dispersion

Theoretical Plate Number

The basic quantity describing dispersion is the variance of a chromatographic zone. If the width of the zone is initially negligible, the variance after some time is a measure of the dispersion to which the zone was exposed during that time. In column elution chromatography, the flux of sample out of the column c(t) is taken as a function of time t, and we obtain the variance σ^2 of the output curve, the "peak," as the second statistical moment $\sigma^2 = m_2$ [see Eq. (3)].

To express the degree of dispersion, the dimensionless parameter n– the "number of theoretical plates"—is commonly used. Its definition is

$$n = \frac{(m_1')^2}{m_2} \tag{7}$$

The term "theoretical plates" is obsolete and is kept only as a historical artefact resulting from early theories of chromatography, in analogy to the theory of fractional distillation (where real plates, not "theoretical," occur).

The evaluation of n from an experimental chromatographic peak strictly according to Eq. (7) can be made only by numerical integration methods using a computer. If the peak is gaussian (which is often approximately, but never exactly, true) several simple formulas for the calculation of n can be deduced [14]:

$$n \approx 4\left(\frac{t_R}{w_i}\right)^2 \approx 5.55\left(\frac{t_R}{w_h}\right)^2 \approx 16\left(\frac{t_R}{w_b}\right)^2 \approx 6.28\left(\frac{t_R h_p}{A}\right)^2 \tag{8}$$

Here, t_R is the retention time, w_i is the width of the peak at the inflection points (at 60.7% of the height), w_h is the width at half-height, w_b is the width at the base, h_p is the height of the peak, and A is the peak area. Usually the formula with w_h is preferred.

Equations (8) should be applied with great caution, especially if absolute measurement of n is attempted. The assumption of a gaussian peak shape is imperative: even a moderate tailing results in a large error, with the result that the obtained value of n does not reflect the true variance of the peak. The proper use of Eq. (7) circumvents this problem and is thus more reliable. This has been strongly advocated by Kaiser [15].

Another way to calculate n was presented by Foley and Dorsey [16]. Based on the exponentially modified gaussian peak model (see p. 90) it also seems to handle somewhat skewed peaks with a fair degree of accuracy.

The plate number n can also be written

$$n = \left(\frac{L}{\sigma_L}\right)^2 \tag{9}$$

where L is the length of the chromatographic column and σ_L is the variance in length units of the column profile when it has reached the column end. The definitions of n from Eqs. (7) and (9) are not exactly equal; a slight difference occurs from the elution effect arising from the conversion of column profile to elution curve. This is discussed in pages 45–47. The difference between Eqs. (7) and (9) is usually negligible.

It should be noted that if a peak is asymmetric, this is often due to nonlinear retention effects, such as adsorption or overloading. In such cases the broadening of the peak is partly due to differences in retention at different solute concentrations, and the equations in this section are not applicable. This type of chromatographic system (nonideal, nonlinear chromatography) is mathematically complex, and its theoretical description is still incomplete. It is discussed briefly in Sec. VI. The more simple theories of dispersion to be presented in this chapter are applicable only to linear systems (i.e., systems in which retention is independent of concentration).

Plate Height

The "height equivalent to a theoretical plate" (HETP), denoted h, is related to n by

$$h = \frac{L}{n} \tag{10}$$

or, after combination with Eq. (9),

$$h = \frac{\sigma_L^2}{L} \tag{11}$$

From statistics we recall that a total variance can be written as a sum of variances arising from independent causes. Therefore, if the various dispersion mechanisms operate independently of each other, we can express the total observed h as a sum of contributions from individual sources of dispersion:

$$h = \sum h_i \tag{12}$$

These contributions can then be individually treated, which is the usual way of attacking the theory of dispersion in chromatography.

The van Deemter Equation

In a pioneering work of utmost importance [17], van Deemter and his co-workers derived an equation in the form of Eq. (12) for dispersion in chromatography. The van Deemter equation expresses h as a function of the average mobile phase velocity u and has the general form

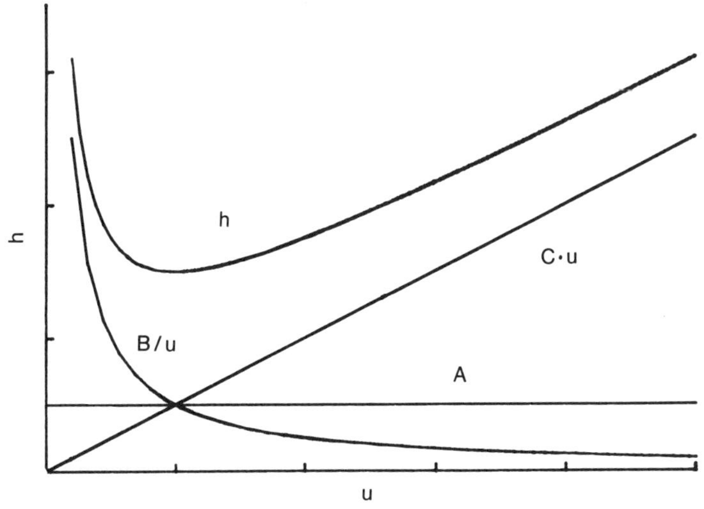

Figure 2. Graph of the van Deemter equation [Eq. (13)].

$$h = A + \frac{B}{u} + Cu \tag{13}$$

where A, B, and C are constants, independent of u.

According to the derivation [17] and the classical interpretation of Eq. (13), the terms are attributed to the following mechanisms of dispersion: flow inequalities in the column packing (A), axial diffusion in the mobile phase (B/u), and resistance to mass transfer (Cu). Over the years, the van Deemter approach has also been extended to incorporate other dispersive mechanisms, as will be discussed below.

In its general form, the equation can be applied to both gas and liquid chromatography, both in open tubular columns and in packed columns. This will be discussed in some detail below. The flow dependence of the three terms gives rise to the characteristic curve, which is schematically shown in Fig. 2. It can usually be fit excellently to experimental data.

The van Deemter equation can also be written in a dimensionless form [18],

$$h_r = \frac{h}{d_p} = a + \frac{b}{\nu} + c\nu \tag{14}$$

where h_r is the reduced plate height and a, b, and c are constants, The reduced velocity ν is defined by

$$\nu = \frac{ud_p}{D_M} \tag{15}$$

where D_M is the diffusion coefficient of the solute in the mobile phase and d_p is the diameter of the column packing particles (or, for open tubular columns, the column inner diameter). We will use either this representation or the dimensional form [Eq. (13)] according to which is convenient in the actual context. The main advantage of expressing the plate height in terms of reduced parameters is that data obtained with different particle sizes and different mobile phases can be directly compared.

Mobile-Phase Velocity and Holdup Volumes

The average mobile-phase velocity u can be calculated from the volumetric flow rate F and the column dimensions

$$u = \frac{FL}{V_c} \tag{16}$$

where V_c is the volume of the mobile phase in the column.

If t_c is the retention time for a nonsorbed tracer compound, we obtain $V_c = Ft_c$, and thus,

$$u = \frac{L}{t_c} \tag{17}$$

These simple expressions are subject to several additional considerations. In liquid chromatography with packed columns, the parameter V_c is not unequivocally determined. Sample molecules may penetrate the pores of the packing particles to different extents depending on the physical size of the molecules, and will thus have different holdup volumes. In steric exclusion chromatography (Chap. 8) this effect is exploited for the separation of macromolecules, but otherwise this effect is of little importance for retention. The matter was clarified by Horváth and Lin [19], who distinguish among three different holdup volumes V_e, V_0, and V_M, where V_e is the interstitial volume (the volume of mobile phases outside the particles of the packing), V_0 additionally incorporates the volumes of the pores in the particles, and V_M is the volume available to a given solute, so $V_e \leqslant V_M \leqslant V_0$. For a nonporous packing (or with open tubular columns), all these volumes are equal. The quantity k_0, defined as

$$k_0 = \frac{V_M - V_e}{V_e} = \frac{V_M}{V_e} - 1 \tag{18}$$

is the ratio of the intraparticle pore volume to the interstitial volume. In liquid chromatography, the volume V_e should be used to define the holdup volume and the average mobile-phase velocity. Thus t_c in Eq. (17) should be defined as the retention time for a nonsorbed compound, the molecules of which are totally excluded from the pores of the packing. In the following text, the velocity is defined in this way and usually termed u_e, when related to liquid chromatography.

In gas chromatography, on the other hand, the solute molecules are typically small compared with the pore dimensions. Then $V_M = V_0$ and V_e is difficult to measure. In this case, t_c refers to small molecules completely penetrating the pores.

The time t_c should *not* be corrected for gas expansion [6]. We want u to be compressibility averaged, and thus $u = ju_o$, where u_o is the velocity at the column outlet and j is the James-Martin compressibility factors, [see Eq. (92) below and Chap. 3, page 777]. Now, $\mu_0 = FL/V_0$, where F is the flow at outlet pressure and V_0 is the geometrical void volume including pores, equal to jt_cF. When these relations are combined, Eq. (17) results.

Of course, the considerations for u in this section are equally valid for the reduced flow velocity ν defined in Eq. (15).

C. Laplace Transform

The technique of Laplace transformation is an important means for solving differential equations. It involves transforming the differential equation into the Laplace domain, where it is simpler and often possible to solve. The solution, still in the Laplace domain, must then, however, be back-transformed, which may be very difficult or even impossible. Fortunately there are ways to obtain useful information from a function in the Laplace domain.

The definition of the Laplace transform of some function f(t) (which must be zero for $t < 0$) is

$$L[f](s) = \int_0^\infty e^{-st} f(t)\, dt \tag{19}$$

Here s is the Laplace variable (complex). For a thorough discussion on the properties of Laplace transforms, see the mathematical literature (for example, Ref. 20).

An important circumstance that is of great importance in our context is the possibility of calculating statistical moments from an expression of the Laplace transform of the distribution function under consideration. Following the nomenclature of Yamaoka and Nakagawa [10], we define a "moment vector" M:

$$M = \begin{vmatrix} m_1' \\ m_2 \\ m_3 \\ m_4 - 3m_2^2 \end{vmatrix} \tag{20}$$

where the elements of M are termed cumulants. This arrangement makes calculations with the moments especially simple [see Eqs. (21) and (27)].

From Eq. (19) and the definition of moments, it is not difficult to prove that

$$M = \lim_{s \to 0} \begin{vmatrix} \dfrac{-\partial}{\partial s} \\ \dfrac{\partial^2}{\partial s^2} \\ \dfrac{-\partial^3}{\partial s^3} \\ \dfrac{\partial^4}{\partial s^4} \end{vmatrix} \ln L[c](s) \tag{21}$$

Using this relationship, it is possible to calculate statistical moments for solutions to differential equations, even if it is not possible to find an explicit solution in the time domain by inversion of the Laplace transform.

D. Construction of Peak Profiles

To visualize the shape of a peak expressed in an indirect way, such as a set of statistical moments or a Laplace transform, it is necessary to construct pictures. Some methods for doing this are discussed in this section.

Gram-Charlier A Series

If the statistical moments of a density distribution function are known, the function is completely specified.* It can be described as a Gram-Charlier A series (GCAS) [21,22], which is written

$$f(z) = \frac{1}{\sigma\sqrt{2\pi}} \exp^{(-z^2/2)} \left(1 + \sum_{i=3}^{\infty} \frac{A_i}{i!} H_i(z) \right) \tag{22}$$

where $\sigma = \sqrt{m_2}$, $z = (t - m_1')/\sigma$, A_i are constants, and $H_i(z)$ are Hermite polynomials, with $A_3 = S$, $A_4 = E$, $H_3 = z^3 - 3z$, and $H_4 = z^4 - 6z^3 + 3$. Usually, the series is truncated after $i = 4$.

This approximation has been used with good results to reproduce chromatographic peaks from calculated statistical moments [23]. This same work also shows a striking example of the difficulty in obtaining accurate moments by numerical integration. The GCAS is in one case

*This is not generally true but holds for those distribution functions encountered in chromatography.

shown to create a completely distorted and grotesque "peak." On the other hand, by least-squares fitting of Eq. (22) to the experimental data, other values were found for the moments, widely different from those obtained by integration. These provided an excellent fit, as shown in Fig. 3. It seems that this procedure could be a better procedure for calculating moments from experimental data than the direct numerical integration. However, the necessary truncation of the infinite series causes moments that are determined in this way to include components from higher moments, which are not considered in the truncated series.

Edgeworth-Cramér Series

A slightly different series with mathematical advantages over the GCAS is the Edgeworth-Cramér series (ECS) [22]. This is in a certain way a more straightforward asymptotic expansion of the normal distribution than Eq. (22), and it might be expected to give a better representation of a chromatographic peak when truncated.

The ECS, truncated after the terms containing moments up to the fourth order, can be written

$$f(z) = \frac{1}{\sigma\sqrt{2\pi}} e^{-z^2/2}\left(1 + \frac{S}{3!}H_3(z) + \frac{E}{4!}H_4(z) + \frac{10S^2}{6!}H_6(z)\right) \qquad (23)$$

where $H_6(z) = z^6 - 15z^4 + 45z^2 - 15$. The application of this series corresponds to the addition of an extra term, containing S^2, to the GCAS. The complete formula for the Edgeworth-Cramér series is given in Ref. 22 (p. 229).

Dondi and co-workers [24] compared the two types of series. They found that the ECS offers significant advantages over the GCAS for fitting to experimental peaks. They also used the ECS series to calculate statistical moments for experimental peaks [25].

Numerical Inversion of Laplace Transforms

Laplace transforms, often difficult or impossible to invert with exact mathematical methods, can be inverted numerically using computer calculations. Algorithms for this inversion have been presented in the literature [26]. The computer procedures needed are rather involved, and the problem is not trivial. Only in rare cases [27,28] have such techniques been applied to chromatographic problems. A further development in this field would be desirable, since it would provide a general means for visualization of the results obtained by theory. Section VIII shows how this technique has been utilized.

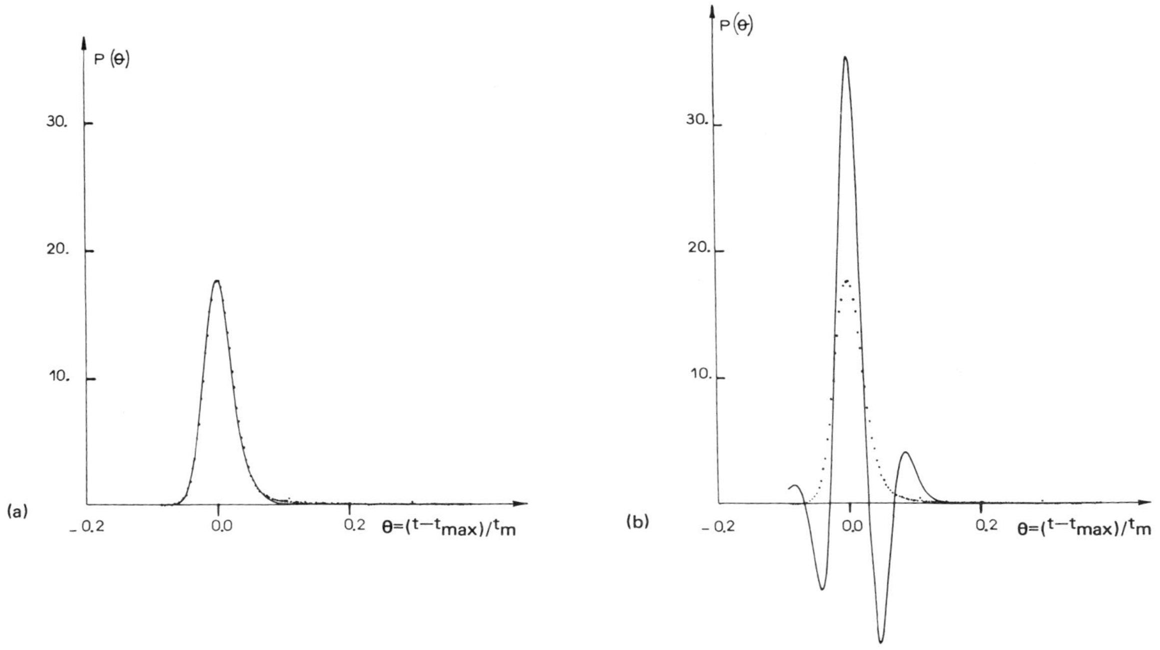

Figure 3. Chromatogram of methane. (a) Best Gram-Charlier theoretical model obtained by least-squares fit of the experimental data: $m_0 = 2970$; $m_1' = 29.46$ sec; $m_2 = \sigma^2 = 0.47$ sec^2; S = 0.45; E = 3.36. (b) Gram-Charlier series corresponding to the moments obtained by integration: $m_0 = 3028$; $m_1' = 29.53$ sec; $m_2 = \sigma^2 = 0.77$ sec^2; S = 2.23; E = 15.86. (After Vidal-Madjar and Guiochon [22], with permission.)

Simulation Methods

Methods for the study of chromatography quite different from the purely mathematical methods discussed here involve the construction of a computer model of the column. This model is then used for the production of simulated chromatograms.

One such technique [29] uses a modified "plate" description of the column. This provides a simple program, which is nevertheless very flexible. The simulated chromatograms agree well with theory, and the method has given useful results [30].

A more mathematically advanced simulation technique [31,32] uses numerical methods for solving the partial differential equations arising from mass-balance considerations (see pages 49–50).

E. Some Concepts from System Theory

System theory [33] deals with the description of systems in a broad sense, and some of the concepts in this discipline of science are of value when describing the behavior of a chromatographic system.

A chromatographic peak is a result of several factors, each of which contributes in a characteristic way to the final peak shape. The process of creating the peak can be viewed as the transformation of an initial concentration signal (the input profile) to a final electrical signal appearing on the recorder. This process can be assumed to occur in several independent steps: the input profile is first changed by dispersion in the injector and in the lines leading to the column, is then retarded and further dispersed in the column, and is again dispersed in new connecting lines and in the detector. The concentration signal is converted to an electrical signal by the detector and its associated electronics. Here electrical noise is added to the electrical signal, forming the final, observed, result.

Mathematically, each of these processes can be characterized by an impulse-response function (IRF). Such a function describes how a particular system responds to an infinitely narrow input function $\delta(t)$, the Dirac delta function. The delta function has the following properties:

$$\delta(t) = 0 \quad \text{for all } t \neq 0$$

$$\int_{-\infty}^{\infty} \delta(t)\, dt = 1 \tag{24}$$

$$L[\delta] = 1$$

The independent variable for the delta function is usually time, as in Eq. (24). It can also be defined with other independent variables, such as distance [see Eq. (34)].

The Laplace transform of an IRF is called a transfer function, which is a central concept in system theory. By the process of convolution, described below, different IRF can be joined together to describe a more complex process. A prerequisite for this combination is that the processes to be combined be independent of each other and linear. (A process is linear if the response to an input signal $x_1 + x_2$ is the sum of the response to the inputs x_1 and x_2, respectively.) These prerequisites do not always exist and care must be exercised not to use the convolution approach when it is not valid.

Much of the rest of this chapter is devoted to the problem of finding the impulse-response function of the chromatographic column in various cases.

F. Convolution

If f(t) and h(t) are two IRF, the combined result of the corresponding processes is given by the convolution integral, defining a new IRF

$$f*h(t) = \int_0^\infty f(t_1)h(t - t_1)\, dt_1 \tag{25}$$

The convolution theorem [20) states that the Laplace transform of the convolution product $f*h$ is found by multiplication of the Laplace transforms of the functions f and h. An analogous relation holds for Fourier transforms. Thus,

$$L[f*h] = L[f]L[h] \tag{26a}$$

$$F[f*h] = F[f]F[h] \tag{26b}$$

The convolution product of Dirac's delta function [see Eq. (24)] and any function f is $\delta * f = f$.

By means of the convulution theorem, it is possible to combine different processes by multiplication of their transfer functions. In chromatography, multiplication of (1) the Laplace transform of the input concentration profile, (2) the transfer functions for various extra-column dispersion processes, (3) the transfer function of the column, and (4) the transfer function for the readout system yields the global transfer function, describing the final, observed peak.

Another implication of the convolution theorem is the possibility of deconvolution by the use of Fourier transforms [34]. By calculating the Fourier transform of the function $f*h$, the transform of h can be found by division (complex) by F[f], if this is known. Then h can be found as the inverse transform of F[h]. This method has been used to remove instrumental contributions to observed signals in spectroscopy [35], electrochemistry [36], and chromatography [37-39].

Most computer libraries contain efficient fast Fourier transform (FFT) routines. For mini- and microcomputer applications, easily implemented procedures are readily available [40,41] as program listings.

Yet another aspect of convolution, which easily follows from a combination of Eq. (26) with Eq. (21), is that convolution of two functions corresponds to addition of the respective statistical moments (more rigorously, addition of the cumulants). Thus [22],

$$M(f) + M(h) = M(f*h) \tag{27}$$

where M(f) is the moment vector [see Eq. (20)] for the function f, and so forth. A well-known special case of this relation is the possibility of adding variances resulting from various independent dispersion processes [see Eq. (13)].

Equation (27) suggests another possible route to deconvolution [42]. First find the moments of the observed peak, then subtract the moments of the instrumental contribution function, and finally, apply the Gram-Charlier series (page 37), for example, to create the deconvoluted peak. Considering the discussion in Secs. II.C and II.D, this deconvolution precedure seems to be far less accurate than the process using Fourier transforms.

III. CONTRIBUTIONS FROM EXTRACOLUMN PROCESSES

In this section the influences of extracolumn processes on the final peak shape will be discussed, before going into the column processes themselves, which may be more interesting. The extracolumn processes are often neglected in theoretical discussions but may significantly influence the final peak. This matter was studied in detail by Sternberg [43] and by Cram and Glenn [44].

Since the natural goal in most practical cases is the elimination of extracolumn processes, their theoretical treatment does not have to be very exact but is intended only to give an approximate view of the matter.

A. Injection Profile

The ideal case of an infinitely narrow injection profile can generally not be obtained experimentally. Most experimental injection profiles can be approximated by one of two functions (or a combination of them): the rectangular plug function and the exponential decay function. The mathematical definitions, Laplace transforms, and statistical moments of these functions are summarized in Table 1.

Table 1. Properties of Some Impulse-Response Functions, Normalized so $m_0 = 1$

	Rectangular plug	Exponential decay
f(t)	$\begin{cases} 1/\tau_r & 0 \leqslant t \leqslant \tau_r \\ 0 & t < 0;\ t > \tau_r \end{cases}$	$\begin{cases} 1/\tau_e e^{(-t/\tau_e)} & t \geqslant 0 \\ 0 & t < 0 \end{cases}$
L[f](s)	$[1 - e^{(-s\tau_r)}]/s$	$1/(s\tau_e + 1)$
M	$\left\| \begin{matrix} \tau_r/2 \\ \tau_r^2/12 \\ 0 \\ -\tau_r^4/120 \end{matrix} \right\|$	$\left\| \begin{matrix} \tau_e \\ \tau_e^2 \\ 2\tau_e^3 \\ 6\tau_e^4 \end{matrix} \right\|$
S	0	2
E	−6/5	6

Rectangular Plug

The rectangular plug function is in many cases a realistic input function. Properly designed valve injectors in both gas and liquid chromatography should very nearly provide a plug injection. The syringe injection of liquid in gas chromatography will, if the vaporization is quick enough, also give a nearly rectangular sample distribution. The only parameter of the normalized rectangular plug function is the time width τ_r.

Exponential Decay

This function results from several phenomena. Liquid injection in gas chromatography when the rate of vaporization is slow can cause the concentration to vary approximately exponentially. The parameter τ_e is then determined by the temperature, the heat of vaporization, and geometric factors [43].

If the injection is made into a chamber of substantial volume, the dilution causes an exponential variation of the concentration in the effluent from the chamber. The parameter τ_e is then equal to v/F, where v is the volume and F is the volumetric flow rate through the chamber.

If "dead volumes" (i.e., unswept and unstirred volumes in contact with the flow stream) exist, an exponential decay also results, with $\tau_e = l^2/2D$, where D is the diffusion coefficient and l is a distance, the characteristic diffusion distance.

Other Injection Profiles

Injection profiles other than those mentioned above, not intended to approach the Dirac's delta, are sometimes used. In preparative-scale chromatography, wide plugs are usually injected. Frontal chromatography, in which the injection profile is a step function, is used for physicochemical measurements and for studies of environmental gas sampling and trace enrichment methods [45]. These injection profiles and several others were studied by Reilly and co-workers [46].

An interesting modern technique in this context is correlation chromatography [34,47], in which the sample profile is a complicated function of time, corresponding to a "pseudorandom binary sequence," and the resulting chromatogram has little resemblance to the usual chromatogram. By the use of deconvolution and correlation techniques, a "normal" chromatogram can be computed.

Influence on Retention Time and Height Equivalent to a Theoretical Plate

The retention time t_{obs} of an observed peak is, according to the principle of addition of moments, simply the sum of the retention time from the actual column process t_{col} and the first moment t_{inj} of the injection function:

$$t_{obs} = t_{col} + t_{inj}$$

The increase in plate height is calculated in a slightly more complicated way [see Eqs. (7) and (10)]

$$h_{obs} = \frac{L(h_{col} \cdot t_{col}^2/L + \sigma_{inj}^2)}{t_{obs}^2}$$

Usually t_{ing} is relatively small, so $t_{obs} \approx t_{col}$. In that case, the increase in h from injection will be equal to $L\sigma_{inj}^2/t_{obs}^2$.

The quantities to be added are the second moments, which is important to realize. Similarly, skews and excesses may not be added directly, but the third and fourth cumulants may.

B. Extracolumn Dispersion

The influence of connectors, connecting lines, and detector volumes on the resulting peak shape remain to be discussed. This question was summarized and discussed by Sternberg [43]. Important information on this matter can also be found in the theoretical treatment of flow injection analysis reviewed in Ref. 48.

Dispersion in Tubes

A straight, smooth tube through which the sample travels introduces as a first approximation a gaussian-type broadening of the sample band [49]. The IRF will be the same as the elution curve in linear, nonideal equilibrium chromatography with no retention [see Eq.(41) with $k = 0$]. By coiling [50] or bending the column tube into a serpentine shape [51], dispersion can be greatly reduced.

Connectors and Various Other Dead Volumes

It is impossible to accurately describe the action of random imperfections in the chromatographic flow stream. Generally, exponential contributions will be introduced, and the discussion on page 43 will apply.

Detector Volume

A detector always has a finite sensing volume, which can be of widely varying size. Some detectors (e.g., flow-through photometric detectors) have a clearly defined volume, but for gas chromatographic flame-ionization detectors, the bounds of this volume are less obvious. The output of the detector will not be a true measure of the instantaneous concentration of a component, but rather an average over the effective detector volume V_{det}. This introduces a rectangular IRF (Table 1), with the time parameter τ_r equal to V_{det}/F, where F is the volumetric flow rate.

Influence of Electronics

The electrical signal from the detector is electronically amplified and finally presented on a strip-chart recorder or similar device. A finite response rate of this equipment introduces an exponential decay function (Table 1), with a time constant τ_e equal to the RC time constant (where R is resistance and C is capacitance). The time constant is usually stated in manuals and technical descriptions.

C. Columns Profile Versus Elution Curve

The independent variable in most chromatographic techniques is time. In some cases volume is used instead, for different reasons. If the mobile-phase flow rate is constant, as we assume here, it is insignificant for the discusssion whether we use volume or time as the abscissa.

A curve that describes chromatographic peaks as a concentration function of time or volume is termed an *elution curve*, and this is the curve that is experimentally observed. We introduce the notation c(t) for elution curves. This concentration is expressed in units of moles per volume of mobile phase leaving the column, or in units of moles per second (i.e., the flux of sample out of the column). The proportionality factor between these representations is the volumetric flow rate, which is assumed to be constant.

The *column profile*, on the other hand, is a plot of concentration as a function of the distance z from the column inlet. This type of curve is observed directly only in open-bed techniques, such as thin-layer chromatography or paper chromatography. Scanning thin-layer plates with a photometric scanner produces curves that are principally column profiles. Theoretical derivations often yield column profiles, which is not always recognized.

The column profile will be denoted C(z, t). The most convenient unit of concentration is moles per column-length unit. Other, equivalent units, such as moles per unit volume (or weight) of one or both of the phases in a short segment of the column, can also be used. It is often tacitly assumed that the column profile and the elution curve have the same shape. This is valid for long columns only. Generally, the elution curve will be more skewed than the column profile because the column profile is continuously broadened as it passes through the column. Thus the first part of a peak to be eluted "belongs to" a more narrow column profile than parts that elute later. This results in a skewed elution curve, even if the column profile is symmetrical. A treatment of this effect in the simple case of a gaussian column profile has been published [52].

Kreft and Zuber [53] clarified these concepts. They did not treat chromatography but the more simple problem of diffusion in a fluid moving in a tube. Their views on the different types of concentration curves and injection and detection modes are most instructive.

Sometimes it is assumed that the elution curve is equal to C(L, t), where L is the column length. This is not rigorously true. The proper relation between column profile and elution curve is [53,54]

$$C(t) = \int_L^{\infty} \frac{\partial C(z, t)}{\partial t}\, dz = -\int_0^{L} \frac{\partial C(z, t)}{\partial t} \qquad (28)$$

or, in the Laplace domain,

$$L[c](s) = s \int_L^\infty L[C](z, s)\, dz = sL\left[\int_L^\infty C\, dz\right](s) \qquad (29)$$

Equation (28) can be derived in the following way. The integral $I = \int_L^\infty C(z, t)\, dz$ is the amount of sample eluted from the column up to time t. Then the flux of sample out to the detector is the time derivative of I, from which follows Eq. (28), since C(z, t) is a continuous and bounded function.

Observe that the notation C(z, t) refers to the total concentration of sample in the column (i.e., the sum of the concentrations in the mobile and the stationary phases, respectively).

IV. LINEAR CHROMATOGRAPHY

A. The Condition of Linearity

When chromatographic migration is governed by a linear isotherm, the chromatographic system is linear. Also, secondary factors that influence the retention in a concentration-dependent way should be absent. Such factors include the sorption effect, the effect of temperature changes accompanying the zones, and viscosity effects [6].

The basic condition of linearity is that the capacity factor is constant and independent of concentration, so that we can write

$$C_S^V = KC_M^V \qquad C_S = kC_M \qquad (30)$$

where C_S^V and C_M^V are molar concentrations (moles per volume unit), and C_S and C_M are concentrations in moles per column-length unit (called length-based concentrations); K is the partition coefficient, and the capacity factor k is equal to K/α, where α is the fractional void volume (V_M/V_S). For the definition of V_M, see page 35.

Adsorption Chromatography

In cases in which chromatographic retention is caused by adsorption principally characterized by a nonlinear isotherm, the condition of linearity is approximated as the concentration of the sample approaches zero. Often a marked curvature of the adsorption isotherm persists to very low concentrations, making the assumptions of linearity invalid in many practical cases. This is especially the case for parasitic adsorption phenomena in gas-liquid chromatography (see Chap. 5). For systems that purposely exploit adsorption as the retention mechanism, such as gas chromatography using porous polymer adsorbents, nonlinearity is minimized by using adsorbents with large surface areas and relatively weak adsorption properties.

Partition Chromatography

In partition chromatography, the situation is different from that in adsorption chromatography. To a good approximation, partition isotherms are linear up to relatively high concentrations (see Chap. 3, pages 135–141). As has been pointed out [55], however, the partition isotherm is not exactly linear even if the solution (in GLC) or solutions (in LLC) involved are ideal in the sense of Raoult's or Henry's laws. This is because molar concentrations, included in the capacity, are not exactly proportional to mole fractions. Deviations from linearity for this reason are, however, relatively slight. At higher concentrations, when solute-solute interactions play a significant role, more important deviations from linearity result.

B. Axial Diffusion: Nonideal Linear Equilibrium Chromatography

Model

Here we first consider dispersion due to axial diffusion (along the column). This corresponds to the B term in the van Deemter equation [Eq. (13)]. This type of dispersion can be treated by introducing a constant diffusion coefficient and by assuming complete equilibrium between the mobile and the stationary phases. Other causes of dispersion, such as finite attainment to equilibrium, cannot be accurately modeled in this simple way and will be treated in subsequent sections.

The following general mass-balance equation was stated by Lapidus and Amundsen [56], and has been used by many authors (for a derivation, see the next paragraph):

$$\gamma_M D_M \frac{\partial^2 C_M}{\partial z^2} + \gamma_S D_S \frac{\partial^2 C_S}{\partial z^2} = u \frac{\partial C_M}{\partial z} + \frac{\partial C_M}{\partial t} + \frac{\partial C_S}{\partial t} \tag{31}$$

Here C_M and C_S signify concentrations in the mobile and stationary phases, respectively, u is the linear flow rate of the mobile phase, and z is the distance from the column inlet; D_M and D_S are diffusion coefficients for the sample in the two phases. In a packed column, diffusion is obstructed by the porous structure of the packing, and a correction factor γ_M is applied to adjust the value of the diffusion coefficient. Usually, γ_M is set to about 0.8. Furthermore, the stationary phase is not continuous but broken into separate pools, at least one for each support particle, effectively hindering the diffusion of sample from one particle to another. Consequently, the correction factor γ_S is in this case close to zero. In open tubular columns, these corrections do not apply, and γ_M and γ_S are both close to unity. If we assume a linear isotherm with the capacity factor k and complete equilibrium (i.e., $C_S = kC_M$), we obtain

$$D \frac{\partial^2 C}{\partial z^2} = u \frac{\partial C}{\partial z} + \frac{\partial C}{\partial t} (1 + k) \tag{32}$$

Here, C is $C_S + C_M$ and D is a conditional diffusion constant describing the axial diffusion in both phases:

$$D = \gamma_N D_M + k\gamma_S D_S \tag{33}$$

Suitable initial and boundary conditions for the solution of Eq. (32) are

$$C(z, 0) = \delta(z)$$
$$C(\pm\infty, t) = 0 \tag{34}$$

where $\delta(z)$ is Dirac's delta function [see Eq. (24)], which describes an idealized sharp pulse. The column is assumed to be infinite (i.e., $-\infty < z < \infty$), and the injection is made at z = 0 as a sharp plug at zero time. Strictly, we should write, in Eq. (34), $M\delta(z)$, where M is the number of moles injected, but since we consider peak profiles to be normalized (see page 29), this is unnecessary.

Derivation of the Mass-Balance Equation

The mass-balance equation can be derived by studying a portion of the column with the infinitesimal length dz and phase volumes dV_S and dV_M. Let us first consider the mobile phase. The flux of sample through the column at position z can be written

$$J(z) = -\gamma_M D_M \frac{\partial C_M^v}{\partial z} + uC_M^v \tag{35}$$

The first term is caused by diffusion according to Fick's first law, with an effective diffusion coefficient equal to $\gamma_M D_M$ (see page 48). The second term is due to convection (i.e., the transport of sample by the flow of the mobile phase). Now, the increase in sample concentration with time, in the volume being studied, is equal to the number of molecules flowing into the volume, $J(z)\,dV_M/dz$, minus the number flowing out, $J(z + dz)\,dV_M/dz$, minus the number of molecules that are lost to the stationary phase, $J_S\,dV_S/df$, all divided by dV_M. Thus,

$$\frac{\partial C_M^v}{\partial t} = \frac{J(z) - J(z + dz)}{dz} - \frac{J_S}{\alpha d_f} \tag{36}$$

But

$$J(z + dz) = J(z) + dz \frac{\partial J(z)}{\partial z}$$

Considering this and Eq. (35), we obtain

$$\frac{\partial C_M^V}{\partial t} = \frac{\partial J(z)}{\partial z} - \frac{J_S}{\alpha d_f} = \gamma_M D_M \frac{\partial^2 C_M^V}{\partial z^2} - u \frac{\partial C_M^V}{\partial z} - \frac{J_S}{\alpha d_f} \tag{37}$$

In the stationary phase we have similar conditions, with a zero convection term:

$$\frac{\partial C_S^V}{\partial t} = \gamma_S D_S \frac{\partial^2 C_S^V}{\partial z^2} + \frac{J_S}{d_f} \tag{38}$$

Owing to the assumption of equilibrium, J_S is here the same as in Eq. (37). Combining Eqs. (37) and (38), J_S is eliminated. After changing to length-based concentrations, Eq. (31) results.

Column Profile

The column-profile solution to Eq. (32), considering Eq. (34), can be written [57,58]

$$C(z, t) = (4\pi D_{eff} t)^{-1/2} \exp\left[\frac{-(z-u_{eff} t)^2}{4D_{eff} t}\right] \tag{39}$$

Here $u_{eff} = u/(1 + k)$ and $D_{eff} = D/(1 + k)$. If this equation is viewed as a function of z, it shows a peak-shaped gaussian column profile with the mean $u_{eff}t$ and variance $2D_{eff}t$. Thus the peak moves through the column with a constant velocity $u_{eff}t$ and at the same time widens, the width being proportional to t and the height to $1/\sqrt{t}$. The maximum of this peak reaches the end of the column when $t = L/u_{eff}$ or $t = L/u(1 + k)$. This is the definition of retention time in linear chromatographic systems.

The Laplace transform of Eq. (39) is [59]

$$L[C](z, s) = \frac{\exp[\frac{u}{2D}(1-\rho)z]}{\rho u} \tag{40}$$

with

$$\rho = \left[1 + \frac{4Ds(1 + k)}{u^2}\right]^{1/2}$$

Elution Curve

By applying Eq. (28) to the column profile in Eq. (39), we can derive the following expression for the elution curve [53]:

$$c(t) = \left(\frac{u_{eff}}{2} + \frac{L}{2t}\right) C(L, t) \tag{41}$$

where C(L, t) is given by Eq. (39).

The same equation was also derived in a slightly different way [52]. It constitutes the impulse-response function of a chromatographic column under the assumptions of linearity and equilibrium between the two phases.

The transfer function of the column in this casc of chromatography is the Laplace transform of c(t), which is

$$L[c](s) = \frac{2D(1 + k)s \exp\left[\frac{u}{2D}(1-\rho)L\right]}{u^2\rho(\rho - 1)} \tag{42}$$

with ρ as in Eq. (40).

The moment vector M [see Eq. (20)] for the elution curve c(t) is [54]

$$M = \begin{vmatrix} \left(\frac{L}{u} + \frac{D}{u^2}\right)(1 + k) \\ \left(\frac{2LD}{u^3} + \frac{5D^2}{u^4}\right)(1 + k)^2 \\ \left(\frac{12LD^2}{u^5} + \frac{44D^3}{u^6}\right)(1 + k)^3 \\ \left(\frac{120LD^3}{u^7} + \frac{558D^4}{u^8}\right)(1 + k)^4 \end{vmatrix} \tag{43}$$

If the column profile is viewed as a function of time at the end of the column, as was done by several authors [58,60], the following simple expression for the elution curve results:

$$c_1(t) = C(L, t) \tag{44}$$

As can be shown by intuitive reasoning [52,54,61], this is not the true elution curve, but a good approximation. The moment vector M_1 for the function $c_1(t)$ is given in Ref. 59, nonequilibrium parameters being neglected. The difference between M and M_1 is relatively small and appears only in the form of different numerical constants in the second terms of the moments.

Initial and Boundary Conditions

An equation that has been given by several authors [10,62,63] as a solution to the mass balance equation [Eq. (32)] is

$$C_2(z, t) = \frac{z}{t} C(z, t) \tag{45}$$

with C(z, t) as in Eq. (39). For the corresponding "elution curve" ($c_2(t) = C_2(L, t)$), the moments are equal to the first terms of M [Eq. (43)] and M_1. When deriving Eq. (45), initial and boundary conditions slightly different from those in Eq. (34) are used, namely

$$\begin{aligned} C_M(0, t) &= \delta(t) \\ C_M(z, 0) &= 0 \\ C_M(\infty, t) &= 0 \end{aligned} \tag{46}$$

Using these conditions, Eq. (45) is readily obtained as a solution to Eq. (32). It might be surprising that the assumption of an infinitely narrow (in width) sample input function, as in Eq. (34), should lead to a final elution curve equation distinctly different from the assumption of an infinitely short (in time) input [Eq. (46)]. The difference is, however, actually more important [54]: under the conditions given in Eq. (46), it is implied [owing to the definition of $\delta(t)$] that $C_M(0, t) = 0$ for all $t > 0$ (i.e., that the concentration in the very first part of the column is always zero). This is in fact not an initial condition but a boundary condition. Furthermore, the column is assumed to be "semi-infinite" (i.e., $0 < z < \infty$, instead of $-\infty < z < \infty$ as before). The question of whether these boundary conditions or the conditions in Eq. (34) are best is important and cannot be answered definitely. However, the assumptions in Eq. (46) bring about strange behavior in the case in which the flow rate is small or zero. In Fig. 4, column profiles from Eqs. (39) and (45) are shown after some time, when $u = 0$. We see that Eq. (39) shows a gaussian profile centered around $z = 0$, which is an intuitively probable picture. In contrast, Eq. (45) shows a peak that has migrated along the column, and C(0, t) is zero, as required by the boundary conditions.

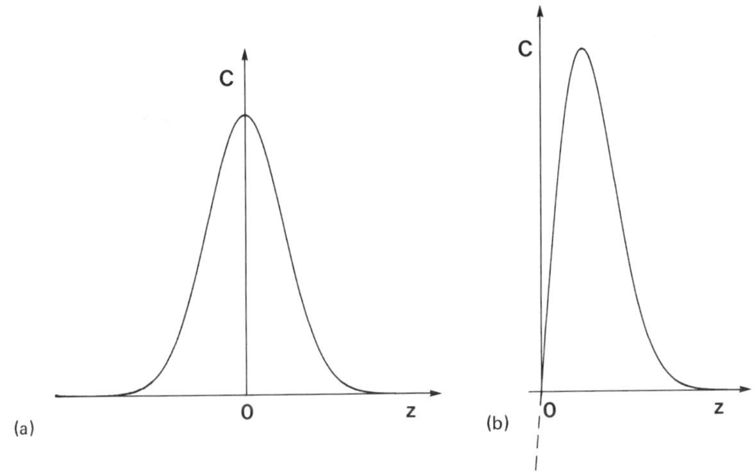

Figure 4. Plots of the column concentration profile for linear nonideal chromatography, calculated with u = 0 under two different types of initial and boundary conditions. (a) Equation (39). (b) Equation (45), (After Jönsson [54].)

We also see that the assumption of a semi-infinite column is essential: Eq. (45) predicts negative concentrations when $z < 0$.

In practice, neither of these pictures is correct, since normally the column would start at $z = 0$. At $z < 0$, we would have a connecting tubing with a situation that is different from the column. Thus the uniform diffusion backward predicted by Eq. (39) certainly is idealized. Nevertheless, backward diffusion is acceptable. A realistic physical arrangement creating the behavior predicted by Eq. (45) can hardly be imagined. A plug or a lid would produce $\partial C(z, t)/\partial z = 0$ at $z = 0$, which gives quite a different picture. Thus the conditions with an infinite column and a $\delta(z)$ injection are preferable.

One question remains: If we assume $\delta(z)$ as our input function, is the solution to the mass-balance equation in this case a true IRF (which by definition is the response to a $\delta(t)$ input)? If not, the convolution with other true IRF is questionable. The problem is solved in the following way [54,64]. We add to the left-hand side of Eq. (31) a new quantity, $I(z, t)$, which is the increase in sample in the actual column element due to injection. Then $I(z, t) = 0$ for all $z \neq 0$, and $I(0, t)$ is the injection time function, which can be set equal to $\delta(t)$. Thus $I(z, t) = \delta(z)\,\delta(t)$. This condition is mathematically equivalent to $C(z, 0) = \delta(z)$ and produces, accordingly, the results presented in Eqs. (39) through (43).

Retention Time Measurements

By definition, the retention time t_R in linear chromatography is

$$t_R^* = \frac{L}{u}(1 + k) \tag{47}$$

This is also, as was seen on page 50, the time for the maximum of the symmetrical column profile in Eq. (39) to reach the end of the column. It is of more interest to consider elution curves, and it can be proved that t_R^* coincides with the median of the elution curve c(t) as defined in Eq. (41). Thus the median, instead of the maximum—which is most often used, or the center of gravity, often stated to be more correct [5,6,60,65]—is the preferred point of the elution curve for retention time measurements. A more detailed discussion of this question is outside the scope of this chapter but can be found elsewhere [61].

The B Term in the van Deemter Equation

Under the simple assumption stated on page 48, the variance of the observed peak is given by the second line of Eq. (43). We can now calculate the corresponding contribution to the plate height h_{ax}, arising from the effect of axial diffusion (see page 33). According to Eqs. (7) and (10).

$$h_{ax} = \frac{Lm_2}{(m_1')^2} = \frac{2D}{u} + \frac{D^2}{Lu^2} - \cdots \tag{48}$$

The first term in Eq. (48) is the classic B term in the van Deemter equation [17]. The other terms are small correction terms arising from the conversion from column profile to elution curve. Neglecting them, we obtain from Eq. (33)

$$h_{ax} = \frac{2\gamma_M D_M}{u} + \frac{2k\gamma_S D_S}{u} \tag{49}$$

where the second term is usually negligible, especially in gas chromatography.

C. Mass Transfer Effects: General Nonideal Linear Chromatography

Model

We now continue the treatment in order to include another mechanism of dispersion, incomplete equilibrium between the two phases. This can be considered to have two components: a finite rate of mass transfer over the phase boundary and a finite rate of diffusion in the bulk of the stationary phase. Lateral diffusion in the mobile phase is still not taken into account; it is best treated in conjunction with flow effects (see Sec. IV. D, pages 59–64).

To model mass transfer, we use a model that was described by Grushka [59]. This model is a reasonable compromise between tractability and accuracy. Grushka considered the column as two infinitely long and wide parallel slabs, constituting the mobile and stationary phases, respectively. Solute molecules are allowed to diffuse into the stationary phase. Only diffusion in the direction normal to the phase interface is considered. For this diffusion, Fick's second law holds:

$$\frac{\partial C_S^v}{\partial t} = D_S \frac{\partial^2 C_S^v}{\partial x^2} \tag{50}$$

where D_S is the diffusion coefficient in the stationary phase and x is the lateral coordinate.

The mass transfer flux J_S from the mobile to the stationary phase over the phase boundary is characterized by

$$J_S = k_f[KC_M^v - C_S^v(x = d_f)] \tag{51}$$

according to Lapidus and Amundsen [56]. Here J_S is the flow of solute to the stationary phase (see page 50), d_f is the thickness of the stationary phase, and k_f is the mass transfer coefficient.

Equation (51) can also be written

$$J_S = d_f[k_s C_M^v - k_d C_S^v(x = d_f)] \tag{52}$$

Here k_s and k_d are the rate constants for sorption and desorption respectively, and Eq. (52) is simply the expression of first-order kinetics. The concentration change inside the brackets in Eq. (52) is converted to flux by multiplication by V_S and division by A_S, the surface area of the stationary phase. If the thickness of the stationary phase is even, then $d_f = V_S/A_S$. Comparing Eqs. (51) and (52), we see that $K = k_s/k_d$ (as required) and $k_f = k_d d_f$. Inserting J_S from Eq. (52) into Eq. (36) and converting to lenght-based concentrations, the mass balance in the mobile phase is given by the equation

$$D \frac{\partial^2 C_M}{\partial z^2} = u \frac{\partial C_M}{\partial z} + \frac{\partial C_M}{\partial t} + k_d[kC_M - C_S(x = d_f)] \tag{53}$$

Suitable boundary conditions [59] are as follows (the last condition was incorrectly written in Ref. 59):

$$
\begin{aligned}
&C_M(z, 0) = \delta(z) \\
&C_M(\infty, t) = 0 \\
&C_S(z, 0) = 0 \\
&\frac{\partial C_S}{\partial x} = 0 \qquad \text{at } x = 0 \\
&\frac{D_S}{d_f}\frac{\partial C_S}{\partial x} = k_d[kC_M - C_S(x = d_f)]
\end{aligned}
\tag{54}
$$

Similar models were presented by other authors, notably by Kučera [60], whose model is still more detailed. It is intended for gas-solid chromatography and in principle takes into account a third "phase," the stagnant mobile phase, in addition to the conventional two. Kučera's model also takes into account various "grain shapes": spherical and cylindrical, in addition to the "infinite slabs" discussed above. Several parameters that are inacessible to experimental determination are introduced, and the results from this model ar very untractable.

Kubín [62] presented a model similar to that used by Grushka. Unfortunately, he used the initial condition $C_M(0, t) = \delta(t)$ which was shown on pp. 52–53 to be unrealistic.

A different way of modeling nonequilibrium is to assume a time delay between the column profiles for C_M and C_S [66]. This method cannot distinguish between the two types of nonequilibrium, and does not seem to give more tractable results than the model used here.

Solution in the Laplace Domain

For the system of equations consisting of Eqs. (50), (53), and (54), no explicit solution for $C_M(z, t)$ is known. It is, however, possible to obtain the Laplace transform of the solution [59], which has a form similar to that of Eq. (40):

$$
L[C_M](z, s) = \frac{\exp[\frac{u}{2D}(1\text{-}R)z]}{uR} \tag{55}
$$

where

$$R = \left(1 - \frac{4AD}{u^2}\right)^{1/2}$$

and

$$A = s + \frac{kk_d(sD_S)^{1/2} \sinh [d_f(s/D_S)^{1/2}]}{(sD_S)^{1/2} \sinh [d_f(s/D_S)^{1/2}] + d_f k_d \cosh [d_f(s/D_S)^{1/2}]}$$

The Elution Curve

Since no explicit equation is known for the column profile, the same is true for the elution curve, and it is necessary to describe it by statistical moments or as numerical inversions of the Laplace transform.

Grushka [59] calculated the statistical moments for the curve $C_M(L, t)$ from Eq. (55) using the method outlined in Sec. II.C(page 32). (A minor error in his calculations was later corrected [23].) For reasons that were discussed above (pages 51–52), this is not a recommended procedure. Instead, the true eleution curve (in the Laplace domain) must be calculated using E. (29) before the moments are calculated. This was recently accomplished [54]. The Laplace transform of the elution curve $c_3(t)$ is given by

$$L[c_3](s) = \frac{2DB \exp[\frac{uL}{2D}(1-R)]}{u^2R(R - 1)} \tag{56}$$

with

$$B = s + k \frac{sd_f k_d \cosh [d_f(s/D_S)^{1/2}]}{(sD_S)^{1/2} \sinh[d_f(s/D_S)^{1/2}] + d_f k_d \cosh[d_f(s/D_S)^{1/2}]}$$

and with R as in Eq. (55).

The moment vector M_3 for the elution curve is found to be

$$M_3 = \begin{vmatrix} \left(\frac{L}{u} + \frac{D}{u^2}\right)(1 + k) - \frac{k}{1 + k} f_2 \\ \left(\frac{2DL}{u^3} + \frac{5D^2}{u^4}\right)(1 + k)^2 + 2\left(\frac{L}{u} + \frac{D}{u^2}\right) k(f_1 + f_2) \\ \left(\frac{12D^2L}{u^5} + \frac{44D^3}{u^6}\right)(1 + k)^3 + 6\left(\frac{2DL}{u^3} - \frac{5D^2}{u^4}\right) k(1 + k)(f_1 + f_2) \\ \left(\frac{120D^3L}{u^7} + \frac{558D^4}{u^8}\right)(1 + k)^4 + 12\left(\frac{12D^2L}{u^5} + \frac{44D^3}{u^6}\right) k(1 + k)^2(f_1 + f_2) \end{vmatrix} \tag{57}$$

Here, $f_1 = 1/k_d$ and $f_2 = d_f^2/(3D_S)$. Only first-order terms in f_1 and f_2 are considered. The parameter f_1 describes the finite mass transfer rate over the phase boundary, and f_2 is due to diffusion in the stationary phase. In the case of equilibrium, both f_1 and f_2 are zero, in which case M_3 is equal to M_1 [see Eq. (43)].

The retention time to the center of gravity of the peak, equal to the first element (m'_1) in the moment vector, depends on the nonequilibrium parameter f_2 (but not on f_1). This contrasts with the conclusions reached by Kucera [60], Grushka [59], and Villermaux [65], who found m'_1 to be independent of nonequilibrium kinetics. An intuitive way to rationalize the expression for m'_1 in Eq. (57) is to imagine that the finite diffusion rate prevents the sample form occupying the entire stationary phase volume, leading to decreased retention. The appearance of the term containing f_2 in m'_1 can be traced to consideration of the stationary-phase concentration in the derivation of the elution curve from the column profile.

The C Terms in the van Deemter Equation

In a manner similar to that on page 54, we can extract the classic C term in the van Deemter equation from the moment expressions in Eq. (57). Thus, the contribution of plate height h_{kin} arising from the slow attainment to equilibrium is found to be

$$h_{kin} = 2u \frac{k}{(1 + k)^2} (f_1 + f_2) + \frac{2Dk}{L(1 + k)^2} (f_2 - f_1) + \cdots \tag{58}$$

Disregarding minor terms and introducing the definitions of f_1 and f_2, we obtain

$$h_{kin} = h_S + h_{neq} = \frac{2}{3} \frac{k}{(1+k)^2} \frac{d_f^2}{D_S} u + 2 \frac{k}{(1+k)^2} \frac{u}{k_d} \qquad (59)$$

The first term, h_S, expresses the contribution to plate height from the finite rate of diffusional mass transfer in the stationary phase. The second term, h_{neq}, shows the effect of slow attainment to equilibrium. In the derivation presented here, a partition equilibrium between two phases is assumed. In this case, h_{neq} is related to interfacial resistance to mass transfer, which is probably not a significant effect, at least not in gas-liquid chromatography [67]. The expression for h_{neq} might also refer to an adsorption equilibrium (i.e., to gas-solid or liquid-solid chromatography). In such cases it may be important.

An alternative formulation of the nonequlilibrium term was given by Khan [68], comprising a "desorption rate constant,: which is in fact a mass-transfer coefficient like k_f in Eq. (51). It was shown on page 55 that $k_f = k_d d_f$. With this relation, Khan's equation agrees with Eq. (59).

D. Mobile Phase Dispersion in Open Tubular Columns

Dispersion in the mobile phase must be treated in different ways according to on whether packed columns or open tubular columns are considered. In this section, the relatively straightforward conditions in the latter case are described.

Laminar and Turbulent Flow

If the velocity of a fluid flowing through a smooth tube is not too high, the flow will be laminar, giving rise to a simple parabolic flow profile. For a straight tube with a circular cross section, the velocity u(r) will vary with radial position as

$$u(r) = 2u\left[1 - \left(\frac{r}{R}\right)^2\right] \qquad (60)$$

where r is the radial coordinate (the distance from the center of the tube), R is the tube radius, and u is the mean velocity. With laminar flow, a certain particle in the fluid will ideally follow the same stream line throught the entire tube; there is no radial convective mixing. This leads to a peak broadening that arises from the different velocities of sample molecules that happen to follow stream lines at different distances from the column tube center. The effect is counteracted by diffusion, which tends to level out radial concentration differences. At higher flow rates, the flow becomes turbulent and the parabolic profile breaks down, which leads to an effective mixing and no corresponding peak-broadening effect.

The onset of turbulence is governed by the Reynolds number Re. If Re in a straight open tube is no greater than about 2000, the flow is essentially laminar. Re is defined as

$$\mathrm{Re} = \frac{2\rho u R}{\eta} \tag{61}$$

where ρ and η are the density and viscosity of the fluid, respectively. For gases, the Reynolds number is approximately equal to the reduced velocity ν [see Eq. (15)]; for liquids it turns out that $\mathrm{Re} \approx \nu \times 10^3$ [1]. Considering the normal working range of open tubular columns, where ν seldom exceeds 100, it is clear that we must consider the peak-broadening effect of the parabolic flow profile.

The Golay Equation

In a famous pioneering work [69], Golay developed the theory for band broadening in open tubular columns. In that work, the contribution to plate height (i.e., the the second statistical moment) from the parabolic profile was developed. Here we also want to calculate the contributions to the other moments, which can be readily deduced from Golay's theory.

The starting point is essentially the mass balance equation for the case of nonideal equilibrium chromatography [see Eq. (32)]. Golay also treated the nonequilibrium case (in a way similar to that in Sec. IV.C and with the same result), but this unnecessarily complicates the treatment here. In cylindrical coordinates (r = radial coordinate and z = axial coordinate), considering the parabolic flow profile, the mass balance equation will read

$$D_M\left(\frac{\partial^2 C}{\partial r^2} + \frac{1}{r}\frac{\partial C}{\partial r} + \frac{\partial^2 C}{\partial z^2}\right) = 2u\left[1 - \left(\frac{r}{R}\right)^2\right]\frac{\partial C}{\partial z} + \frac{\partial C}{\partial t} \tag{62}$$

where C is the mobile-phase concentration as a function of z, r, and t, and all other symbols are defined as before.

At the tube wall, it must hold that

$$2\frac{D}{R}\frac{\partial C}{\partial r} = -k\frac{\partial C}{\partial t} \qquad r = R \tag{63}$$

To facilitate the solution, Golay assumed a small concentration difference across the tube. Thus, we can write

$$C = \overline{C} + \Delta C \tag{64}$$

where $\overline{C}$ is the concentration averaged over the tube cross section, (a function of z and t only). The corresponding average of ΔC is zero. Then we can assume that $\Delta C \ll C$. Accepting this, r can be eliminated from Eqs. (62) and (63) and we obtain

$$(a - bu) \frac{\partial^2 \overline{C}}{\partial z^2} = u \frac{\partial \overline{C}}{\partial x} + b \frac{\partial^2 \overline{C}}{\partial z \, \partial t} + (1 + k) \frac{\partial C}{\partial t} \tag{65}$$

where

$$a = D_M + \frac{1 + 6k + 11k^2}{48(1 + k)^2} \frac{u^2 R^2}{D_M}$$

$$b = \frac{k(1 + 4k)}{24(1 + k)} \frac{uR^2}{D_M}$$

Before discussing the solution to this problem, we should note that, if $k = 0$ (i.e., if the sample molecules do not enter the stationary phase), the situation is equivalent to band broadening in an uncoated tube, a standard problem in chemical engineering. In that case, $b = 0$ and Eq. (65) is analogous with Eq. (32), leading to a gaussian peak profile [as in Eq. (39)], the only difference being an increase in the effective diffusion coefficient (i.e., in the width of the peak). Referring to Eq. (39), we obtain (with $k = 0$ and $b = 0$)

$$D_{eff} = D + \frac{u^2 R^2}{48D} \tag{66}$$

This is in complete agreement with the standard treatment of this problem by Taylor [49]. It should be observed that the second term in Eq. (66) depends inversely on the diffusion coefficient—rapid diffusion decreases the lateral concentration difference, leading to decreased effects of the velocity differences.

With $k > 0$, the problem is not only quantitatively but also qualitatively different because of the $\partial^2 C/\partial z \, \partial t$ term that appears when $b > 0$. Equation (65) is then no longer satisfied by a gaussian column profile. This can be intuitively rationalized by noting that the exchange between the mobile and the stationary phase takes place at the column wall only (i.e., in the most slowly moving fluid). In the "beginning" of a peak, where sample flows into the stationary phase, a relatively lower concentration occurs at the wall than at the "end" of the peak. Thus the leading part of the peak should move somewhat faster than the tail. This leads to a negative influence on the skew of the elution curve, counteracting the skew arising from the "elution effect" (Sec. III.C page 45).

Elution Curve and Moments

By standard produces, the Laplace transform of the elution curve solution to Eq. (62), with initial and boundary conditions as in Eq. (33), is found to be

$$L[\bar{C}](s) = \frac{2s(1+k)(a-bu)\exp\left[\frac{bs+u}{2(a-bu)}(1-\varphi)L\right]}{u(bs+u)\varphi(\varphi-1)} \tag{67}$$

with

$$\varphi = \left[1 + \frac{(a-bu)s(1+k)}{(bs+u)^2}\right]^{1/2}$$

and a and b are as in Eq. (65).

This expression is very similar to Eq. (42), the corresponding expression for the case in which the laminar flow effect is not considered. However, Eq. (67) cannot be inverted to the time domain, as can Eq. (42), because the Laplace variable s enters Eq. (67) in another (and more complicated) way. What we can do is calculate statistical moments using Eq. (21). The result is

$$M_{fp} = \left|\begin{matrix} \left(\frac{a}{u^2} + \frac{L}{u}\right)(1+k) \\ \left(\frac{5a^2}{u^4} + \frac{2aL}{u^3}\right)(1+k)^2 \\ \left[\frac{44a^3}{u^6} + \frac{12a^2L}{u^5} - \frac{6abL}{u^4(1+k)}\right](1+k)^3 \\ \left[\frac{558a^4}{u^8} + \frac{120a^3L}{u^7} - \frac{120a^2bL}{u^6(1+k)} + \frac{24ab^2L}{u^5(1+k)^2}\right](1+k)^4 \end{matrix}\right| \tag{68}$$

where a and b are given in Eq. (65) and some minor terms are neglected. The first term in each moment expression is, for conditions typical for open column chromatography, also very small. These terms are kept here only for comparison with Eq. (43).

We can see that the first two moments (giving the retention time and peak width) are identical to those in Eq. (43), if a is treated as a diffusion coefficient. Thus the effect of the parabolic profile on the peak width, and therefore on h, is still simply equivalent to an increase in the diffusion coefficient, as stated in the definition of a.

However, for the third and fourth moments, and consequently for the skew and excess, terms containing b appear. These terms are, with typical values of the parameters, not negligible. The flow profile effect on the skew and excess is relatively smaller than what should be caused by a simple increase in the diffusion coefficient.

Flow Profile Term in the van Deemter Equation

By combining Eqs. (7) and (10) with the expressions for m_1' and m_2 in Eq. (68) and eliminating terms that arise from longitudinal diffusion only [i.e., those present in Eq. (43)], we arrive at the following expression for the increment in the plate height arising from the effect of the parabolic flow profile in open tubular columns:

$$h_{fp} = \frac{(1 + 6k + 11k^2)}{24(1 + k)^2} \frac{R^2}{D_M} u \tag{69}$$

This term was not included in the original van Deemter equation [17] as it applies to open tublular columns only. It was derived by Golay [69] in the previously cited article. With common nonmenclature (see page 33), this is a C term (i.d., it increases proprtionaly with the flow rate).

Effects of Column Coiling

The treatment above applies strictly to straight tubes only. For obvious reasons, open tubular columns are in practice always coiled. The coiling introduces, by centrifugal force, a secondary flow component in the direction of the coil radius, influencing the parabolic flow profile. The effect increases with increasing flow rate and decreasing coil radius. The parabolic profile may eventually completely break down, leading to effective lateral mixing. These phenomena were discussed in detail by Tijssen [50,70]. From a chromatographic point of view, this is mainly an advantage, since to some extent it decreases the peak-broadening effects of the flow profile.

Looking at the problem semiquantitively, the expression in Eq. (69) will be influenced by coiling in two ways. First, the diffusion coefficient will appear to increase by the addition of an apparent radial diffusion coefficient D_r (a flow-dependent phenomenon) arising from mass transport by the secondary flow. Second, when the flow particle is no longer parabolic, the constant 24 and the part of Eq. (69) containing the capacity factor will change, also in a flow-dependent way. These two phenomena will work in the same direction, toward a decrease in h_{fp}. The main effect arises from the influence on D_M.

It was found [50] that only if

$$\mathrm{Re}^2 \frac{R}{R_h} \frac{\rho}{\eta D_M} > 10 \qquad (70)$$

is the chromatographic effect of coiling noticeable. The terms Re, ρ, and η were defined in conjunction with Eq. (61), and R_h is the radius of the helical coil. The effect of coiling will usually be negligible in common practice. For example, if R = 0.01 cm and R_h = 8 cm, the left side of Eq. (70) will exceed 10 when u exceeds about 180 cm/sec (for gas chromatography, with nitrogen as the mobile phase and D_M = 0.01 cm^2/sec). As the matter is fairly complicated, no further detailed theory will be given here.

E. Diffusion in the Mobile Phase in Packed Columns

The complicated geometry of packed chromatographic columns leads to several band-broadening mechanisms:

1. There is a velocity difference between various flow paths that the mobile phase follows through the porous bed. This process is usually (but slightly improperly) termed "eddy diffusion."
2. There will be a stagnant layer of mobile phase close to the particle surfaces, owing to the laminar flow profile in the highly irregular channels between particles. This introduces a similar band-broadening mechanism, as the one discussed in Sec. IV.D (page 59).
3. Stagnant mobile-phase regions will also be present in pores of the packing material. Additional band broadening is then caused by diffusion between these regions and the moving mobile phase.

Several more or less sophisticated and successful theories explaining this problem complex have been presented, some of which will be reviewed below. The results are given in terms of plate height only. Accepting the approximate view that these plate height increments can be considered increments in an effective diffusion coefficient, higher moments may be calculated with the aid of Eq. (43).

Eddy Diffusion

The simplest approach to point 1 is the oldest, arising from the original work of van Deemter et al. [17]. As is discussed in more detail below, this approach is in many cases quite sufficient.

Assuming that the effects of eddy diffusion are random and symmetrical, they can be formally expressed as an increase in the diffusion coefficient of sample in the mobile phase. In analogy to Eq. (33) (neglecting diffusion in the stationary phase), we obtain

$$D = \gamma_M D_M + \lambda d_p u \tag{71}$$

where λ is an empirical parameter, d_p is the particle diameter, and u is the mobile-phase linear velocity. The form of the second term in Eq. (71) was merely suggested by van Deemter [17] but was later rationalized by Chilcote and Scott [71]. It was also derived by Giddings using a stochastic method [72].

Accepting Eq. (71), the effect of eddy diffusion will parallel that of axial diffusion, giving a problem already solved in Sec. IV.B. Thus, D from Eq. (71) can be used in Eqs. (39) through (48). We obtain from Eq. (48) the following contribution from eddy diffusion to the total plate height:

$$h_e = 2\lambda d_p = A \tag{72}$$

This is the classic A term in the van Deemter equation, predicting that h_e is independent of flow rate.

The Coupling Theory of Giddings

In the late 1950s, many reports presented experimental data that were inconsistent with the simple concept of eddy diffusion and the original van Deemter equation. These reports were critically summarized by Giddings and Robison [73]. Zero or even negative values of A were found and, at large velocites, a downward bending of the plot of h versus u was sometimes observed. To account for these discrepancies, Giddings [74,75] suggested the following expression:

$$h_{e1} = \frac{1}{1/A + 1/E_1 u} \tag{73}$$

where A is the eddy diffusion term from Eq. (72). The expression E_1u arises [75] from concepts similar to those described in Sec. IV.D concerning the effects of the parabolic flow profile in cylindrical tubes, and it is intended to include mechanism 2 above. If the space between the particles of the packing were cylindrical, E_1u would be identical to h_{fp} in Eq. (69). Giddings originally wrote it for the case of parallel plates (i.e., with other numerical coefficients). Of course, the real geometry is much more irregular, and thus no exact form for E_1 can be given. The terms A and E_1u are not added straightforwardly but in the same way as resistances connected in parallel. This is based on the assumption that the two peak-broadening mechanisms 1 and 2 are not statistically independent, but coupled. The term "coupling theory" is often used when referring to Eq. (73).

The same problem was attacked using the random walk method [74], with the same results as in Eq. (73), again without details regarding the exact form of Eu.

For large values of u, h_{e1} is approximately equal to A, but for small flow rates the E_1 term becomes more important and the overall h_{e1} decreases. This dependence was considered to correspond more closely to experimental findings [71] than the constant A term alone.

Giddings [1] also developed an extended version of the coupling theory, which was essentially a summation of five terms similar to that in Eq. (73), each of them describing a certain type of packing or flow irregularity. The resulting equation, although describing in detail all related processes, is hardly useful, since there seems to be no way for experimentally checking it or generally using it in practice.

Other Coupling Theories

Several authors, notably Huber [76,77] and later Horváth and Lin [19, 78], have elaborated the concept of eddy diffusion and arrived at equations similar to Giddings' coupling equation.

Following the reasoning of Horváth and Lin [19], we assume that the second term in Eq. (71) correctly describes the phenomenon of eddy diffusion. Then, the contribution to plate height will, according to Einstein's diffusion law and Eq. (11), read

$$h_e = \frac{2\lambda d_p u t^*}{L} \tag{74}$$

where t^* is the time that the solute spends in the moving regions of the mobile phase. If all of the mobile phase moves, then t^* will be equal to L/u and Eq. (74) will degenerate to the classic A term, as in Eq. (72). However, a fraction of the mobile phase forms a stagnant layer surrounding each particle. In reality there will be no sharp division between this nonmoving layer and the moving fluid, but nevertheless it is possible to define a flow-dependent parameter ζ, which is the fraction of the column void space occupied by the stagnant layer. Then,

$$t^* = \frac{L(1 - \zeta)}{u} \tag{75}$$

since $u/(1 - \zeta)$ is the velocity in the free stream. Thus we obtain

$$h_e = 2\lambda d_p(1 - \zeta) \tag{76}$$

Horváth and Lin [19] estimated ζ from the theoretical work of Pfeffer [79] and obtained

$$h_{e3} = \frac{2\lambda d_p}{1 + \omega(D_M/ud_p)^{1/3}} \tag{77}$$

where ω is a parameter that depends only on the porosity of the packing. Equation (77) is more conveniently expressed in reduced variables:

$$h_{r,e3} = \frac{2\lambda}{1 + \omega\nu^{-1/3}} \tag{78}$$

In analogy with Eq. (73), we may also write

$$h_{e3} = \frac{1}{1/A + 1/E_3u^{1/3}} \tag{79}$$

Huber and Hulsman [77] used experimental correlations by Hiby [80] to estimate ζ and obtained a very similar result:

$$h_{e2} = \frac{1}{1/A + 1/E_2u^{1/2}} \tag{80}$$

which is another alternative to the original coupling equation [Eq. (73)].

The differences among Eqs. (73), (79), and (80) are limited to the power of u only. It is not easy to decide unequivocally which is correct. However, Knox and Parcher [81] have experimentally studied expressions of the form of Eq. (80) with the exponent for u treated as am empirical parameter n'. It was found [82] that, to a good approximation, a simplification is possible:

$$h_k = E_k u^{n'} \tag{81}$$

Values of n' ranging from 0.25 to 0.35 were found. As a compromise that fit reasonably well with a large number of experimental data, $n' = 1/3$ was chosen. This is the basis for the so-called Knox equation, an alternative to the classic van Deemter equation. (See page 74 and Sec. V.D.)

According to these results, it is reasonable to prefer Eq. (79) to Eqs. (73) and (80).

Diffusion in Stagnant Mobile-Phase Regions

In a packed column, there are two different regions of stagnant mobile phase: the film surrounding each particle and the pores inside the particles. These regions separate the region of stationary phase, where sorption takes place, from the moving mobile phase. Solute molecules must thus perpetually be transported by diffusion across the stationary regions during the chromatographic process. The resistance to this transport is an often important cause of band broadening.

The *film-resistance contribution* term as derived by Horváth and Lin [70] (written in reduced parameters) is

$$h_{rf} = \frac{\kappa(k_0 + k + k_0 k)^2}{(1+k)^2(1+k_0)^2} \nu^{2/3} \tag{82}$$

Here κ is a constant, the packing structure parameter, having the approximate value of 1/15, and k_0 is the ratio of the available part of the intraparticulate pore volume to the interstitial void space in the column [see Eq. (18)].

In Eq. (82) we notice that k_0 enters the expression in the same way as the capacity factor k. Indeed, k_0 is the capacity factor related to retention by steric exclusion effects (see Chap. 8).

Huber [76,77] developed an expression analogous to Eq. (82) but with the power of ν equal to 1/2 instead of 2/3. Apparently, these powers are related to the powers 1/2 and 1/3 in Eqs. (80) and (79), respectively. Again referring to the work of Knox and coworkers (see pages 67), it is reasonable to prefer the power 2/3 for the flow dependence of the film-resistance term. This term is usually considered negligible even if by calculation (see p. 78) it might be significant.

The contribution from *diffusional mass transport in pores* is, especially in liquid chromatography, a major cause of band broadening. Consideration of the corresponding plate-height increment was first made by van Deemter [17], who, however, assumed it to be negligible in gas-liquid chromatography. Questioning that assumption, Jones [83] later derived the following equation:

$$h_m = C_2\left(\frac{k}{1+k}\right)^2 \frac{d_M^2}{D_M} u \tag{83}$$

This expression is analogous to the plate-height contribution from diffusion in the stationary phase [h_s in Eq. (59)]. Here, C_2 is a numerical constant and d_M is the diffusion path length in the mobile phase, both of which are obviously related to pore characteristics and the diameter d_p of the particles. These terms were linked by several authors who presented equations in the following form:

$$h_p = f \frac{d_p^2}{D_M} u \tag{84}$$

or, in reduced parameters,

$$h_{r,p} = f\nu \tag{85}$$

The factor f depends on the capacity factor k of the solute and on the porosity of the packing. It always involves empirical parameters.

A relatively general but still tractable expression for f was presented by Horváth and Lin [78]. Their equation reads

$$h_p = \frac{\theta(k_0 + k + k_0 k)^2}{30k_0(1 + k_0)^2(1 + k)^2} \frac{d_p^2 u_e}{D_M} = \frac{\theta}{30k_0} \left(\frac{k_e}{1 + k_e}\right)^2 \frac{d_p^2 u_e}{D_M} \tag{86}$$

where θ describes the tortuosity of the porous particles and k_0 is related to the pore volume as described above, in conjunction with Eqs. (82) and (18). Here k_e is the capacity factor based on the interstitial volume (see page 35). It holds that

$$k_e = (1 + k)(1 + k_0) - 1 \tag{87}$$

The flow velocity is given in terms of the interstitial flow velocity u_e, where $u_e = u(1 + k_0)$ and u is the overall average flow velocity. Essentially the same equation as Eq. (86) was developed by Giddings [84], Grubner [8], and Huber and Hulsman [76].

Another expression for h_p [85] reads

$$h_p = \frac{(a + bk + ck^2)}{24(1 + k)^2} \frac{d_p^2}{D_M} u_e \tag{88}$$

Here, a, b, and c are numerical constants, given with surprising precision as 0.37, 4.69, and 4.04, respectively. The constants a, b, and c were determined by the curve fitting of experimental results obtained from measurements of 69 compounds on a single column (Partisil 20, silica gel). It seems reasonable to believe that for other types of column packings these numerical factors will be different. However, this was not mentioned; instead, Eq. (88) was recommended for general use with the coefficients given. We may easily see by comparison with Eq. (86) that, for large values of k_e, $c \approx 0.8\theta/k_0$, obviously dependent on the properties of the column packing.

An interesting extension was made by Knox and Scott [86]. They considered, in addition to diffusion in regions of stagnant mobile phase in the pores of a column packing, diffusion in the stationary phase. This should not be confused with the largely gas-liquid chromatographic problem of mass transfer between the bulk of the stationary phase and the surface, leading to Eq. (59). Here we are dealing with liquid chromatography with a thin film of bonded stationary phase in which the solutes may diffuse in a direction more or less parallel to the surface. The diffusion coefficients in that layer and in the stagnant liquid mobile phase may very well be of comparable magnitude. Knox and Scott use a terminology that is rather different from that used above. To facilitate comparison, their equation is rewritten

$$h_p = \frac{(k_0 + k + k_0 k)^2 d_p^2 u_e}{30(1 + k_0)^2 (1 + k)^2 (k_0 \gamma_{SM} D_M + k(1 + k)\gamma_S D_S)} \tag{89}$$

γ_{SM} and γ_S are obstructive factors in the stagnant mobile phase and in the stationary phase, respectively. It is seen that if the contribution from the stationary phase is neglected (i.e., when $\gamma_S = 0$) and $\theta = 1/\gamma_{sm}$, Eqs. (89) and (86) are identical. Note that the consideration of D_S decreases h_p by permitting a more efficient mass transfer in pores. In gas chromatography, this effect is negligible. For more details about this concept in liquid chromatography, see pages 77–79.

V. DETAILED DISPERSION EQUATIONS

After the separate and general treatment of the various dispersion mechanisms, it is now possible to join them together into detailed descriptions, typically in the form of the van Deemter equation, for each of the common chromatographic techniques, starting with the most straightforward.

A. Gas Chromatography with Open Tubular Columns

By a combination of Eqs. (49), (59), and (69) describing axial diffusion, liquid phase mass transfer, and mobile phase dispersion, respectively, we end up with the so called Golay equation describing the plate height for a gas chromatographic open tubular column.

$$h = h_{ax} + h_{fp} + h_S = \left[\frac{2D_{M,o}}{u_o} + \frac{(1 + 6k + 11k^2)}{24(1 + k)^2} \frac{R^2}{D_{M,o}} u_o \right] j_1 + \frac{2}{3} \frac{k}{(1 + k)^2} \frac{d_f^2}{D_S} u_o j \quad (90)$$

It is necessary to take into account the effects of pressure on the flow velocity [87]. This is done by noting that the mean velocity u is given by

$$u = u_o j \quad (91)$$

where u_o is velocity at the column outlet and j is the James-Martin pressure correction factor [88].

$$j = \frac{3(p^2 - 1)}{2(p^3 - 1)} \quad (92)$$

where p is the ratio of inlet pressure to outlet pressure ($p = P_i/P_o$). This accounts for the pressure correction for the last term in Eq. (90).

In the first terms of Eq. (90), the quotient u/D_M appears. Both u and D_M are inversely dependent on pressure (i.e., $D_M = D_{M,o}P_o/P$, where $D_{M,o}$ is D_M at the outlet pressure). Thus, u/D_M is independent of pressure and equal to $u_o/D_{M,o}$.

The factor j_1 was set equal to unity by Bohemen and Purnell [89] and given by Giddings [87] as

$$j_1 = \frac{9(p^4 - 1)(p^2 - 1)}{8(p^3 - 1)^2} \quad (93)$$

where j_1 varies only between 1 and 9/8, and may often be neglected. It is caused by the effects of decompression, by which the width of a sample zone increases when the carrier gas successively expands while traveling through the column.

For porous layer open tubular columns, Eq. (90) was modified by Golay [90]. This adds substantially to the complexity of the equation by introducing two additional terms.

As is seen in Eq. (90), the value of h depends on several independent parameters, each of which influences the result in different and sometimes complex ways. In order to get a clear picture and an understanding of the process, it is necessary to calculate h in a large number of typical cases. This was done by Cramers and co-workers

[91] and by Ingraham and co-workers [92]. The latter authors also presented a large number of insturctive computer plots of Eq. (90) in different cases. Only a few examples of the results obtained in these studies will be reviewed here; for a closer study the original works are recommended.

In Fig. 5 [92], the influence of column radius R and column length (which affects P_i) is shown for two different carrier gases (affecting D_M). Both d_f and k are kept constant. We see, for example, that the minimum value of h increases when R is increased, that the flow rate u_{opt}, which corresponds to that minimum, is shifted to higher values when the column length is decreased, and that u_{opt} is generally larger when hydrogen rather than helium is used as a carrier gas (i.e., when D_M is increased and P_i decreased).

Cramers et al. [91] largely compared several simplified version of Eq. (90). Their main conclusion is that the stationary phase term [the last term in Eq. (90)] may be neglected only in certain cases, especially if the thickness of the stationary-phase film is not uniform.
not uniform.

B. Liquid Chromatography with Open Tubular Columns

In principle, the Golay equation [Eq. (90)] is also applicable to liquid chromatography. Written in reduced form [see Eqs. (14) and (15)] and without pressure corrections, it reads

$$h_r = \frac{2}{\nu} + \frac{(1 + 6k + 11k^2)}{96(1 + k)^2}\nu + \frac{2}{3}\frac{k}{(1 + k)^2}\left(\frac{d_f}{d_c}\right)\frac{D_M}{D_S}\nu \tag{94}$$

where d_c is the diameter of the column. Comparing liquid and gas chromatography, a main difference is that diffusion coefficients in liquids are roughly 10^4 times smaller than in gases. Therefore, the last term in Eq. (94) may be neglected in liquid chromatography. This leaves us with two simple terms, neither of them containing any column parameters. The reduced plate height obtainable with a liquid chromatographic open tubular column will be comparable to that obtainable in gas chromatography with the same reduced velocity. But from the definition of ν [Eq. (15)], it can be seen that it is necessary to decrease the product ud_c 10^4 times in order to obtain comparable values of ν. It turns out [93] that in order to gain an advantage with respect to the speed of separation with open tubular columns compared with packed columns, d_c should be around 10 μm. This means that it is necessary to work in the nanoliter or even picoliter range, which poses hitherto unsolved practical equipment problems. However, the possibility of very efficient separations is so great [94] that efforts devoted to the development of this technique are justified.

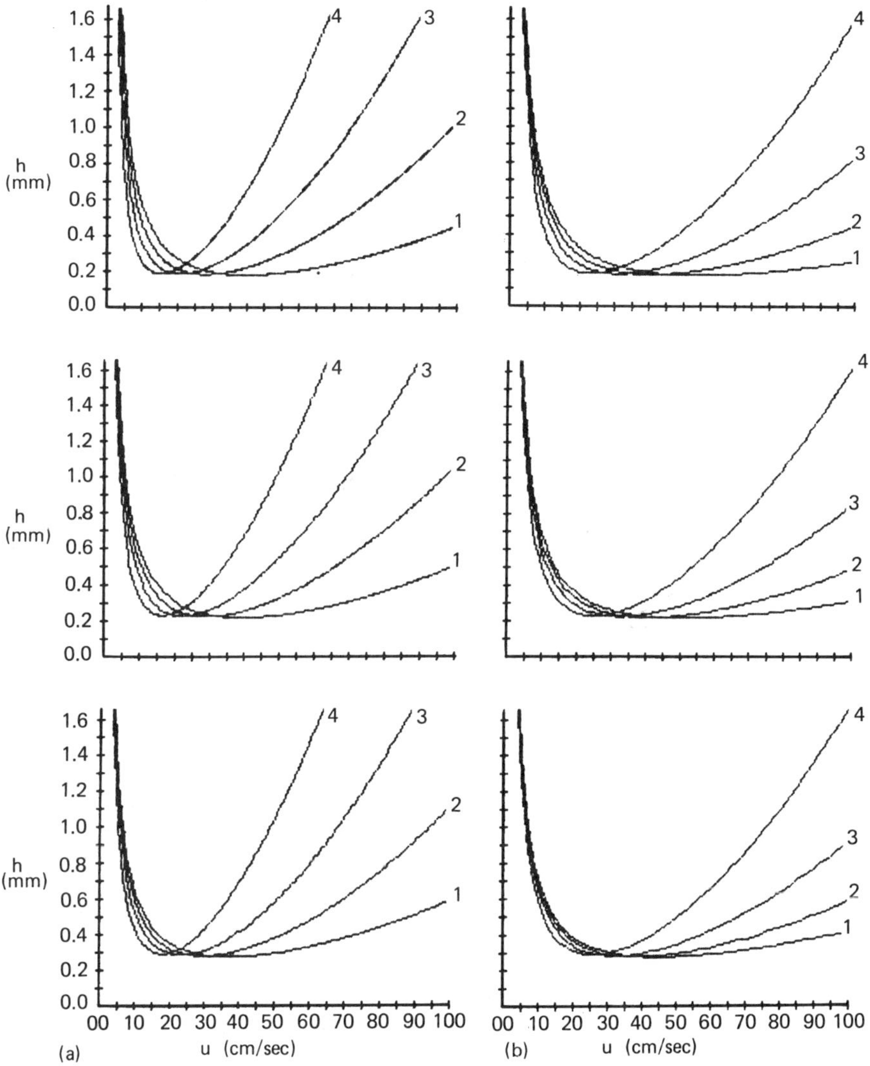

Figure 5. Interrelationships of column diameter, column length, and carrier gas choice. For all graphs, $d_f = 0.25$ μm, k = 5.0. Column lengths: (1) 10 m; (2) 25 m; (3) 50 m; (4) 100 m. (a) Helium; (b) hydrogen. Column diameters: upper graphs, 0.20 mm; center graphs, 0.25 mm; lower graphs, 0.32 mm. (After Ingraham, Shoemaker, and Jennings [92], with permission.)

C. Gas Chromatography with Packed Columns

Compared with open tubular columns, the involved geometry of packed columns adds to the complexity of the detailed dispersion equations. The van Deemter equation was originally developed for packed gas chromatographic columns, and during the 1950s and 1960s a vivid debate on its applicability and detailed formulation was conducted. The evolution of open tubular columns and modern liquid chromatography has during the past decade largely superseded that discussion, and the evaluation of the dispersion theory of packed column gas chromatography has ebbed. Currently, the van Deemter equation in the following general form seems to be generally accepted:

$$h = \left(A + \frac{B}{u_o} + C_G u_o\right) j_i + C_L u_o j \qquad (95)$$

At the flow velocities normally used, the coupling concept (see pages 65–67) appears to be unnecessary to account for experimental results [95]. At high flow rates, above those commonly used in gas chromatography, a lower slope of the curve of h versus u_o is sometimes observed. This behavior can be explained by the coupling concept, by introducing the expression given in Eq. (73) [or, alternatively, those in Eqs. (79) or (80)] instead of the first, constant term in Eq. (95). Also, the first term could be exchanged for $Au_o^{1/3}$. Thereby, Eq. (95) is transformed into the so-called Knox equation. This type of equation was found [96] to closely fit their experimental data. It has mainly been applied to liquid chromatography, even if it can be used in gas chromatography. A more detailed discussion of the Knox equation will be postponed until Sec. V.D.

Choosing from the expressions given in the preceding sections, the following detailed dispersion equation may be written for the case of packed column gas chromatography with a liquid stationary phase [Eqs. (49), (59), (72), and (86)].

$$h = h_e + h_{ax} + h_p + h_S$$

$$= \left[2\lambda d_p + \frac{2\gamma_M D_{M,o}}{u_o} + \frac{\theta(k_0 + k + k_0 k)^2 d_p^2 u_o}{30k_0(1 + k_0)(1 + k)^2 D_{M,o}}\right] j_1 + \frac{2}{3}\frac{k}{(1 + k)^2}\frac{d_f}{D_S} u_o \qquad (96)$$

The factors j and j_1, correcting for gas compressibility, were described in Eqs. (91) through (93). Because D_M is typically about 10^4 times greater than D_S, longitudinal diffusion in the stationary phase is neglected. Thus, Eq. (86) is preferred over Eq. (89), and only one term in Eq. (49) is retained. The term h_P, describing slow diffusion in pores, was written in Eq. (86) for liquid chromatography, in which the flow velocity is properly related to the interstitial column volume. Here it is written with the flow velocity related to the entire volume of the mobile phase in the column. The last term in Eq. (96), h_S, refers to finite rate of mass transfer in a liquid stationary phase. In the case of gas-solid adsorption chromatography, this term is not applicable and should be replaced with a similar term describing the kinetics of the adsorption equilibrium [see Eq. (59)]:

$$h_{neq} = 2 \frac{k}{(1+k)^2} \frac{u_o}{k_d} j \tag{97}$$

In principle, this term is also applicable to gas-liquid chromatography, in which it describes a finite rate of mass transfer across the gas/liquid interface. This effect is, however, negligible, except in very special cases [67] (see page 59).

D. Liquid Chromatography with Packed Columns

Liquid chromatography with packed columns has evolved rapidly during the past decade, to a great extent owing to recent theoretical advances. There has been considerable discussion in the literature of the details of the dispersion equation, and several alternative formulations have been presented. The matter seems not yet finally settled. In a simplified form, the main candidate equations for the dispersion equation are (in chronological order)

$$h = A + \frac{B}{u} + Cu \tag{98}$$

$$h = \frac{A}{1 + E/u} + \frac{B}{u} + Cu \tag{99}$$

$$h = \frac{A}{1 + E/u^{1/2}} + \frac{B}{u} + Cu + Du^{1/2} \tag{100}$$

$$h = Au^{1/3} + \frac{B}{u} + Cu \tag{101}$$

$$h = \frac{A}{1 + E/u^{1/3}} + \frac{B}{u} + Cu + Du^{2/3} \tag{102}$$

Of these, Eq. (98) is the original van Deemter equation [17]. Equation (99) is Giddings' modification [74,75], introducing the "coupling" concept (see page 65). Equation (100) is the equation of Huber and Hulsman [76,77], introducing a term for film resistance (see pages 67–68). Equation (101) is the Knox equation (96). Finally, Eq. (102) is due to Horváth and Lin [78] [see Eqs. (79) and (82)].

The main debate is whether a coupling term $E/u^{n'}$ exists and , in such a case, the value of the exponent. Also, the existence of a film-resistance term $Du^{n''}$ is questioned. Finally, the detailed expression of the separate terms, especially C, is discussed.

The question of choosing among Eqs. (98) through (102) can be solved in a pragmatic way. Obtain a representative set of data describing h as a function of u, fit the equations to these data, and decide which equation provides the best fit. This was attempted by Katz et al. [85]. There are several problems related to this approach, most of them realized by these authors. These equations are in practice very similar, and all of them will fit reasonably well to most experimental data. The additional criterion that all constants must be nonnegative partly facilitates the decision. In any case, very accurate data are needed to obtain clear distinctions. In their study, Katz and co-workers based their calculations on peak-width measurements, assuming a gaussian peak shape. Although this probably is a very reasonable assumption in this case, calculation of the second statistical moment will generally lead to more accurate results (see page 32).

On the basis of data obtained with porous silica columns, Katz and co-workers preferred the original van Deemter equation (98). The Knox equation (101) also fit the data in a relevant way, as expected, but the parameters obtained were found not be consistent with their detailed formulation as adapted by Katz et al. In several cases, the other equations provided negative or zero values for the constants D and E.

Do these results mean that nearly 30 years of development of the chromatographic theory has been useless? Certainly not. Even if it strongly emphasizes the importance of the work of van Deemter and his co-workers, it is necessary to realize that the work cited here refers to one type of column packing only. Packings with other pore characteristics, especially pellicular, might necessitate the inclusion of a coupling term and perhaps also a film-resistance term. Only uncoated silica packing was studied, although the majority of applications use bonded-phase columns. Thus the effects of the bonded layer, which might be important (see below), are neglected in these comparisons. Furthermore, the Knox equation has been shown by other authors to give consistent results (see below).

For the ensuing discussion, we write the detailed dispersion equation in the following form, comprising both the van Deemter and the Knox formulations of the first term:

$$h = \begin{Bmatrix} h_e \\ h_k \end{Bmatrix} + h_{ax} + h_p + h_{neq} = \begin{Bmatrix} 2\lambda d_p \\ A'u_e^{1/3} \end{Bmatrix} + \frac{2(\gamma_M D_M + k\gamma_S D_S)}{u_e}$$

$$+ \frac{(k_0 + k + k_0 k)^2}{30(1 + k_0)^2(1 + k)^2} \frac{d_p^2 u_e}{[k_0 \gamma_{SM} D_M + k(1 + k_0)\gamma_S D_S}$$

$$+ \frac{2ku_e}{(1 + k)^2(1 + k_0)k_d} \qquad (103)$$

The flow velocity used, u_e, is the interstitial flow velocity, i.e., the velocity of the liquid flowing between the grains of the packing (see Sec. II.B). Equation (103) could also be written in reduced form [see Eqs. (14) and (15)]:

$$h_r = \begin{Bmatrix} 2\lambda \\ A\nu_e^{1/3} \end{Bmatrix} + \frac{2(\gamma_M + k\gamma_S D_S/D_M}{\nu_e} + \frac{(k_0 + k + k_0 k)^2}{30(1 + k_0)^2(1 + k)^2}$$

$$\times \frac{\nu_e}{[k_0 \gamma_{SM} + k(1 + k)\gamma_S D_S/D_M]} + \frac{2kD_M \nu_e}{(1 + k)^2(1 + k_0)k_d d_p^2} \qquad (104)$$

where ν_e is the reduced velocity corresponding to u_e.

The second and third terms, accounting for axial diffusion and for mass transfer resistance in the pores, also include—following Knox and Scott [86]—a contribution from diffusion in the stationary phase. This would be applicable to columns with a bonded stationary phase, as mentioned on page 70. For liquid-solid adsorption columns (i.e., uncoated silica columns) this contribution is probably negligible. Applying Eq. (104) to data obtained with a bonded octadecyl silica packing and water-methanol as eluent, Knox and Scott [86] found consistent values of $\gamma_S D_S/D_M$ around 0.5, indicating that diffusion in the bonded layer is significant. This is also supported by recent spectroscopic data by Bogar and co-workers [97].

The work of Stout et al. [98], presented simultaneously with that of Knox and Scott, complements and supports these findings. Similar experiments with uncoated silica columns yield, as expected, comparatively low values for $\gamma_s D_s / D_M$. Both these studies assumed the first term to be of the Knox type.

The last term in Eq. (103) refers to the kinetics of the sorption equilibrium and should be applicable to both uncoated and coated columns. This term was not included in the studies referred to above [86,98], but it was claimed by Horváth and Lin [78] to be very important, especially with small particle size.

To get an overall impression of the relative magnitude of the terms in Eq. (104) (which is by no means claimed to be conclusive), some simple numerical calculations were performed, starting from the following values: $\lambda = 1$, $\gamma_{sM} = \gamma_M = 0.65$, $\gamma_s D_s / D_M = 0.5$ [86], $k_0 = 0.66$ [85], $D_M = 10^{-5}$ cm^2/sec, $k_d = 2.5 \times 10^2$ sec^{-1} [78], and $d_p = 10$ μm. This applies to a modern, bonded-phase "HPLC" column.

For k = 1, the results, as functions of ν_e, are given in row 1 of Table 2. By differentiation, the position of the minimum value of h is found to be $\nu_e = 8.5$. The values of the respective terms for this ν_e value and for the double value are found in rows 2 and 3. Increasing k to 5, we find that h_{min} is now obtained at $\nu_e = 22$ and the relevant terms are listed in rows 4-6.

We see that the A term, which depends mainly on the quality of the packing, is highly significant, even dominating. It is the term about which there is still the most debate. If we assume a Knox-type A term with A = 1 (which is a favorable value), the 2 in the first column would be succesively replaced with $\nu_e^{1/3}$, 2.0, 2.6, $\nu_e^{1/3}$, 2,8, 3.5, further increasing the dominance of this term, but otherwise not causing much change in the general picture.

In this example, the influence of stationary phase diffusion (included in $h_{r,ax}$ and $h_{r,p}$) is significant. If we assume an uncoated column, $h_{r,ax}$ decreases while $h_{r,p}$ increases. In rows 7-12, the same examples are recalculated with $\gamma_s D_s / D_M = 0$. The reduced velocities giving a minimum h are now lower and decrease with k. The kinetic term $h_{r,neq}$ is strongly dependent on particle diameter; with 5-μm particles it will be four times larger than stated in the table.

The film-resistance term $h_{r,f}$ [Eq. (82)], usually assumed to be negligible, was also calculated and found to be of about the same magnitude as $h_{r,neq}$. Therefore it may be erroneous to neglect it. On the other hand, in the study in which it was derived [78], its significance was not experimentally proved. The method used for evaluation of the different terms was based on a calculation of $h_{r,f} + h_{r,p}$ from measured and assumed quantities.

Generally, from these and several other similar calculations and experimental studies [19,76-78,81-82,85-86,95-99], the following conclusions can be drawn about packed column liquid chromatography:

Table 2. Calculated Values of the Terms in the van Deemter Equation for Packed Column Liquid Chromatography [Eq. (104)][a]

Row	$h_{r,e}$	$h_{r,ax}$	$h_{r,p}$	$h_{r,neq}$	h_r	Comment
1	2	$2.3/\nu_e$	$0.020\nu_e$	$0.012\nu_e$		k = 1, bonded phase
2	2	0.27	0.17	0.10	2.54	$\nu_e = 8.5$ (at h_{min})
3	2	0.14	0.34	0.20	2.68	$\nu_e = 17$
4	2	$6.3/\nu_e$	$0.006\nu_e$	$0.007\nu_e$		k = 5, bonded phase
5	2	0.29	0.13	0.15	2.57	$\nu_e = 22$ (at h_{min})
6	2	0.15	0.26	0.30	2.71	$\nu_e = 44$
7	2	$1.3/\nu_e$	$0.038\nu_e$	$0.012\nu_e$		k = 1, uncoated
8	2	0.25	0.19	0.06	2.50	$\nu_e = 5.1$ (at h_{min})
9	2	0.13	0.39	0.12	2.64	$\nu_e = 10.2$
10	2	$1.3/\nu_e$	$0.063\nu_e$	$0.007\nu_e$		k = 5, uncoated
11	2	0.30	0.27	0.029	2.60	$\nu_e = 4.3$ (at h_{min})
12	2	0.15	0.54	0.057	2.75	$\nu_e = 8.6$

[a]Reduced plate height h_r as a function of reduced interstitial velocity ν_e. Parameters as listed in text.

- None of the A, B, or C types of terms should be neglected.
- The influence of diffusion in a bonded-phase layer might significantly alter the relation between the B and C terms, thereby shifting the position of the minimum plate height.
- The kinetic term (h_{neq}) may be significant and will increase in importance as the particle diameter is decreased.
- Despite great theoretical and experimental efforts, many details of the dispersion equation for liquid chromatography in packed columns are still unknown: there is no strict evidence for (or against) the Knox type of the first term, and there is uncertainty about the formulation and importance of the other terms, especially the film-resistance and kinetic terms.

E. Steric Exclusion Chromatography

Steric exclusion chromatography is a special case of liquid chromatography in a packed column with the following characteristics:

There is (ideally) no interaction between the stationary phase material and the analyte. Thus, $k = 0$.

The retention observed is caused by the ability of analyte molecules to penetrate the pores of the packing to various extents. Thus, the parameter k_0 defined in Eq. (18) governs the retention according to the equation

$$V_R = V_e(1 + k_0) \tag{105}$$

where V_e is the interstitial void space in the column. The term k_0 is analogous with k in "common" liquid chromatography.

The analytes are usually macromolecules. Thus, D_M is small.

The samples chromatographed are often polydisperse, introducing yet another type of band broadening, h_{poly}.

The following equation is based on Eq. (103), which is modified and extended as a result of the above characteristics:

$$h = h_e + h_{ax} + h_p + h_{poly} = 2\lambda d_p + \frac{\gamma_M D_M}{u_e} + \frac{k_0}{30(1 + k_0)^2} \frac{d_p^2 u_e}{\gamma_M D_M} + \frac{L \ln P}{D_2^2 V_R^2} \tag{106}$$

The first and second terms describe eddy diffusion and axial dispersion, respectively, according to the original van Deemter treatment [see Eqs. (49) and (72)]. The third term represents diffusional mass transfer in the pores [99]. It can be readily derived from Eq. (86) with $k = 0$. The last term describes the influence of polydispersity. It is due to Dawkins and Yeadon [100]. An essentially equivalent expression was given by Knox and McLennan [101]. In this term, P is the polydispersity and D_2 is the slope of the calibration curve (for a definition of these, see the cited references), L is the column length, and V_R is the retention volume.

For totally excluded molecules all terms but the first vanish, as $k_0 = 0$, $D_2 = \infty$, and D_M is small. For totally penetrating molecules, D_M is large, leading to a comparatively small third term, but the second term is still small provided the flow rate is not unusually low.

The last term vanishes as $D_2 = \infty$. Thus, only for partially penetrating molecules are the third and fourth terms significant and introduce a marked dependence of h on u_e. These general trends agree with experimental observations [100] (see Fig. 6).

VI. NONLINEAR CHROMATOGRAPHY

A. Nonlinear Isotherms

The concept of nonlinear chromatography takes into account a nonlinear sorption isotherm. Let us recall that the isotherm is an expression of the concentration of solute in the stationary phase (on the sorbent, in the case of adsorption chromatography) as a function of the concentration of solute in the mobile phase:

$$C_S = f(C_M) \tag{107}$$

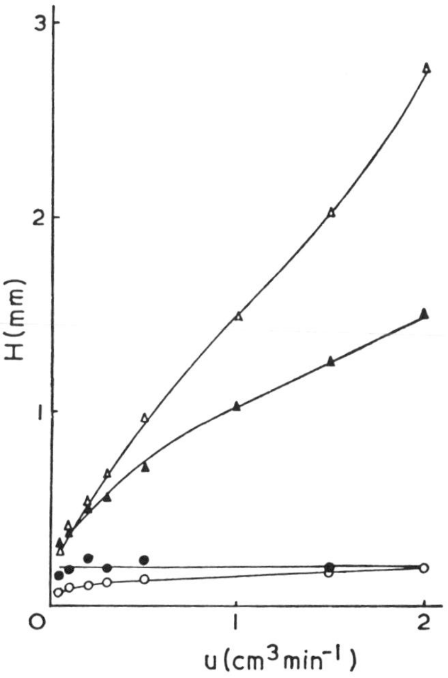

Figure 6. Dependence of plate height on flow rate for H4 silica: ●, PS-1987000; △, PS-35000; ▲, PS-9800; ○, toluene (numbers signify molecular weight of polystyrene standards). (After Dawkins and Yeadon [100], with permission.)

The isotherm, when plotted, can be convex toward the C_S axis (i.e., "Langmuir-like"), concave, or a combination of these. Examples of common types of isotherms are shown in Fig. 7. The curvature typically results from interactions in one or both of the phase. When the retention mechanism is adsorption, interactions between molecules on the adsorbing surface usually cause a convex isotherm (see Chaps. 4 and 5). Partition isoterms are, to a good approximation, linear up to relatively high concentrations (see Chap. 3). The direction of curvature then depends on the type of interactions and on the phase in which the interactions occur. In GLC, isotherm curvature is for this reason usually in the concave direction. A combination of several retention mechanisms (see Chap. 5) can create isotherms that have other forms, either s-shaped or more complicated.

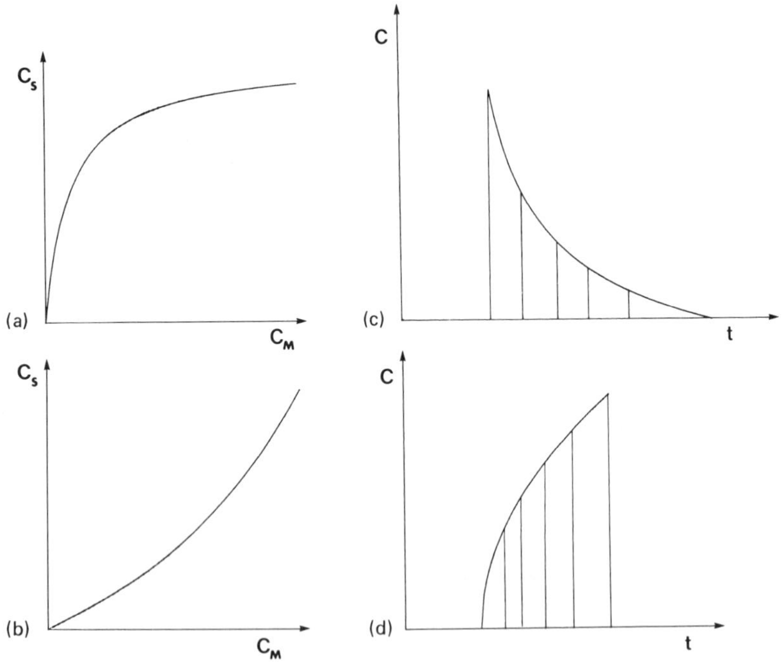

Figure 7. (a) Typical adsorption isotherm (Langmuir type); (b) Partition isotherm ("anti-Langmuir" type); (c) and (d) corresponding peak shapes according to Eq. (110).

B. Ideal Nonlinear Chromatography

Assumption of Ideality

The introduction of nonlinearity causes considerable mathematical difficulty making a general treatment of nonideal nonlinear chromatography impossible. Thus we must make simplifications and start by neglecting all dispersion processes, such as eddy diffusion, axial diffusion, and nonequilibrium. In the case of linear chromatography these drastic assumptions of ideality result in an infinitely narrow peak with a retention time of $L/u(1 + k)$ and all other moments equal to zero. This can easily be seen from Eq. (57), with $D = f_1 = f_2 = 0$.

Model

The mass-balance equation [Eq. (31)], under assumptions of ideality, reads

$$u \frac{\partial C_M}{\partial z} + \frac{\partial C_M}{\partial t} + \frac{\partial C_S}{\partial t} = 0 \tag{108}$$

which is easily transformed to

$$u \frac{\partial C_M}{\partial z} + \left(1 + \frac{\partial C_S}{\partial C_M}\right) \frac{\partial C_M}{\partial t} = 0 \tag{109}$$

This is the more than 40-year-old equation first presented by de Vault [102].

Solution

The solution to the mass-balance equation (108)(assuming a Dirac delta function injection) is most conveniently written

$$t = \frac{L}{u}\left(1 + \frac{\partial C_S}{\partial C_M}\right) \tag{110}$$

This peak-shape equation has a different form than those obtained for linear chromatography. It expresses the time required for a portion of the sample with a specified concentration to reach the column outlet. This gives rise to a peak with one vertical flank and one curved (often called a diffuse) flank (see Fig. 7). If we compare peaks with different areas (i.e., containing different amounts of sample), we find that the diffuse flanks of these peaks coincide, the area of the peak being determined by the position of the vertical flank. This is not a bad approximation of the peak shape actually observed in chromatographic systems with a strongly curved isotherm. The diffuse flank, having a simple relation to the isotherm, can be used for experimental mapping of adsorption isotherms [6,103].

Consequences of Nonlinearity

It is neither convenient nor necessary for us to derive an elution curve in the form c(t) for a general isotherm function. Such elution curves have been presented in the literature for special types of isoterms [102] but add nothing to the form of Eq. (110). We will also not calculate transfer functions or statistical moment expressions, because these concepts are not applicable in this context. The introduction of a nonlinear isotherm also makes the system nonlinear in a mathematical and system-theoretical sense, which is not always recognized but must be strictly stressed.

This means that the convolution theorem [Eq. (25)] is not valid. Even if we could derive transfer functions and statistical moments for nonlinear peaks, the combination of different injection profiles and instrumental contributions to the peak transfer function, as was described in Sec. II, is not permitted and gives erroneous results.

Retention in Nonlinear Chromatography

From Eq. (110) it follows that the retention time for a zone of a specified concentration is governed by the derivative $\partial C_S/\partial C_M$ of the isotherm at that concentration. On the other hand, the velocity of the individual molecules is still given by the ratio C_S/C_M (or $C_S^V/C_M^V \times V_S/V_M$)—the number of molecules in the stationary phase divided by the number in the mobile phase [see Eqs. (30), (39), and (47)]. In linear chromatography, the derivative and the ratio are equal, which is not true in the nonlinear case. This apparently leads to a paradox: How can the velocity of the peak maximum, for example, be different from that of the individual molecules?

This was discussed by Helfferich [104] and reviewed by Conder and Young [6]. The solution of the paradox can be summarized as follows. The molecules in the peak maximum do indeed travel at the velocity given by the value of the ratio at the actual concentration, but as the peak height continuously decreases, this velocity is not constant throughout the whole run. The retention time of the maximum is given by the mean velocity of the molecules, which cannot be equal to their velocity at the outlet concentration (i.e., the concentration observed).

C. Nonideal Nonlinear Chromatography

This is the most complicated case of chromatography, and a purely mathematical treatment of it, with a general isotherm equation, has never been and probably will never be presented.

First, let us again rule out the tempting possibility of convoluting a peak profile resulting from the ideal nonlinear case (page 83) with some (i.e., gaussian) diffusion operator. This is not allowed, since the processes of diffusion and nonequilibrium are not independent of the peak-shaping process, owing to nonlinearity. The system is nonlinear in the mathematical sense, thereby prohibiting us from taking this approach.

Simplified Isotherm Model

The most useful results are obtained when an isotherm of the following kind is assumed:

$$C_S = k_1 C_M - k_2 C_M^2 \tag{111}$$

This expression is a virial expansion of the isotherm function (see Chap. 4), and it describes small variations from linearity. Depending on whether k_2 is positive or negative, the isotherm is concave or convex toward the C_S axis; if $k_2 = 0$, the isotherm is linear.

For the case of complete equilibrium and axial diffusion (analogous to that treated in Sec. IV.B), an explicit solution to Eq. (31), considering Eq. (111) was given by Houghton [105]. It is necessary to assume that the deviation from linearity is sufficiently small [$k_2 C_M << (1 + k_1)/2$]. The solution is in the form of a column profile $C_H(x, t)$, and the elution curve is assumed (as is often done) to be equal to $C_H(L, t)$. (See pages 57–52 for a discussion of this assumption.)

The equation given by Houghton is realtively complex and is not reproduced here. It is readily found in Refs. 105 or 106. The injection profile assumed by Houghton is not a Dirac delta function but a rectangular plug. As was stated on page 43, this is a realistic assumption and its inclusion into the derivation of the column profile is perfectly relevant, since the combination of an impulse-response function with an injection profile using the convolution theorem is not allowed in this nonlinear case. Houghton's equation correctly reduces to Eq. (39) when $k_2 = 0$ and the width of the injection plug approaches zero [107].

Sorption Effect

The presence of sample molecules in the mobile phase increases its local velocity in the sample zone, thereby reducing the retention time. This effect, known as the "sorption effect," is significant in those cases (rare in anlytic applications of chromatography) in which the concentration of sample in the mobile phase is relatively high. The extent to which the retention volume is reduced is proportional to the molar fraction of sample in the mobile phase. The sorption effect is thus a cause of nonlinearity other than isotherm curvature. For a more detailed discussion, see Refs. 6 and 107.

It was shown by several authors [108-110] that the sorption effect could be included into Houghton's treatment. Following de Clerk and Buys]110], the nonlinearity parameter k_2 (Eq. (46)] should be modified in the following way:

$$k_2^* = k_2 - \frac{k_1 RT}{2P} \tag{112}$$

where P is the total column pressure. The corrected nonlinearity parameter k_2^* can then be used in e.g. Houghton's equation. Qualitatively, we see that the effect of a Langmuir-like isotherm (i.e., $k_2 < 0$) is amplified by the sorption effect. This is qualitatively correct (Ref. 6, p. 46).

Recently, Jaulmes and co-workers [107] further extended Houghton's treatment, including the sorption effect and development of a peak profile equation that could be successfully fitted to experimental peaks [111].

General Systems

The problems inherent in a nonlinear isotherm of a general type, as well as dispersion from various causes, defy conventional mathematical treatment. It is therefore necessary to resort to numerical methods for the solution of the differential equations. The solutions will appear in the form of graphs, providing visual comprehension of the influence of the various parameters. Funk and Houghton [112,113] made such calculations assuming an isotherm as in Eq. (111) but without the condition that the deviation from linearity be small. They introduced the nonequilibrium parameters in a way similar to that discussed in Sec. IV.C.

Smit and co-workers [32] presented numerical solutions to a system with a Langmuir isotherm:

$$C_S = \frac{AC_M}{1 + BC_M} \tag{113}$$

assuming axial diffusion and instantaneous equilibrium. Some of their curves are shown in Fig. 8. They made a careful analysis [31] of the errors and convergence situation for the numerical solutions. However, their treatment does not include nonequilibrium parameters.

A complete numerical treatment for general isotherms, taking into account all dispersion processes, has not been undertaken. Probably,

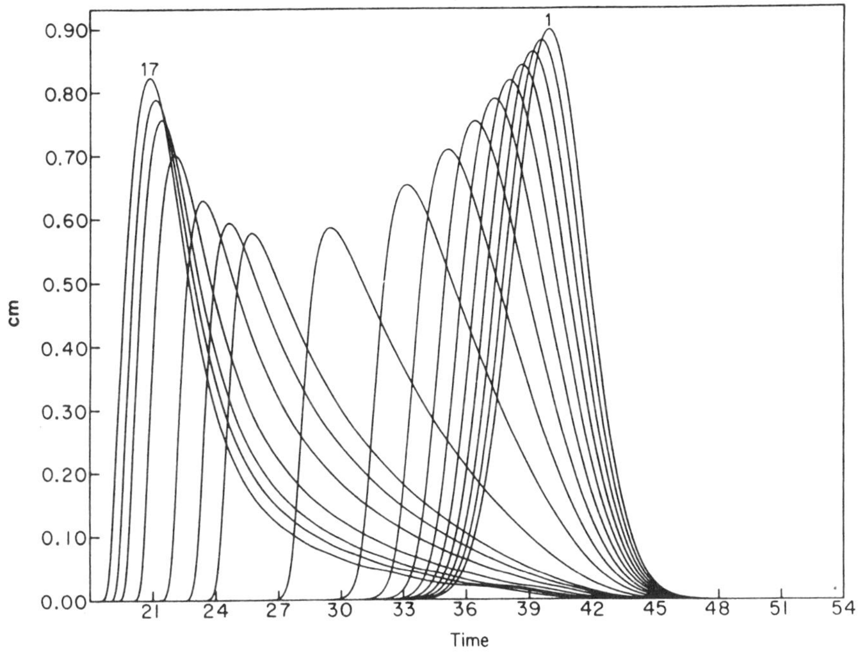

Figure 8. Family of elution profiles, calculated by the simulation program SAM, showing the influence of the isotherm curvature on the elution profiles; D = 0.5 cm^2/sec; u = 5 cm/sec; L = 100 cm; isotherm of the Langmuir type with A = 1 and different B values (for peak 1, B = 0, and for peak 17, B = 9.9). (After Smith and coworkers [32], with permission.)

the simplified description of dispersion as a "lumped" diffusion coefficient assumed by Smit et al [31,32] is sufficient for applications to practical problems.

The comparison of calculated and experimentally obtained peak shapes, in order to extract nonlinearity and dispersion parameters, is a very difficult task, demanding very high accuracy in both theoretical modeling and the necessary experimental work.

For example, consider the asymmetry of an observed peak. It can be caused by a variety of reasons, such as nonlinearity, axial diffusion, slow mass transfer, and the sorption effect. To relate an observed asymmetry effect to a particular combination of these processes, and to extract relevant information from this, is indeed a delicate task. Such studies must be done in simplified systems, in which proper experimental design minimizes the influence of all but one or two causes of asymmetry, permitting the application of more simple models than the general nonideal, nonlinear model.

VII. EMPIRICAL PEAK-SHAPE EQUATIONS

The difficulty in obtaining, by mathematical means, an explicit equation describing the shape of a chromatographic peak has motivated the construction of equations that do not have a rigorous physicochemical foundation but acceptably fit experimental peaks. Such equations find application in the area of computerized handling of chromatographic data. The fitting of an empirical equation to a peak facilitates the calculation of its area and other parameters. It also permits enhancement of resolution [34] and noise reduction, storage of the peak (e.g., in a computer memory) in an effective way, and other operations.

We have already (on page 38) mentioned the Edgeworth-Cramer series. With this series, equations for the peak can be derived, the parameters of which are the statistical moments. These equations can therefore to a great extent be connected with theory and are probably the best general approach to the problem of fitting experimental peaks. Unfortunately, the equations [Eqs. (22) and (23)] are relatively complicated and the fitting procedure is not trivial. Other equations, considerably more simple, have been developed by various authors, and a brief review of these follows.

A. Gaussian-like Equations

The Gaussian Model

The most simple approach to curve fitting is to use the gaussian function:

$$c_G(t) = \frac{a}{\sigma\sqrt{2\pi}} e^{[-(t-t_m)^2/2\sigma^2]} \tag{114}$$

This equation has three parameters, the area a, the width (standard deviation) σ, and the center position t_m. It has been used [114,115] mainly for the resolution of overlapped peaks. As is well known, the moment vector [see Eq. (20)] is given by

$$M_G = \begin{vmatrix} t_m \\ \sigma^2 \\ 0 \\ 0 \end{vmatrix} \tag{115}$$

The inherent asymmetry of any chromatographic peak makes the simple and symmetrical gaussian model inadequate for more demanding applications, and several modified equations have been presented in the literature.

The Bigaussian Model

By dividing the peak with a vertical line through its maximum and assuming "half-gaussian" shape for each of the parts of the peak, we obtain the bigaussian peak model [116,117] (see Fig. 9). It contains four parameters, h, σ_1, σ_2, and t_m, the meanings of which are analogous to those of the gaussian case (h is the height).

$$c_B(t) = \begin{cases} he^{[-(t-t_m)^2/2\sigma_1^2]} & 0 < t < t_m \\ he^{[-(t-t_m)^2/2\sigma_2^2]} & t_m \leqslant t < \infty \end{cases} \tag{116}$$

Expressions for the first three statistical moments were given by Buys and de Clerk [117]:

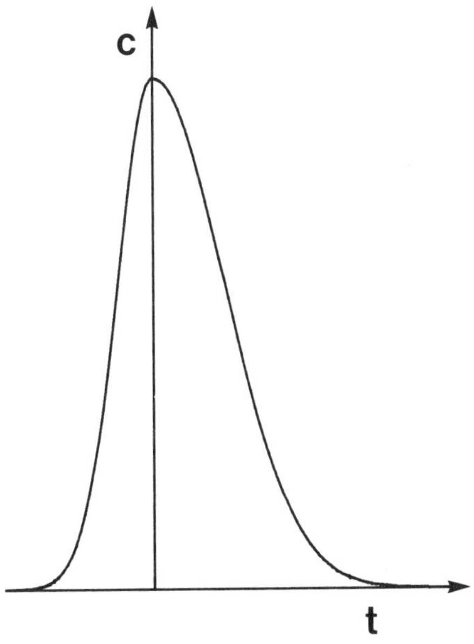

Figure 9. Bigaussian peak.

$$M_B = \begin{vmatrix} t_m + 0.798\Delta\sigma \\ \bar{\sigma}^2 + 0.113(\Delta\sigma)^2 \\ 0.798\bar{\sigma}^2 \, \Delta\sigma + 0.0415(\Delta\sigma)^3 \end{vmatrix} \tag{117}$$

with

$$\Delta\sigma = \sigma_2 - \sigma_1 \quad \text{and} \quad \bar{\sigma} = \frac{\sigma_1 + \sigma_2}{2}$$

The bigaussian distribution has no rigorous theoretical background. In spite of this, it has been successfully used in different applications [114,116].

Exponentially Modified Gaussian Equation

By combining a gaussian equation with an exponential decay function, the exponentially modified gaussian peak model arises. The model is more justified than the bigaussian model, if we assume that column processes create a gaussian peak as the response to an impulse input. With a long column and a linear isotherm, this is a fair approximation. Then extracolumn effects, such as dead volumes, detector characteristics, and, in some cases, imperfect injection profiles introduce an exponential modifier (see page 43) with a certain time constant τ_e. The combination of the gaussian and the exponential contributions to the peak shape is made by convolution [Eq. (25)]. The final result is [118]

$$c_E(t) = \frac{a}{\tau_e \sigma\sqrt{2\pi}} \int_0^\infty e^{\{-[(t-t_m-t')^2/2\sigma^2]-(t'/\tau_e)\}} \, dt' \tag{118}$$

This equation has four independent parameters, the peak area a, the retention time t_m, the standard deviation σ of the gaussian constituent, and the exponential time constant τ_e.

Delley [119,120] showed that it is possible to simplify Eq. (118) in the following way:

$$C_E(t) = \frac{a}{\sigma\sqrt{2\pi}} \frac{SZ(S-T)}{(S-T)} e^{-T^2/2} \tag{119}$$

Here, $S = \sigma/\tau_e$, $T = (t - t_m)/\sigma$, and Z is a function that can be approximated by

$$\frac{Z(x)}{x} = \left(\frac{\pi}{2}\right)^{1/2} e^{x^2/2} - x - \frac{x^3}{3} - \frac{x^5}{3 \cdot 5} - \frac{x^2}{3 \cdot 5 \cdot 7} - \cdots \tag{120}$$

if $x < 4$. For other values of x, tabulations and the definition of Z(x) can be found in the original works.

The statistical moments for the exponentially modified gaussian distribution are easily found [21,118]:

$$M_E = \begin{vmatrix} t_m + \tau_e \\ \sigma^2 + \tau_e^2 \\ 2\tau_e^3 \\ 6\tau_e^4 \end{vmatrix} \tag{121}$$

where M_E is simply the sum of M_G and the moment vector of the exponential decay function (Table 1) according to Eq. (27). The exponentially modified gaussian model appears to be the most often used model for the fitting of chromatographic peaks. This is due both to its good ability to acceptably fit experimental peaks and to its theoretical justifications.

The fitting of this equation to experimental peaks was discussed by many workers [16,115,121,122]. van Rijswick [114] constructed a curve-fitting program that could automatically choose between four peak models: the gaussian, the bigaussian, exponentially modified gaussian, and the Edgeworth-Cramér series.

B. The Chesler-Cram Equation

The peak shape models described in Sec. VI.A, as well as the GCAS and ECS models (pages 37–38), assume a peak shape that is fairly defined. The closeness by which an experimentally obtained, "real" peak can be fitted by the equations depends on how well the peak conforms to the corresponding mathematical model. Consequently, a perfectly general, mathematical model is sometimes desired—an equation that fits every possible type of chromatographic peak without systematic bias. As has been pointed out [114], however, such an equation is not always desired; in curve fitting with unresolved chromatograms, the use of a perfectly general model will provide an unlimited number of peak combinations.

The most successful general equation for peak shape was given by Chesler and Cram [12]. It is a combination of several components: a gaussian and an exponential function are joined by a hyperbolic tangent function. A similar idea was presented by Sternberg [43], who suggested a triangular joining function. Also, the kinetic tailing model [23,123] is related to this approach. The Chesler-Cram equation (CCE) reads

$$c_c(t) = C_1[e^{[-(t-C_4)^2/2C_5]} + (1 - 0.5\{1 - \tanh[C_2(t - C_3)]\}) \times C_6 \exp\{-0.5C_7[|t - C_8| + (t - C_8)]\}] \tag{122}$$

This equation has eight constants, C_1-C_8, which can be partly graphically interpreted as different characteristics of the peak shape. For example, C_1 is the height, C_4 the position of the maximum, C_7 the rate of decrease of the exponential, and so forth. (See Ref. 12 for a complete description.) The CCE, having many degrees of freedom, can be fitted to peaks of very varying types, but there are no theoretical connections between the constants C_1-C_8 and physicochemical theory; the CCE is entirely empirical.

In principle it is possible to derive expressions for the statistical moments of the CCE by applying Eqs. (1) through (4). This was attempted [124], but the result is not tractable, and expressions for the moments in terms of C_1-C_8 cannot be given. By numerically integrating Eq. (122) after the constants have been determined by fitting, the moments can be found. In this way, also, the third and forth moments can be determined with satisfying precision.

A simplified version of the CCE, permitting the derivation of reasonable expressions for the moments has also been given [12]. This equation can also successfully be used for moments calculation. It is not known whether the moments calculated in this way are unbiased or if the assumption of a particular equation introduces systematic errors.

Much effort has been put into comparisons of the CCE parameters with chromatographic dispersion parameters, such as diffusion coefficients and rate constants [125,126]. Reliable correspondences were not found, only more or less significant correlations. Owing to the lack of a theoretical foundation for the model, nothing better can be expected, and thus the use of the CCE for the measurement of physicochemical quantities seems to be limited.

VIII. CONCLUDING EXAMPLES

To illustrate the system-theoretical approach to the chromatographic peak shape theory involving the convolution of various peak components, some numerical examples are given. Most of these calculations involve numerical inversions of Laplace transforms. They were done on a tabletop computer using the techniques published in Refs. 26 and 40. A more detailed description of the calculation procedures can be found in Ref. 54. In all examples, the nonideality is exaggerated in order to make the effects clearly visible.

Figure 10, curve 1, shows a peak generated by using Eq. (41). This is an example of a peak from equilibrium, nonideal, linear chromatography. Now let us assume a rectangular sample introduction function (see page 43). The resulting peak (Fig. 10, curve 2) is seen to be of a very similar shape to that of Curve 1 but displaced with half the width of the rectangular function. The increase in width, skew, and excess is slight, as can also be seen from the moment vector in Table 1. The curve was generated by numerical inversion of the product of Eq. (42) and the Laplace transform of the rectangular plug function (Table 1), corresponding to a convolution of the rectangular plug function with impulse-response function form Eq. (41), according to the convolution theorem [Eq. (26)].

To continue, we assume an exponential decay component, arising from dead volumes or slow vaporization, for example (page 43). The resulting peak (Fig. 10, curve 3), created analogously to that above, is considerably lower and wider than curve 2. A comparison of the moment vectors in Table 1 shows that, if the time constant has the same value for the two functions, the influence of an exponential contribution to the peak shape will be more pronounced than the influence of a rectangular contribution.

Finally, curve 4 shows the combined effects of both a rectangular and an exponential peak shape contribution. This curve was produced

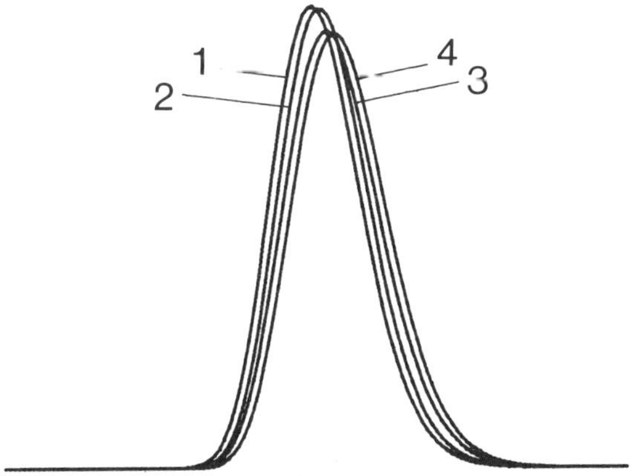

Figure 10. Elution curves in linear nonideal equilibrium chromatography, calculated as described in the text. For all curves, L = 1, u = 1, k = 1, D = 0.01, $f_1 = 0$, $f_2 = 0$. (1) $\tau_r = 0$, $\tau_e = 0$. (2) $\tau_r = 0.1$, $\tau_e = 0$. (3) $\tau_r = 0$, $\tau_e = 0.1$. (4) $\tau_r = 0.1$, $\tau_e = 0.1$.

by numerical inversion of the product of the corresponding three Laplace transforms.

To see the effects of nonequilibrium, Eq. (56) was numerically inverted with various values of f_1 and f_2 inserted. In Fig. 11, some of these curves can be studied. More of them can be found elsewhere [54]. Curve 1 in Fig. 11 is the same as curve 1 in Fig. 10, the equilibrium case. In curve 2, f_2 has a nonzero value, corresponding to slow diffusion in the stationary phase. The curve is wider, lower, and more skewed. It is also slightly displaced to the left, according to the moments expression [Eq. (57)].

Also taking into consideration a finite rate of mass transfer over the phase boundary by letting $f_1 > 0$, we obtain curve 3, which is still wider, lower, and more skewed than curve 2. However, it is not more displaced.

Finally, an exponential contribution (curve 4) and a rectangular plug function (curve 5) are added, having effects similar to those in the equilibrium case.

In these examples, various independent peak shape elements are implicitly or explicitly considered: The impulse-response function of the column is influenced by several parameters modeling axial diffusion, a finite mass transfer rate, and a finite diffusion rate in the stationary phase, respectively. The injection process can be modeled with rectangular or exponential functions (or a combination of them). The effects of dispersion in tubes and in the detector, as well as the electrical characteristics of the readout system, can also be modeled,

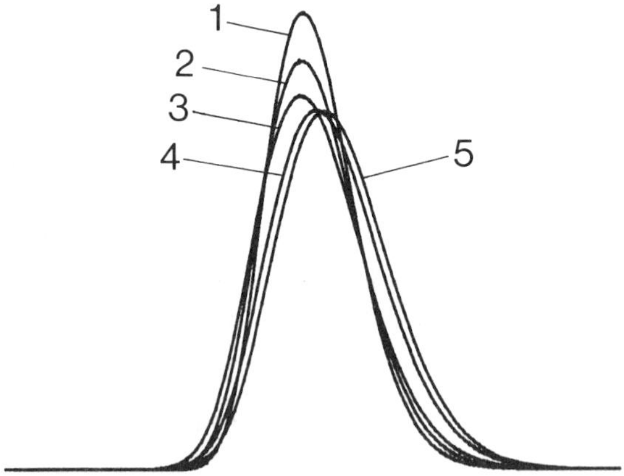

Figure 11. Elution curves in linear nonideal chromatography, calculated as described in the text. For all curves, $L = 1$, $u = 1$, $k = 1$, $D = 0.01$. (1) $f_1 = 0$, $f_2 = 0$, $\tau_r = 0$, $\tau_e = 0$. (2) $f_1 = 0$, $f_2 = 0.1$, $\tau_r = 0$, $\tau_e = 0$. (3) $f_1 = 0.1$, $f_2 = 0.1$, $\tau_r = 0$, $\tau_e = 0$. (4) $f_1 = 0.1$, $f_2 = 0.1$, $\tau_r = 0$, $\tau_e = 0.1$. (5) $f_1 = 0.1$, $f_2 = 0.1$, $\tau_r = 0.1$, $\tau_e = 0.1$.

again using exponential and rectangular contributions. This convolution approach offers the most versatile description of linear nonideal chromatography known.

As was noted above, the convolution approach is not valid for nonlinear systems. The description of these systems is considerably more complicated and difficult. Fortunately, in chromatographic practice, strongly nonlinear systems are rare, and when they do occur, the nonlinearity is subject to considerable elimination.

REFERENCES

1. J. C. Giddings, *Dynamics of Chromatography. Part I*, Marcel Dekker, New York (1965).
2. Cs. Horváth and W. R. Melander, Theory of chromatography, *Chromatography, Part A: Fundamentals and Techniques* (E. Heftmann, ed.), Elsevier, Amsterdam, pp. A27-A135 (1983).
3. J. A. Perry, *Introduction to Analytical Gas Chromatography. History, Principles, and Practice*, Marcel Dekker, New York (1981).
4. J. H. Knox, Kinetic factors influencing column design and operation, *Techniques in Liquid Chromatography* (C. F. Simpson, ed.), Wiley Heyden, Chichester (1982).
5. J. R. Conder, Peak distortion in chromatography, Part 1, Concentration-dependent behavior, *J. HRC&CC*, 5:341 (1982), Part 2, Kinetically controlled factors, *J. HRC&CC*, 5:397 (1982).
6. J. R. Conder and C. L. Young, *Physicochemical Measurement by Gas Chromatography*, John Wiley & Sons, Chichester (1979).
7. G. Guiochon and A. Siouffi, Study of the performance of thin layer chromatography. II: Band broadening and plate height equation, *J. Chromatogr. Sci.*, *16*:470 (1978).
8. O. Grubner, Statistical moments theory of gas-solid chromatography: Diffusion-controlled kinetics, *Advances in Chromatography, Vol. 6* (J. C. Giddings and R. A. Keller, eds.), Marcel Dekker, New York, pp. 173-209 (1968).
9. E. Grushka, M. N. Myers, P. D. Schettler, and J. C. Giddings, Computer characterization of chromatographic peaks by plate height and higher central moments, *Anal. Chem.*, *41*:889 (1969).
10. K. Yamaoka and T. Nakagawa, Statistical moments in linear equilibrium chromatography, *J. Chromatogr.*, *93*:1 (1974), Moment vectors in linear gas chromatography, *J. Chromatogr.*, *100*:1 (1974).
11. S. N. Chesler and S. P. Cram, Effect of peak sensing and random noise on the precision and accuracy of statistical moment analysis from digital chromatographic data, *Anal. Chem.*, *43*:1922 (1971).

12. S. N. Chesler and S. P. Cram, Iterative curve fitting of chromatographic peaks, *Anal. Chem.*, *45*:1354 (1973).
13. W. W. Yau, Characterizing skewed chromatographic band broadening, *Anal. Chem.*, *49*:395 (1977).
14. B. A. Bidlingmeyer and F. V. Warren, Jr., Column efficiency measurements, *Anal. Chem.*, *56*:1583A (1984).
15. R. E. Kaiser and E. Oelrich, *Optimization in HPLC*, Hüthig Verlag, Heidelberg (1981).
16. J. P. Foley and J. G. Dorsey, Equations for calculation of chromatographic figures of merit for ideal and skewed peaks, *Anal. Chem.*, *55*:730 (1983).
17. J. J. van Deemter, F. J. Zuiderweg, and A. Klinkenberg, Longitudinal diffusion and mass transfer as causes of nonideality in chromatography, *Chem. Eng. Sci.*, *5*:271 (1956).
18. J. C. Giddings, Reduced plate height equation: A common link between chromatographic methods, *J. Chromatogr.*, *13*:301 (1964).
19. Cs. Horváth and H.-J. Lin, Movement and band spreading of unsorbed solutes in liquid chromatography, *J. Chromatogr.*, *126*:401 (1976).
20. G. Doetsch, *Einführung in Theorie und Anwendung der Laplace-Transformation, 2 Aufl.*, Birkhäuser Verlag, Basel und Stuttgart (1970).
21. D. A. McQuarrie, On the stochastic theory of chromatography, *J. Chem. Phys.*, *38*:437 (1963).
22. H. Cramér, *Mathematical Methods of Statistics*, Princeton University Press, Princeton (1946).
23. C. Vidal-Madjar and G. Guiochon, Experimental characterization of elution profiles in gas chromatography using central statistical moments, *J. Chromatogr.*, *142*:61 (1977).
24. F. Dondi, A. Betti, G. Blo, and C. Bighi, Statistical analysis of gas chromatographic peaks by the Gram-Charlier series of Type A and the Edgeworth-Cramér series, *Anal. Chem.*, *53*:496 (1981).
25. F. Dondi and F. Pulidori, Applicability limits of the Edgeworth-Cramér series in chromatographic peak shape analysis, *J. Chromatogr.*, *284*:293 (1984).
26. J. W. Cooley, P. A. W. Lewis, and P. D. Welch, The fast Fourier transform algorithm: Programming considerations in the calculation of sine, cosine and Laplace transforms, *J. Sound Vib.*, *12*:315 (1970).
27. J. Kühne, Bemerkungen zur Anwendung der numerischen Umkehrung der Laplace-Transformation aufgrund von Legendre-Polynomen im Falle der Chromatographie, *Chromatographia*, *11*:403 (1978).
28. J. Villermaux, Theory of linear chromatography, *NATO Adv. Study Inst. Ser. Ser E.*, *33*:83 (1981).

29. J. Å. Jönsson, Computer simulation of the chromatographic process, *Chromatographia, 13*:273 (1980).
30. J. Å. Jönsson, Non-ideal effects in linear chromatography, *Chromatographia, 13*:729 (1980).
31. J. C. Smit, H. C. Smit, and E. M. DeJager, Computer implementation of simulation models for non-linear, non-ideal chromatography, Part 1, Fundamental mathematical considerations, *Anal. Chim. Acta, 122*:1 (1980).
32. J. C. Smit, H. C. Smit, and E. M. DeJager, Computer implementation of simulation models for non-linear, non-ideal chromatography, Part 2, Numerical experiments and results, *Anal. Chim. Acta, 122*:151 (1980).
33. P. Faurré and M. Depeyrot, *Elements of System Theory*, North-Holland, Amsterdam (1977).
34. R. Annino, Signal and resolution enhancement techniques in chromatography, *Advances in Chromatography, Vol. 15* (J. C. Giddings, E. Grushka, J. Cazes, and P. R. Brown, eds.), Marcel Dekker, New York, pp. 33-67 (1977).
35. G. Horlick, Fourier transform approaches to spectroscopy, *Anal. Chem., 43*(8):61A (1971).
36. D. E. Smith, Data processing in electrochemistry, *Anal. Chem., 48*:517A (1976).
37. R. W. Dwyer, Jr., Isolation of column phenomena in gas chromatography, *Anal. Chem., 45*:1380 (1973).
38. T. A. Maldacker, J. E. Davis, and L. B. Rogers, Applications of Fourier transform techniques to steric-exclusion chromatography, *Anal. Chem., 46*:637 (1974).
39. S. G. Welser, Minimizing flow-associated noise in electrochemical detectors for liquid chromatography, *Anal. Chem., 54*:2126 (1982).
40. D. M. Monro, Complex discrete fast Fourier transform, algorithm AS 83, *Appl. Stat., 24*:153 (1975).
41. D. M. Monro, A portable integer FFT in Fortran, *Comp. Prog. Biol. Med., 7*:267 (1977).
42. K. Yamaoka and T. Nakagawa, Moment analysis for isolation of intrinsic column efficiencies in gas chromatography, *Anal. Chem., 47*:2050 (1975).
43. J. C. Sternberg, Extracolumn contributions to chromatographic band broadening, *Advances in Chromatography, Vol. 2* (J. C. Giddings and R. A. Keller, eds.), Marcel Dekker, New York, pp. 205-270 (1966).
44. S. P. Cram and T. H. Glenn, Jr., Instrumental contributions to band broadening in gas chromatography, *J. Chromatogr., 112*:329 (1975).
45. O. Grubner and W. A. Burgess, Simplified description of adsorption breakthrough curves in air cleaning and sampling devices, *Am. Ind. Hyg. Assoc. J., 40*:169 (1979).

46. C. N. Reilly, G. P. Hildebrand, and J. W. Ashley, Jr., Gas chromatographic response as a function of sample input profile, *Anal. Chem.*, *34*:1198 (1962).
47. H. C. Smit, R. P. J. Duursma, and H. Steigstra, A microprocessor-based instrument for correlation chromatography and data processing, *Anal. Chim. Acta*, *133*:283 (1981).
48. J. Růžička and E. H. Hansen, *Flow Injection Analysis*, John Wiley & Sons, New York (1981).
49. G. Taylor, Dispersion of soluble matter in solvent flowing slowly through a tube, *Proc. R. Soc. London*, *A219*:186 (1953).
50. R. Tijssen, Axial dispersion and flow phenomena in helically coiled tubular reactors for flow analysis and chromatography, *Anal. Chim. Acta*, *114*:71 (1980).
51. E. D. Katz and R. P. W. Scott, Low-dispersion connecting tubes for liquid chromatography systems, *J. Chromatogr.*, *268*:169 (1983).
52. J. Å. Jönsson, Shape of the chromatographic elution curve derived from the normal distribution of sample in the column, *J. Chromatogr.*, *150*:11 (1978).
53. A. Kreft and Z. Zuber, On the physical meaning of the dispersion equation and its solutions for different initial and boundary conditions, *Chem. Eng. Sci.*, *33*:1471 (1978).
54. J. Å. Jönsson, Elution curves and statistical moments in non-ideal, linear chromatography, *Chromatographia*, *18*:427 (1984).
55. A. J. B. Cruickshank and D. H. Everett, The shape of the elution peaks in gas chromatography, *J. Chromatogr.*, *11*:289 (1963).
56. L. Lapidus and N. R. Amundsen, Mathematics of adsorption in beds, VI, The effect of longitudinal diffusion in ion-exchange and chromatographic columns, *J. Phys. Chem.*, *56*:984 (1952).
57. H. Röck, Gaschromatographie und extraktive Destillation, *Chem. Ing. Tech.*, *28*:489(1956).
58. J. J. Carberry, Determination of heats of adsorption by transient-response techniques, *Nature*, *189*:391 (1961).
59. E. Grushka, Chromatographic peak shapes. Their origin and dependence on the experimental parameters, *J. Phys. Chem.*, *76*:2586 (1972).
60. E. Kučera, Contribution to the theory of chromatography. Linear non-equilibrium elution chromatography, *J. Chromatogr.*, *19*:237 (1965).
61. J. Å. Jönsson, The median of the chromatographic peak as the best measure of retention time, *Chromatographia*, *14*:653 (1981).
62. M. Kubín, Beitrag zur Theorie der Chromatographie, *Coll. Czech. Chem. Commun.*, *30*:1104 (1965).
63. O. Grubner and D. Underhill, Comparison of equations describing mass transfer in packed beds, *J. Chromatogr.*, *73*:1 (1972).

64. P. Lövkvist and J. Å. Jönsson, Description of sample introduction in chromatography as a separate term in the mass balance equation, *J. Chromatogr.*, *356*:1 (1986).
65. J. Villermaux, Relations entre la forme des pics chromatographiques et les paramètres physiques et opératoires de la colonne, *J. Chromatogr.*, *83*:205 (1973).
66. J. P. Muth, D. J. Wilson, and K. A. Overholser, Effect of non-equilibrium in gas chromatography, *J. Chromatogr.*, *87*:1 (1973).
67. M. R. James, J. C. Giddings, and H. Eyring, Contribution of interfacial resistance to theoretical plate height in gas chromatography, *J. Phys. Chem.*, *68*:1725 (1964).
68. M. A. Khan, Non-equilibrium theory of capillary columns and the effect of interfacial resistance on column efficiency, *Gas Chromatography 1962* (M. van Swaay, ed.), Butterworths, London, pp. 3-17 (1962).
69. M. J. E. Golay, Theory of chromatography in open and coated tubular columns with round and rectangular cross-sections, *Gas Chromatography 1958* (D. H. Desty, ed.), Butterworths, London, pp. 36-55 (1958).
70. R. Tijssen, Liquid chromatography in helically coiled open tubular columns, *Sep. Sci. Technol.*, *13*:681 (1978).
71. D. D. Chilcote and C. D. Scott, A theoretical derivation of the axial diffusion coefficient in chromatography, *J. Chromatogr.*, *87*:315 (1973).
72. J. C. Giddings, The random downstream migration of molecules in chromatography, *J. Chem. Educ.*, *35*:558 (1958).
73. J. C. Giddings and R. A. Robison, Failure of the eddy diffusion concept of gas chromatography, *Anal. Chem.*, *34*:885 (1962).
74. J. C. Giddings, "Eddy" diffusion in chromatography, *Nature*, *184*:357 (1959).
75. J. C. Giddings, Lateral diffusion and local nonequilibrium in gas chromatography, *J. Chromatogr.*, *5*:61 (1961).
76. J. F. K. Huber and J. A. R. J. Hulsman, A study of liquid chromatography in columns. The time of separation, *Anal. Chim. Acta*, *38*:305 (1967).
77. J. F. K. Huber, Stofftransport und Stoffverteilung bei chromatographischen Prozessen, *Ber. Bunsenges. Phys. Chem.*, *77*:179 (1973).
78. Cs. Horváth and H.-J. Lin, Band spreading in liquid chromatography. General plate height equation and a method for the evaluation of the individual plate heigh contribution, *J. Chromatogr.*, *149*:43 (1978).
79. R. Pfeffer, Heat and mass transport in multiparticle systems, *Ind. Chem. Fund.*, *3*:380 (1964).
80. J. W. Hiby, Longitudinal and transverse mixing during single-phase flow through granular beds, *Symposium on Interaction between Fluids and Particles*, Institute of Chemical Engineers, London, p. 312 (1962).

81. J. H. Knox and J. F. Parcher, Effect of column to particle diameter ratio on the dispersion of unsorbed solutes in chromatography, *Anal. Chem.*, *41*:1599 (1969).
82. G. J. Kennedy and J. H. Knox, The performance of packings in high performance liquid chromatography (HPLC), I, Porous and surface layered supports, *J. Chromatogr. Sci.*, *10*:549 (1972).
83. W. L. Jones, Modifications to the van Deemter equation for the height equivalent to a theoretical plate in gas chromatography, *Anal. Chem.*, *33*:829 (1961).
84. J. C. Giddings, Plate height contributions in gas chromatography, *Anal. Chem.*, *33*:962 (1961).
85. E. Katz, K. L. Ogan, and R. P. W. Scott, Peak dispersion and mobile phase velocity in liquid chromatography: The pertinent relationship for porous silica, *J. Chromatogr.*, *270*:51 (1983).
86. J. H. Knox and H. P. Scott, B and C terms in the van Deemter equation for liquid chromatography, *J. Chromatogr.*, *282*:297 (1983).
87. J. C. Giddings, Role of column pressure drop in gas chromatographic resolution, *Anal. Chem.*, *36*:741 (1964).
88. A. T. James and A. J. P. Martin, Gas-liquid partition chromatography. A technique for the analysis of volatile materials, *Analyst*, *77*:915 (1952).
89. J. Bohemen and J. H. Purnell, Diffusional band broadening in gas chromatographic columns, Part I, The elution of unsorbed gases, *J. Chem. Soc.*, *1961*:360.
90. M. J. E. Golay, Height equivalent to a theorical plate of an open tubular column lined with a porous layer, *Anal. Chem.*, *40*:382 (1968).
91. C. A. Cramers, F. A. Wijnheymer, and J. A. Rijks, Optimum gas chromatographic conditions in wall-coated capillary columns. Extended and simplified forms of the Golay equation, *J. HRC & CC*, *2*:329 (1979).
92. D. F. Ingraham, C. F. Shoemaker, and W. Jennings, Computer comparisons of variables in capillary gas chromatography, *J. HRC & CC*, *5*:227 (1982).
93. J. H. Knox and M. T. Gilbert, Kinetic optimization of straight open-tubular liquid chromatography, *J. Chromatogr.*, *186*:405 (1979).
94. G. Guiochon and H. Colin, Narrow-bore and micro-bore columns in liquid chromatography, *Micro Column High-Performance Liquid Chromatography* (P. Kucera, ed.), Elsevier, Amsterdam, pp. 1-38 (1984).
95. J. H. Knox, Evidence for tubulence and coupling in chromatographic columns, *Anal. Chem.*, *38*:253 (1966).
96. J. N. Done, G. J. Kennedy, and J. H. Knox, The role of particle size and configuration in high-speed liquid chromatography, *Gas Chromatography 1972* (S. G. Perry and E. R. Adlard, eds.), Applied Science, Barking, pp. 145-155 (1973).

97. R. G. Bogar, J. C. Thomas, and J. B. Callis, Lateral diffusion of solutes bound to the alkyl surface of C_{18} reversed-phase liquid chromatographic packings, *Anal. Chem.*, *56*:1080 (1984).
98. R. W. Stout, J. J. DeStefano, and L. R. Snyder, High-performance liquid chromatographic column efficiency as a function of particle compositions and geometry and capacity factor, *J. Chromatogr.*, *282*:263 (1983).
99. J. C. Giddings and K. L. Mallik, Theory of gel filtration (permeation) chromatography, *Anal. Chem.*, *38*:997 (1966).
100. J. V. Dawkins and G. Yeadon, High-performance gel permeation chromatography of polystyrene with silica microspheres, *J. Chromatogr.*, *188*:333 (1980).
101. J. H. Knox and F. McLennan, Allowance for polydispersity in the true plate height in GPC, *Chromatographia, 10*:75 (1977).
102. D. de Vault, The theory of chromatography, *J. Amer. Chem. Soc.*, *65*:532 (1943).
103. J. F. K. Huber and R. G. Gerritse, Evaluation of dynamic gas chromatographic methods for the determination of adsorption and solution isotherms, *J. Chromatogr.*, *58*:137 (1971).
104. F. Helfferich, Travel of molecules and disturbances in chromatographic columns. A paradox and its resolution, *J. Chem. Educ.*, *41*:410 (1964).
105. G. Houghton, Band shapes in non-linear chromatography with axial dispersion, *J. Phys. Chem.*, *67*:84 (1963).
106. H. Brusset, D. Depeyre, and J. P. Petit, Résolution de l'équation de bilan-matière d'une colonne chromatographique, *Chromatographia, 11*:287 (1978).
107. A. Jaulmes, C. Vidal-Madjar, A. Landurelli, and G. Guiochon, Study of peak profiles in nonlinear gas chromatography, 1, Derivation of a theoretical model, *J. Phys. Chem.*, *88*:5379 (1984).
108. G. V. Yeroshenkova, S. A. Volkov, and K. I. Sakodynskii, Influence of sorption on the shape of chromatographic elution curves, *J. Chromatogr.*, *198*:377 (1980).
109. J. R. Conder and J. H. Purnell, Gas chromatography at finite concentrations, Part 2, A generalized retention theory, *Trans. Faraday Soc.*, *64*:3100 (1968).
110. K. De Clerk and T. S. Buys, The effect of sorption on retention times in non-linear chromatography, *J. Chromatogr.*, *84*:1 (1973).
111. A. Jaulmes, C. Vidal-Madjar, M. Gaspar, and G. Guiochon, Study of peak profiles in nonlinear gas chromatography, 2, Determination of the curvature of isotherms at zero surface coverage on graphitized carbon black, *J. Phys. Chem.*, *88*:5385 (1984).
112. J. E. Funk and G. Houghton, A mathematical model for gas-liquid partition chromatography, *Nature, 188*:389 (1960).

113. J. E. Funk and G. Houghton, A lumped-film model for gas-liquid partition chromatography, Part I, Numerical methods of solution, *J. Chromatogr.*, *6*:193 (1960).
114. M. H. J. van Rijswick, Adaptive program for high precision off-line processing of chromatograms, *Chromatographia*, *7*:491 (1974).
115. A. B. Littlewood, A. H. Anderson, and T. C. Gibb, Computer analysis of unresolved digitized chromatograms, *Gas Chromatography 1968* (C. L. A. Harbourn, ed.), Institute of Petroleum, London, pp. 297-316 (1969).
116. E. Grushka, M. N. Myers, and J. C. Giddings, Moments analysis for the discernment of overlapping chromatographic peaks, *Anal. Chem.*, *42*:21 (1970).
117. T. S. Buys and K. de Clerk, Bi-Gaussian fitting of skewed peaks, *Anal. Chem.*, *44*:1273 (1972).
118. E. Grushka, Characterization of exponentially modifed Gaussian peaks in chromatography, *Anal. Chem.*, *44*:1733 (1972).
119. R. Delley, The peak width of nearly Gaussian peaks, *Chromatographia*, *18*:374 (1984).
120. R. Delley, Series for the exponentially modified Gaussian peak shape, *Anal. Chem.*, *57*:388 (1985).
121. R. E. Pauls and L. B. Rogers, Band broadening studies using parameters for an exponentially modified Gaussian, *Anal. Chem.*, *49*:625 (1977).
122. K.-H. Jung, S. J. Yun, and S. H. Kang, Characterization of peak shape parameters with normal and derivative chromatograms, *Anal. Chem.*, *56*:457 (1984).
123. J. C. Giddings, Kinetic origin of tailing in chromatography, *Anal. Chem.*, *35*:1999 (1963).
124. S. D. Mott and E. Grushka, Chromatographic peak shape, II, The use of shape models for the generation of moments, *J. Chromatogr.*, *148*:305 (1978).
125. S. D. Mott and E. Grushka, Chromatographic solute identification using peak shape analysis, *J. Chromatogr.*, *126*:191 (1976).
126. E. Grushka and S. D. Mott, Chromatographic peak shape, III, Influence of kinetics parameters on solute profiles, *J. Chromatogr.*, *186*:117 (1979).

3

Retention in Gas-Liquid Chromatography

Josef Novák†
Institute of Analytical Chemistry
Czechoslovak Academy of Sciences
Brno, Czechoslovakia

I. BASIC CONCEPTS

A. Mechanism of the Distribution of Solute Between Bulk Liquid and Gas

The mechanism of the chromatographic process is rather complex. In order to properly understand this mechanism, it is necessary to take into account the dynamic nature of the equilibrium between the solute concentrations in the coexisting phases [1]; in a more rigorous conception of this problem, however, equilibrium between the solute concentrations is understood as a consequence of equality between the chemical potentials of the solute in the two phases of the system. Even if the system is stationary and at equilibrium, the solute molecules constantly pass from one phase to the other, remaining in either phase for a certain time after each transition. Since the process is quite random on the molecular level, the individual times for which the solute molecules persist in a given phase are also random and, consequently, very different. However, the average times for which all of the solute molecules are present in either phase are constant under given conditions, and the ratio of these average times constitutes the basic factor of chromatographic retention. Hence, the ratio at which a given quantity of solute distributes itself at equilibrium between the phases of the system is determined by the probability with which the

†Deceased.

solute molecules will occur in these phases, rather than by their mere static presence in the phases. If under these circumstances one phase is moving relative to the other, the solute molecules move along with the moving phase during their presence in it, whereas they are stagnant while in the stationary phase. Owing to statistical fluctuations, some molecules of a given solute will, during a certain space of time, cover distances that are longer or shorter than the distance corresponding to the average dwelling time of the solute molecules in a given phase. Together with longitudinal diffusion, this leads to spreading of the migrating zone of the solute. However, thanks to its statistical nature, this spreading increases only as the square root of the average length of the zone migration, so that the zones of different solutes, migrating at different velocities, may separate from each other after having migrated a certain distance.

B. Definition of the Problem

The preceding qualitative description of the chromatographic process shows that there are two basic general problems of chromatography, those of retention and zone broadening. These constitute the basis of almost all theoretical as well as practical problems of chromatographic separation. In this chapter, only the problem of chromatographic retention is considered, and only under conditions in which it is quite easy to describe retention quantitatively. In every case we consider only isothermal elution chromatography in a column with a nonvolatile liquid stationary phase. However, since the integral of an elution profile represents the corresponding frontal profile in linear chromatography, the rules derived for the movement of the concentration maximum of an elution zone apply equally to the movement of the concentration inflection of the frontal zone.

In general, chromatographic retention and zone broadening are more or less interdependent; however, we shall confine our discussion only to cases in which both problems can be fairly well distinguished from each other and dealt with independently. This means that, among real models of chromatography, we shall consider only the model of linear chromatography under near-equilibrium conditions—that is, those conditions in which there are only slight deviations from equilibrium in the migrating zone and in which positive deviations in the mobile phase, for example, practically equal negative deviations on an average. Such a situation, illustrated schematically in Fig. 1, can occur at very low solute concentrations in the sorbent (solvent) and at moderate velocities of the mobile phase. It is only under such conditions that the profile of an elution zone may be a symmetrical gaussian peak (provided the solute has been introduced in the form of a very narrow symmetrical pulse), the maximum (center) of which represents sorption equilibrium [2]. If the solute concentration in the migrating zone exceeds that in the Henry's law region, we speak of nonlinear

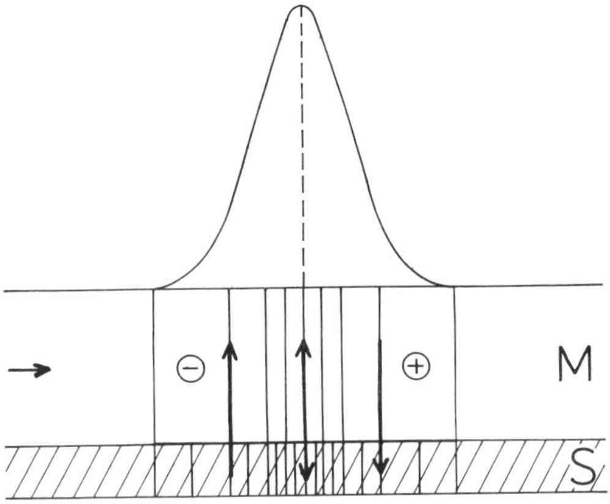

Figure 1. Schematic representation of the situation in an elution chromatographic zone migrating under the conditions of quasi-equilibrium linear chromatography: M, mobile phase; S, stationary phase.

chromatography. In such a case, the zone will be nonsymmetrical and there will be no parameter that would unequivocally represent equilibrium. However, at larger deviations from equilibrium, such as those caused by a large mobile-phase flow velocity, an asymmetric zone is obtained even at solute concentrations well within the Henry's law region. In such a case, equilibrium is better (although not exactly) represented by the center of gravity of the zone [3,4]. Models of chromatography that allow the solute concentration in the zone to exceed that in the Henry's law region are substantially more complicated, and the mathematical description of such models is rather intricate. Except for a brief discussion of the sources of nonlinearity (see page 135), models of this type will not be dealt with in this chapter. The reader interested in this difficult problem is referred to the papers of Conder and Purnell on gas chromatography at finite solute concentrations [5-8], especially to Part II of the series [6], or to the book by Condor and Young [9]. (See also Chap. 2, Sec VI.)

Along with the more or less real model of linear quasi-equilibrium chromatography, we shall deal largely with the hypothetical model of ideal linear chromatography, which is very useful from the viewpoint of the description of chromatographic retention. First, the model of ideal linear chromatography with an incompressible mobile phase will be discussed briefly. This model does not comport with the behavior of real gas-liquid chromatography (GLC) systems. However, it can easily be described mathematically and may be regarded as a suitable reference model. The problem of retention in systems with gaseous mobile phases will then be solved with reference to the fact that the models of linear quasi-equilibrium, linear, and ideal linear chromatography with incompressible mobile phases render, under certain circumstances, the same fundamental retention equation [10-12].

II. IDEAL LINEAR CHROMATOGRAPHY WITH AN INCOMPRESSIBLE MOBILE PHASE

The concept of ideal linear chromatography is derived from a model [10] that should have the following properties: (1) an infinitely quick establishment of equilibrium between the solute concentrations in the mobile and stationary phases; (2) zero longitudinal diffusion of solute in both phases; (3) an absolutely linear sorption isotherm; and (4) plug flow of the mobile phase. Although this model is evidently not real, it is very interesting in that it renders a fairly correct description of chromatographic retention. Clearly, this model does not give any information about zone spreading, since no spreading factors have been considered in its specification. Under the conditions of ideal linear chromatography, the initial concentration profile of solute would, without any change in its shape, move down the column at the same velocity at which the center of a broadening elution zone travels under the conditions of nonideal, quasi-equilibrium linear chromatography [10-12]. However, a more rigorous treatment [13,14] of the model of nonideal linear chromatography shows that the zone maximum retention time is not quite independent of spreading factors.

The solute mass balance in an infinitesimal volume of the chromatographic bed (column), defined by two parallel cross sections of equal areas A situated normal to the mobile-phase flow at distances z and z + dz from the beginning of the bed, leads to the equation

$$\frac{\partial c_{iM}}{\partial t} + \frac{\phi_S}{\phi_M}\frac{\partial c_{iS}}{\partial t} = -u\frac{\partial c_{iM}}{\partial z} + D_M\frac{\partial^2 c_{iM}}{\partial z^2} + \frac{\phi_S}{\phi_M} D_S \frac{\partial^2 c_{iS}}{\partial z^2} \tag{1}$$

where c_{iM} and c_{iS} are the cross-sectional average concentrations (mass/volume) of the solute (i) in the mobile (M) and stationary (S) phases, ϕ_M and ϕ_S are the fractions of area A occupied by the mobile phase and by the stationary phase, respectively, D_M and D_S are the solute diffusion coefficients in the mobile and stationary phases, u is the forward-flow velocity of the mobile phase (averaged over the flow-through cross section i.e., $u = F/A_M$, where F is the volumetric flow rate of the mobile phase and A_M is the cross-sectional area occupied by the mobile phase), and t and z are time and distance from the beginning of the bed (column inlet) in the direction of the mobile phase flow, respectively.

For ϕ_M and ϕ_S, it holds that $\phi_S/\phi_M = A_S/A_M$, where A_S and A_M are the absolute cross-sectional areas occupied by the stationary and mobile phases.

At infinite dilution of the solute in the phases of the system, D_M and D_S are practically constant, and A_S, A_M, ϕ_S, and ϕ_M in a given column cross section are also constant, independently of whether or not a solute zone is merely passing through that cross section. Hence, the contingent presence of the solute in either phase does not appreciably influence the volume of the phase, and the solute concentration is, in practical terms, given by the mass of solute per unit volume of pure phase. Provided the column packing is uniform, it holds that $\phi_S/\phi_M = A_S/A_M^0 = V_S/V_M^0$, with V_S and V_M^0 the total volume of the sorbent and the total free (mobile-phase) volume in the column, respectively. The superscript 0 in V_M^0 is to remind us that we are dealing with a volume expressed at the temperature and mean pressure in the column.

When in Eq. (1) the terms that stand for the longitudinal diffusion of solute in the mobile and stationary phases are ignored and the solute partition coefficient is defined as the ratio of solute equilibrium concentrations in the sorbent and in the mobile phase,

$$K = \left(\frac{c_{iS}}{c_{iM}}\right)_{eq} \tag{2}$$

and using the relation

$$\frac{\partial c_{iS}}{\partial t} = K \frac{\partial c_{iM}}{\partial t} \tag{3}$$

we obtain

$$\frac{\partial c_{iM}}{\partial t}\left(1 + \frac{\phi_S}{\phi_M} K\right) = -u \frac{\partial c_{iM}}{\partial z} \tag{4}$$

which is actually a mathematical definition of ideal linear chromatography. The solution of Eq. (4) for a given input concentration profile, $z = L$, and $t = t_R$, leads to the fundamental retention equation, Eq. (6). According to the theorems on the properties of partial differentials, we can write, with regard to Eq. (4),

$$\frac{(\partial c_{iM}/\partial z)_t}{(\partial c_{iM}/\partial t)_z} = -\left(\frac{\partial t}{\partial z}\right)_{c_{iM}} = -\frac{1 + (\phi_S/\phi_M)K}{u} \tag{5}$$

and, provided the migration of a very narrow pulse of invariant c_{iM} is considered,

$$dt = \frac{1 + (\phi_S/\phi_M)K}{u} dz$$

$$\int_0^{t_R} dt = \frac{1 + (\phi_S/\phi_M)K}{u} \int_0^L dz$$

and

$$t_R = \frac{L}{u}\left(1 + \frac{\phi_S}{\phi_M} K\right) \qquad (6)$$

where t_R is the retention time of the pulse. The term $(\phi_S/\phi_M)K$ is the so-called capacity ratio k, representing the ratio of equilibrium masses or amounts of solute in the stationary and mobile phases as follows:

$$k = \left(\frac{m_{iS}}{m_{iM}}\right)_{eq} = \left(\frac{n_{iS}}{n_{iM}}\right)_{eq} = \frac{\phi_S}{\phi_M} K \qquad (7)$$

where m and n denote the mass and amount (mole number) of the solute, respectively.

III. SIMPLIFIED DESCRIPTION OF CHROMATOGRAPHIC RETENTION IN SYSTEMS WITH A GASEOUS MOBILE PHASE

A. Ideal Gaseous Phase and Invariant Capacity Ratio

Application of the Model of Ideal Linear Chromatography

In this section, the procedure used to derive relations describing solute retention in a GLC column is based on the same philosophy as that employed by James and Martin in their fundamental work on gas-liquid chromatography [15]. When limited to the problem of solute retention only, this philosophy leads to the concept of a hypothetical mean-pressure isobaric column as an equivalent of the real pressure-gradient column. This concept has been proven many times, and there is no doubt about its validity for GLC systems with a near ideal carrier gas and practically constant k along the column.

The finding that the center of a broadening elution zone in non-ideal (quasi-equilibrium) linear chromatography migrates in the same manner as an invariant concentration pulse would migrate, according to the concept of ideal linear chromatography under the given experi mental conditions, is very important in view of the theory of chromatographic retention. Since the local forward velocity of migration of the zone center (average velocity of the whole zone) is a defined fraction of the corresponding local forward velocity of the mobile phase (cross-sectional average), under all circumstances in quasi-equilibrium linear chromatography, the concept of ideal linear chromatography can also be applied to systems with a compressible mobile phase (i.e., to gas chromatography). If k is invariant (independent of the pressure in the column), the local zone center velocity is a constant fraction of the corresponding local velocity of the mobile phase throughout the column. Equation (4) will then incorporate the length-dependent velocity u(z) instead of the invariant quantity u. Provided the mobile-phase flow is laminar, the Darcy law states that:

$$u(z) = -\frac{B_0}{\eta\varepsilon_e}\frac{dP}{dz} \tag{8}$$

where B_0 is the specific permeability constant (cm^2) of the chromatographic bed, η is the dynamic viscosity (dyn sec cm^{-2} = g cm^{-1} sec^{-1}) of the mobile phase (carrier gas), ε_e is the interparticle porosity (with packed beds, ε_e is about half the total porosity ε) of the chromatographic bed, and dP/dz is the pressure gradient (pressure change per unit column length in the direction of mobile-phase flow, dyn cm^{-3} = g cm^{-2} sec^{-2}) across the column. For an empty capillary ($\varepsilon_e = 1$) of circular cross section, $B_0 = r^2/8$, where r is the cross-sectional radius of the capillary, and for packed beds it can be supposed, according to the Kozeny-Carman equation [16,17], that $B_0 = d_p^2\varepsilon_e^3/180(1 - \varepsilon_e)^2$, where d_p is the diameter of the packing particles.

Mobile-Phase Flow Velocity at the Mean Column Pressure. If the carrier gas is ideally compressible, we can write, according to Boyle's law, $u(z)P(z) = u_oP_o$, where the quantities on the left denote the local velocity and pressure at the given column cross section, u_o is the velocity at the column outlet pressure and column temperature, and P_o is the column outlet pressure. By substituting for u(z) from this equation into Eq. (8), we obtain

$$P(z)\,dP = -\frac{u_oP_o}{B_0/\eta\varepsilon_e}\,dz \tag{9}$$

and after integration (provided the carrier gas viscosity is independent of pressure and the column packing is uniform), we have

$$[P(z)]^2 = -\frac{2u_o P_o}{B_0/\eta\varepsilon_e} z + C_I \tag{10}$$

where C_I is an integration constant. For z = 0 and z = L, we have $P(z) = P_i$ and $P(z) = P_0$, L and P_i being the column length and column inlet pressure, respectively, so that $C_I = P_i^2$, and Eq. (16) can be rewritten as

$$[P(z)]^2 = P_i^2 - \frac{2u_o P_o}{B_0/\eta\varepsilon_e} z \tag{11}$$

For the column outlet (i.e., for z = L), the relationship

$$P_o^2 = P_i^2 - \frac{2u_o P_o}{B_0/\eta\varepsilon_e} L \tag{12}$$

apparently is true. From Eqs. (11) and (12), we thus obtain for u_o, B_0, and P(z),

$$u_o = \frac{B_0}{\eta\varepsilon_e L} \frac{P_i^2 - P_o^2}{2P_o} \tag{13}$$

$$B_0 = \frac{2P_o u_o \eta\varepsilon_e L}{P_i^2 - P_o^2} \tag{14}$$

$$P(z) = \left[P_i^2 - \frac{z}{L} (P_i^2 - P_o^2) \right]^{1/2} \tag{15}$$

Since, under the above assumptions, $u(z) = u_o P_o / P(z)$, we can write

$$u(z) = \frac{u_o P_o}{[P_i^2 - \frac{z}{L} (P_i^2 - P_o^2)]^{1/2}} \tag{16}$$

Retention in a Hypothetical Isobaric GLC Column at Mean Pressure. By combining Eqs. (4) and (16), we arrive at a mathematical definition of ideal linear chromatography with an ideally compressible gaseous phase:

$$\frac{\partial c_{iM}}{\partial t}(1 + k) = -u_o \frac{P_o}{[P_i^2 - \frac{z}{L}(P_i^2 - P_o^2)]^{1/2}} \frac{\partial c_{iM}}{\partial z} \tag{17}$$

Employing a procedure, analogous with the derivation of the basic retention equation from Eq. (5), we obtain in this case

$$t_R = \frac{1 + k}{u_o P_o} \int_0^L [P_i^2 - \frac{z}{L}(P_i^2 - P_o^2)]^{1/2} \, dz \tag{18}$$

and after integration,

$$t_R = \frac{(1 + k)L}{P_o u_o} \frac{2}{3} \frac{P_i^3 - P_o^3}{P_i^2 - P_o^2} \tag{19}$$

Equation (19) can be rewritten as

$$t_R = \frac{L}{u_o} \frac{\langle P \rangle}{P_o}(1 + k) = \frac{L}{u_o} \frac{1 + k}{j} \tag{20}$$

where j is the James-Martin compressibility factor [15] and <P> is the mean column pressure defined by

$$\langle P \rangle = \frac{1}{L} \int_0^L P(z) \, dz = \frac{2}{3} \frac{P_i^3 - P_o^3}{P_i^2 - P_o^2} = \frac{P_o}{3[(P_i/P_o)^2 - 1]/2[(P_i/P_o)^3 - 1]}$$

$$= \frac{P_o}{j} \tag{21}$$

Equation (20), derived in the above way, can be regarded as evidence of the generally recognized correctness of the James-Martin concept that the properties of a gas chromatographic column with a pressure gradient are equivalent (so far as solute retention is concerned) to those of a hypothetical isobaric column at the corresponding mean pressure under the conditions of linear chromatography (with a k that is independent of pressure and velocity).

When substituting in Eq. (20) $\langle P \rangle / u_o P_o = u_{\langle P \rangle}$, where $u_{\langle P \rangle}$ is the forward velocity of the carrier gas at the mean column pressure and column temperature, we obtain

$$t_R = \frac{L}{u_{<P>}}(1 + k) = t_M(1 + k) \tag{22}$$

and by multiplying Eq. (22) by the volumetric flow rate of the carrier gas at the mean column pressure and column temperature $F_{<P>}$, we have

$$t_R F_{<P>} = V_R^0 = L\phi_M(1 + k) = V_M^0(1 + k) \tag{23}$$

where V_R^0 and V_M^0 are the retention volume of the solute and the dead retention volume, both at the temperature and mean pressure of the column. The quantity V_M^0 apparently equals the geometric free volume in the column. It follows from Eqs. (20), (22), and (23) that the dead retention time and dead retention volume are given by the relations

$$t_M = \frac{L\langle P\rangle}{u_o P_o} = \frac{L}{u_{<P>}} \tag{24}$$

$$V_M^0 = L\phi_M = F_{<P>}t_M \tag{25}$$

From Eqs. (22) and (23), and taking account of the relation $k = KV_S/V_M^0$, we obtain, for k and K,

$$k = \frac{t_R - t_M}{t_M} = \frac{V_R^0 - V_M^0}{V_M^0} \tag{26}$$

$$K = \frac{t_R - t_M}{t_M}\frac{V_M^0}{V_S} = \frac{V_R^0 - V_M^0}{V_S} \tag{27}$$

At this point it is expedient to note that t_M and V_M^0, as defined by Eqs. (24) and (25), concern only the chromatographic column proper (i.e., without contributions from extracolumn volumes). However, the spaces between the sample inlet port and column inlet and between the column outlet and detector sensor cannot be completely eliminated. Since the hydrodynamic conditions in these spaces differ from those in the column, the extracolumn spaces cannot be regarded as parts of the hypothetical isobaric column at the mean pressure. The dead retention time t_M^+, determined experimentally by means of a nonsorbed compound, represents the passage of this compound not only through the geometric free volume V_M^0 in the column at the mean column pressure <P> but also through the extracolumn volumes $\Sigma V_{M,e}$ at the respective pressures P_e, and it is therefore always longer than t_M. When denoting the sum of the times that the center of the zone of a nonsorbed compound spends in the individual extracolumn spaces as $\Sigma t_{M,e}$ we can apparently write $t_M^+ = t_M + \Sigma t_{M,e}$. If t_R' is the adjusted retention time of the solute (the average time the solute molecules have spent, during their passage through the column, in the sorbent), then the following is true, provided the solute is not sorbed in the extracolumn spaces: $t_R = t_R' + t_M^+ = t_R' + t_M + \Sigma t_{M,e}$. For the retention volume V_R^0, we can write $V_R^0 = V_N + V_M^0 + \Sigma V_{M,e}$, where V_N is the net retention volume given by the product $t_R F_{<P>}$ and the apparent dead retention volume (including the extracolumn volumes) $V_M^+ = V_M^0 + \Sigma V_{M,e} = t_M F_{<P>} + \Sigma(t_{M,e} F_{P,e})$. However, in current practice, the volume V_M^+ is usually calculated (incorrectly, in fact) as $V_M^{\S} = t_M^+ F_{<P>} = (t_M + \Sigma t_{M,e}) F_{<P>}$. Hence, if we want to calculate the value of k from data involving substantial contributions of extracolumn volumes, we should use the equation

$$k = \frac{t_R'}{t_M} = \frac{t_R - t_M^+}{t_M^+ - \Sigma t_{M,e}} \tag{28}$$

instead of Eq. (26), where $\Sigma t_{M,e}$ apparently is given by the equation $\Sigma t_{M,e} = \Sigma(V_{M,e}/F_{P,e})$, with $F_{P,e}$ the volumetric carrier gas flow rate (at the actual pressure and temperature) in a given extracolumn space. If K is to be calculated, Eq. (27) is also valid when t_M and V_M^0 are substituted by the time t_M^+ and the apparent volume $V_M^{\S}$. Hence it follows that, when calculating the capacity factor from retention data, it is necessary to know and take into account the extracolumn volumes, whereas in the calculation of the partition coefficient the effect of extracolumn volumes is compensated for. Thus, in chromatographic terms, the partition coefficient K of a given solute in a given GLC system under the given conditions is defined as the net retention volume ($V_N = V_R^0 - V_M^0$) of this solute per unit volume of the sorbent (liquid stationary phase), all the volumes being considered at the temperature and mean pressure in the column. Since the volumes of both phases of the system depend on temperature and pressure, there are practical as well as theoretical reasons that the quantity K is not very suitable. For this reason, the quantity called the specific retention volume, and denoted V_g, was introduced. It is defined as the net retention volume per gram of the liquid stationary phase, measured at the temperature and mean pressure in the column and expressed at a standard temperature of 0°C. According to this definition, it holds that

$$V_g = \frac{V_R^0 - V_M^0}{W_S} \frac{T^0}{T} = K \frac{T^0}{Td_S} \tag{29}$$

where W_S, T, T^0, and d_S are the mass of the liquid stationary phase in the column, temperature of the column, temperature of 273.15 K, and density of the liquid stationary phase, respectively. It can easily be shown that the specific retention volume V_g represents the partition coefficient, defined as

$$V_g = \left(\frac{g_{iS}}{c_{iM}}\right)_{eq} \frac{T^0}{T} \tag{30}$$

where g_{iS} and c_{iM} are the solute mass fraction in the liquid stationary phase (at infinite solute dilution) and the solute concentration (mass/volume) in the gaseous phase at the conditions (T and $\langle P \rangle$) in the column. Recalculation from the column temperature to 273.15 K is merely a correction for the thermal expansivity of the gaseous mobile phase. The quantity V_g is obviously a function of temperature. Calculation of V_g from directly measured experimental data is done with the equation

$$V_g = \frac{t_R - t_M}{W_S} \frac{F_{Pf} P_f T^0}{\langle P \rangle T_f} \tag{31}$$

where, in addition to the symbols already presented, F_{Pf} is the volumetric carrier gas flow rate as measured at pressure P_f and temperature T_f in the flowmeter. The mean pressure $\langle P \rangle$ can be calculated directly by Eq. (21).

In analytical gas chromatography, extensive use is made of relative retention data and retention indices. The relative retention of solute i (i.e., the retention of this solute relative to that of reference solute j, r_{ij}), is defined by the relation

$$r_{ij} = \frac{t'_{R,i}}{t'_{R,j}} = \frac{V_{N,i}}{V_{N,j}} = \frac{V_{g,i}}{V_{g,j}} = \frac{k_i}{k_j} = \frac{K_i}{K_j} = \frac{r_{ik}}{r_{jk}} \tag{32}$$

where r_{ik} and r_{jk} are the retentions of solutes i and j relative to reference solute k. The retention times or volumes of solute i and reference compound j (or k) must be measured on the same column and under the same conditions. Provided such quantities as V_g, k, K, and r are used to express relative retention, it is theoretically possible to use, for solute i and the reference compound, data that have been measured on different columns with the same stationary liquid phase and at the same temperature. However, with respect to the requirement of maximum attainable reproducibility, it is even better in these cases to work with the same column.

According to Kováts' original definition [18], the retention index is a 100-multiple of the carbon number of a hypothetical n-alkane that would display, on a given column and under given conditions, the same t'_R as the solute compound under investigation. Hence, in order to determine the retention index of solute i, it is necessary to measure, on a given column at constant conditions, the t'_R values for solute i and at least two reference n-alkanes with carbon numbers z and z + n. The retention index I_i is then calculated by the equation

$$I_i = 100 \left[z + n \frac{\log (t'_{R,i}/t'_{R,z})}{\log (t'_{R,z+n}/t'_{R,z})} \right] \tag{33}$$

Instead of t'_R (or V_N), however, it is possible to employ V_g, k, K, and r, with the same consequences as mentioned for relative retention.

Application of LeRosen's Model of Chromatography

The precise mathematical solution of a general model of linear chromatography has not yet been found. Therefore, approximate methods were sought that would make it possible to characterize this model by virtue of analysis of the individual mechanisms operating in the chromatographic process [19,20]. Such an approach easily leads to the basic retention equation and provides the description of the individual zone-spreading factors in terms of the physical properties of the system. When limited to the various aspects of chromatographic retention, this approach essentially corresponds to LeRosen's concept of chromatography [21]. Regarding the ease with which this method makes it possible to arrive at retention equations even in cases of more complex retention mechanisms, we shall consider in this section a system in which the adsorption of solute on the surface of bare support and at the gas/stationary liquid and support/stationary liquid phase interfaces takes place along with the dissolution of solute in the bulk stationary liquid. Such a system is shown schematically in Fig. 2. Under the term "support," we shall mean a granular solid material or the wall of a capillary column.

During the time that an "average" solute molecule spends in the mobile phase, it moves forward at the same velocity as the mobile phase at a given location in the column, but during its presence in the sorbent this molecule is stagnant. Hence, molecules of different solutes spend, on an average, the same time in the mobile phase during chromatography under given conditions, independent of their holdup

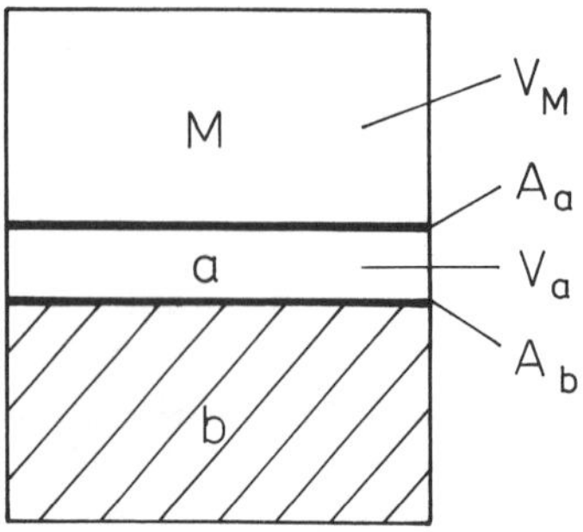

Figure 2. Schematic representation of the system composed of a mobile phase (M), stationary liquid (a), and solid support (b). V_M, volume of the mobile phase; A_a, surface area of the stationary liquid phase/mobile phase interface; V_a, volume of the bulk liquid phase; A_b, surface area of the solid support/stationary liquid phase interface; A_b^*, surface area of the bare solid support surface (not shown in the figure).

time in the sorbent, and the velocity of migration of the center of a solute zone relative to the mobile-phase velocity is given by the average probability of occurrence of the solute molecules in the mobile phase. This probability equals the average fraction of the average total retention time, during which the solute molecules occur in the mobile phase, and the average total retention time is apparently the sum of the average times that the solute molecules spend in the mobile and in the stationary phase(s) during their residence in the column. The average fraction of the average total retention time that the solute molecules spend in either phase of the system during chromatography is equal to the number of solute molecules (moles) in the given phase within a very narrow segment around the center of the migrating elution zone, relative to the total number of solute molecules (moles) in both (all) phases within that segment.

Let t_{iM}, t_{ia}^S, t_{ia}^V, t_{ib}^S, and t_{ib}^{S*} be the average times the molecules of solute i spend, during the migration of the elution zone down the column, in the mobile phase, adsorbed at the surface of the liquid stationary phase (gas/liquid phase interface), dissolved in the bulk liquid phase, and adsorbed at the liquid phase/solid support interface and on the bare support surface, respectively. The corresponding numbers of moles of solute i in the respective phase or at the phase interfaces are n_{iM}, n_{ia}^S, n_{ia}^V, n_{ib}^S, and n_{ib}^{S*}. Then, according to the foregoing concept,

$$\frac{t_{iM}}{\Sigma t_{ij}} = \frac{n_{iM}}{\Sigma n_{ij}} = \frac{m_{iM}}{\Sigma m_{ij}} = \frac{u_i}{u} = R_i \qquad (34)$$

where m is the mass of solute, u_i is the velocity of migration of the center of zone of solute i, and R_i is the so-called retardation factor (which with certain reservations [22] is equal to the quantity R_F, used to express retention in chromatography in planar systems). Since

$$\Sigma t_{ij} = t_{iM} + t_{ia}^{s} + t_{ia}^{V} + t_{ib}^{s} + t_{ib}^{s*} = t_R \tag{35}$$

$$\Sigma n_{ij} = n_{iM} + n_{ia}^{s} + n_{ia}^{V} + n_{ib}^{s} + n_{ib}^{s*} \tag{36}$$

and the time t_{iM} is apparently identical to the dead retention time t_M, we can write, with regard to Eq. (34),

$$t_R = t_M\left(1 + \frac{n_{ia}^{s}}{n_{iM}} + \frac{n_{ia}^{V}}{n_{iM}} + \frac{n_{ib}^{s}}{n_{iM}} + \frac{n_{ib}^{s*}}{n_{iM}}\right) \tag{37}$$

where the superscripts s and V denote adsorption at a surface and dissolution in bulk liquid. It can formally be written for n_{ia}^{V}/n_{iM} and n_{ib}^{s}/n_{iM} that

$$\frac{n_{ia}^{V}}{n_{iM}} = \frac{n_{ia}^{s}}{n_{iM}}\,\frac{n_{ia}^{V}}{n_{ia}^{s}}$$

$$\frac{n_{ib}^{s}}{n_{iM}} = \frac{n_{ia}^{V}}{n_{iM}}\,\frac{n_{ib}^{s}}{n_{ia}^{V}}$$

so that Eq. (37) can be rewritten as

$$t_R = t_M\left(1 + \frac{n_{ia}^{s}}{n_{iM}} + \frac{n_{ia}^{s}}{n_{iM}}\,\frac{n_{ia}^{V}}{n_{ia}^{s}} + \frac{n_{ia}^{s}}{n_{iM}}\,\frac{n_{ia}^{V}}{n_{ia}^{s}}\,\frac{n_{ib}^{s}}{n_{ia}^{V}} + \frac{n_{ib}^{s*}}{n_{iM}}\right) \tag{38}$$

where all the molar ratios on the right side concern situations in adjacent phases. These ratios apparently constitute the respective capacity ratios, according to which it can be written, for a column with a uniform packing,

$$\frac{n_{ia}^{s}}{n_{iM}} = k_{a/M}^{s/V} = K_{a/M}^{s/V}\,\frac{A_a}{V_M^0} \tag{39}$$

$$\frac{n_{ia}^{V}}{n_{ia}^{s}} = k_{a/a}^{V/s} = K_{a/a}^{V/s} \frac{V_a}{A_a} \tag{40}$$

$$\frac{n_{ib}^{s}}{n_{ia}^{V}} = k_{b/a}^{s/V} = K_{b/a}^{s/V} \frac{A_b}{V_a} \tag{41a}$$

$$\frac{n_{ib}^{s*}}{n_{iM}} = k_{b/M}^{s*/V} = K_{b/M}^{s*/V} \frac{A_b^*}{V_M^0} \tag{41b}$$

where the meaning of the superscripts and subscripts with k and K follows from the specification of the respective molar ratios, and A_a, A_b, A_b^*, V_M^0, and V_a denote the total surface areas of the phase interfaces M/a, a/b, and M/b and the total volumes of the phases M and a in the column. Upon substitution from Eqs. (39), (40), and (41) into Eq. (38), we obtain

$$t_R = t_M \left(1 + K_{a/M}^{s/V} \frac{A_a}{V_M^0} + K_{a/M}^{s/V} K_{a/a}^{V/s} \frac{V_a}{V_M^0} + K_{a/M}^{s/V} K_{a/a}^{V/s} K_{b/a}^{s/V} \frac{A_b}{V_M^0} + K_{b/M}^{s*/V} \frac{A_b^*}{V_M^0} \right) \tag{42}$$

Equation (42) can also be rewritten as

$$t_R = t_M \left(1 + K_{a/M}^{s/V} \frac{A_a}{V_M^0} + K_{a/M}^{V/V} \frac{V_a}{V_M^0} + K_{b/M}^{s/V} \frac{A_b}{V_M^0} + K_{b/M}^{s*/V} \frac{A_b^*}{V_M^0} \right) \tag{43}$$

and by comparing the right sides of Eqs. (42) and (43), we have

$$K_{a/M}^{s/V} K_{a/a}^{V/s} = K_{a/M}^{V/V}$$

$$K_{a/M}^{s/V} K_{a/a}^{V/s} K_{b/a}^{s/V} = K_{b/M}^{s/V} = K_{a/M}^{V/V} K_{b/a}^{s/V}$$

The dimensions of the partition coefficients $K^{s/V}$, $K^{V/s}$, and $K^{V/V}$ follow from the conditions that the individual terms on the right sides of Eqs. (42) and (43) must be dimensionless; (e.g., $K^{s/V}_{a/M}$ is defined as the ratio of the surface concentration and volume concentration, $[(m^s_{ia}/A_a)/c_{iM}]_{eq}$, whereas $K^{V/V}_{a/M}$ is a ratio of volume concentrations, namely, $[c_{ia}/c_{iM}]_{eq}$). Upon multiplying Eqs. (42) and (43) by the volumetric carrier gas flow rate $F_{<p>}$, we obtain the corresponding retention equations expressed in terms of retention volumes:

$$V^0_R = V^0_M + K^{s/V}_{a/M} A_a + K^{s/V}_{a/M} K^{V/s}_{a/a} V_a + K^{s/V}_{a/M} K^{V/s}_{a/a} K^{s/V}_{b/a} A_b + K^{s*/V}_{b/M} A^*_b \tag{44}$$

$$V^0_R = V^0_M + K^{s/V}_{a/M} A_a + K^{V/V}_{a/M} V_a + K^{s/V}_{b/M} A_b + K^{s*/V}_{b/M} A^*_b \tag{45}$$

From Eqs. (26) and (45), it follows that

$$kV^0_M = V_N = K^{s/V}_{a/M} A_a + K^{V/V}_{a/M} V_a + K^{s/V}_{b/M} A_b + K^{s*/V}_{b/M} A^*_b \tag{46}$$

By measuring retention data on columns with different defined contents of liquid stationary phase and plotting the V_N values against the corresponding volumes V_a, we obtain a graph from which it is possible to estimate the contributions of the individual retention mechanisms to the total volume of V_N. Such a situation is illustrated in Fig. 3. For the intercept Q on the coordinate V_N, obtained by extrapolating the linear course of the function $V_N(V_a)$ to $V_a = 0$, we find that $Q = K^{s/V}_{a/M} A_a + K^{s/V}_{b/M} A_b$ and the slope of the linear section is $dV_N/dV_a = K^{V/V}_{a/M}$. The value of Q can be interpreted as the contribution of solute adsorption at the surfaces of the liquid stationary phase and support to the value of V_N under conditions of complete coverage of the support with the stationary liquid. The value of $K^{V/V}_{a/M}$, obtained from the differentiation dV_N/dV_a, represents the net partition coefficient of the solute in the bulk liquid-gas system.

The problems of several concurrent retention mechanisms in GLC were dealt with by a number of authors, especially after Martin pointed out that the solute may be significantly adsorbed on the surface of the stationary liquid phase in some cases [23]. The papers on this topic are reviewed in the book by Conder and Young [24]. Relationship (46) represents a generalized retention equation, which is valid, in the present form, only for cases in which the individual retention mechanisms are independent of each other, i.e., at very low solute concentrations. A more detailed discussion of these problems can be found in Chap. 5.

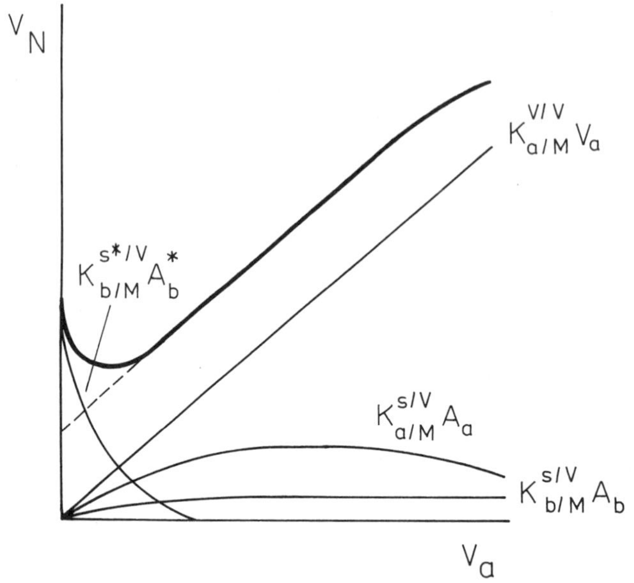

Figure 3. Representation of dependence of the net retention volume V_N on the volume of the stationary liquid V_a in the column. $K^{V/V}_{a/M}$, bulk stationary liquid/mobile phase partition coefficient; $K^{s/V}_{a/M}$, stationary liquid surface/mobile phase partition (adsorption) coefficient; $K^{s/V}_{b/M}$, solid support surface/bulk stationary liquid partition (adsorption) coefficient; $K^{s*/V}_{b/M}$, bare solid support surface/mobile phase partition (adsorption) coefficient. For V_a, A_a, A_b, and A^*_b, see legend to Fig. 2.

B. Real Gaseous Phase and Pressure-Dependent Capacity Ratio

The approximative model of gas chromatography with an ideal gaseous phase and invariant capacity ratio along the column is the model most frequently used. In comparison with other sources of uncertainty in specifying the chromatographic system (variations in the properties of the stationary phase and its true content in the column, solute adsorption at the phase interfaces, errors in the readout of the true retention time, and others), it is quite sufficient for the routine presentation of retention data for analytic purposes. However, in a real GLC column, the gaseous phase is more or less nonideal and the degree of this nonideality varies, owing to the pressure gradient, along the column. The zone maximum velocity is not a constant fraction (although it is still a defined fraction) of the mobile-phase velocity in this case, even when the zone migrates strictly under conditions of linear quasi-equilibrium chromatography. The properties of the stationary liquid phase are also dependent on pressure, but with regard to the relatively small compressibility of liquids, these pressure effects can usually be neglected. Hence, capacity ratios and related quantities determined by virtue of the foregoing idealized model actually represent effective mean values, the magnitudes of which depend to various degrees on the kind of carrier gas being used and the pressure profile in the column. Performing consistent corrections for these nonideality effects, with data precisely measured in properly designed and specified GLC systems, would certainly significantly improve the reproducibility of chromatographic retention data. For physicochemical interpretations of retention data, such specifications and corrections are necessary. A most suitable way to make these corrections is to extrapolate data measured at an actual pressure profile in the column to zero pressure. Since the variability of retention data for a given solute in various GLC systems with an identical liquid stationary phase and at a given temperature stems from a pressure-induced variability in the thermodynamic properties of all the components of the system (disregarding other, pressure-independent factors), it is evident that thermodynamics is an indispensible tool for solving this problem.

The first attempt to take account of the effects of gas-phase nonideality on solute retention in GLC systems was made by Kwantes and Rijnders in 1958 [25]. More rigorous theories were published by Everett and Stoddard in 1961 [26] and, a year later, by Desty et al. [27]. Both these theories involved the concept of a hypothetical mean pressure isobaric column. In 1965, Everett [28] further and substantially refined his former theory. The most rigorous models to date of retention in GLC systems with a real gaseous phase up to now are due to Cruickshank et al. [14,29]. A somewhat different approach to these problems was taken by Martire and Locke [30] and by Schettler et al. [31]. Systems at higher mean pressures and with relatively small pressure gradients were studied by Wičar et al. [32-34]. The treatise presented in this section follows (except for some formal details) Everett's model [28]. The necessary thermodynamics was gathered from the books by Hála et al. [35], Denbigh [36], and Moore [37].

Conventionally Measured K and V_g versus Local K and V_g

In contradistinction to the model with invariant k, a column in which k depends on the pressure profile is, so far as its retention properties are concerned, not equivalent to a hypothetical isobaric column at the mean pressure $\langle P \rangle$ (and mean value of k). However, even in this case the concept of a hypothetical isobaric column may, under certain circumstances, be an acceptable approximation. However, let us now deal with a real column in which k is a function of z. In this case, it is also possible to proceed from the equation for retention time in an infinitesimal column segment, obtained by solving the model of ideal linear gas chromatography, but it is necessary to calculate using a local capacity ratio k(z) instead of an invariant. Thus,

$$dt_R = \frac{1 + k(z)}{u(z)} \, dz \qquad (47)$$

As $dt_M = dz/u(z)$, it is possible to write, for the adjusted retention time,

$$t'_R = t_R - t_M = \int_0^L \frac{k(z)}{u(z)} dz \qquad (48)$$

The local capacity ratio is defined exactly by the relationship $k(z) = K(z)A_S(z)/A_M(z)$, but at this stage of the derivation we introduce the approximation

$$A_S(z) = \frac{1}{L} \int_0^L A_S(z) \, dz = \frac{V_S}{L}$$

This apparently implies the approximation

$$A_M(z) = \frac{V_M^0}{L}$$

Thus, by employing the equation $A_M(z) = F(z)/u(z)$, we obtain:

$$t'_R = \frac{V_S}{L} \int_0^L \frac{K(z)}{F(z)} \, dz \qquad (49)$$

and after substituting for F(z) from the modified real gas equation of state $F(z)P(z)/TZ_M = F_{Pf}P_f/T_fZ_{Mf}$, we have

$$t'_R = \frac{V_S T_f Z_{Mf}}{F_{Pf} P_f T} \; \frac{1}{L} \int_0^L \frac{K(z)P(z)}{Z_M} dz \qquad (50)$$

where, besides the already known symbols, Z_M and Z_{Mf} are the local factor (function of z) of carrier gas nonideal compressibility in the column and carrier gas nonideal compressibility under the conditions of measurement in the flowmeter, respectively.

Equation (50) leads directly to the relationship between the conventional partition coefficient, as calculated from directly measured experimental data, and the true (considering the nonideal compressibility of the carrier gas) local partition coefficient K(z). The conventional partition coefficient K is calculated by

$$K = \frac{t'_R}{V_S} \frac{F_{Pf} P_f T}{\langle P \rangle T_f} \tag{51}$$

so that by combining Eqs. (50) and (51) we obtain, after rearrangement,

$$K = \frac{Z_{Mf}}{\langle P \rangle} \frac{1}{L} \int_0^L \frac{K(z)P(z)}{Z_M} dz = Z_{Mf} \frac{\langle K(z)P(z)/Z_M \rangle}{\langle P \rangle} \tag{52}$$

When disregarding the factors Z_M and Z_{Mf} (which do not differ very much from unity under usual GLC conditions) in Eq. (52), we see that the partition coefficient calculated by Eq. (51) is actually the mean value of the product K(z)P(z) divided by the mean pressure $\langle P \rangle$. Since the mean value $\langle K(z)P(z)/Z_M \rangle$ generally does not equal the product of the individual mean values of the respective functions of z, (i.e., $\langle K(z) \rangle \langle P(z) \rangle / \langle Z_M \rangle$), the last term of relation (52) cannot be reduced to $(Z_{Mf}/\langle Z_M \rangle)\langle K(z) \rangle$. This finding actually explains why it is not justified to use the concept of a hypothetical mean pressure isobaric column with this model. As has already been mentioned, this model can in practice be used when setting appropriate limitations, but it is necessary to reckon with a rougher approximation. For the conventionally defined specific retention volume [see Eq. (31)], we can write, analogically to Eq. (52),

$$V_g = \frac{Z_{Mf}}{\langle P \rangle} \frac{1}{L} \int_0^L \frac{V_g(z)P(z)}{Z_M^0} dz = \frac{\langle V_g(z)P(z)/Z_M^0 \rangle}{\langle P \rangle} \tag{53}$$

where Z_M^0 is the coefficient of nonideal compressiblity of carrier gas at 273.15 K and a local pressure P in the column.

K(z) and $V_g(z)$ as Functions of the Thermodynamic Properties of the GLC System

In order to proceed further, it is necessary to express K(z) and $V_g(z)$ in terms of thermodynamics. The individual pressure-dependent variables occurring in the relationships for K(z) and $V_g(z)$ will also represent certain local values in a given infinitesimal column segment of width dz, but the respective symbols will be quoted without the affix (z) for the sake of brevity. However, for consistency with the previous notation, the affix (z) will continue to be used in symbols for local partition coefficients and specific retention volumes.

At an infinite dilution of the solute in the liquid stationary phase, the local partition coefficient can be expressed as

$$K(z) = \left(\frac{n_{iS}}{n_{iM}}\right)_{eq} \frac{V_M^0}{V_S} = \left(\frac{x_i}{y_i}\right)_{eq} \frac{n_S}{n_M} \frac{V_M^0}{V_S} = \left(\frac{x_i}{Py_i}\right)_{eq} \frac{Z_M RT d_S}{M_S} \tag{54}$$

where $(n_{iS}/n_{iM})_{eq}$, $(x_i/y_i)_{eq}$, V_M^0/V_S, d_S, and Z_M are the ratio of the equilibrium amounts (numbers of moles) of solute in the liquid and gaseous phases, ratio of the equilibrium mole fractions of solute in the liquid and gaseous phases, gaseous-liquid phase volume ratio, density of the liquid phase, and coefficient of nonideal compressibility of the carrier gas, respectively, all at a local P within a column segment defined by two cross sections at distances z and z + dz from the column inlet. The pressure-independent quantities n_S/n_M, R, T, and M_S are the liquid-gas substance amount ratio in the column, molar perfect gas constant, column temperature, and molar mass of the stationary liquid phase, respectively. In a similar way, for the local specific retention volume, we obtain

$$V_g(z) = \left(\frac{x_i}{Py_i}\right)_{eq} \frac{Z_M^0 RT^0}{M_S} \tag{55}$$

And since

$$f_{iM} = \nu_{iM} P y_i = \gamma_i^\theta f_{i,P}^\theta x_i \tag{56}$$

where f_{iM} and ν_{iM} are the fugacity and fugacity coefficient of gaseous solute in the carrier gas-solute mixture, and γ_i^θ and $f_{i,P}^\theta$ are the Raoult's law solute activity coefficient in the liquid phase-solute solution and fugacity of pure liquid solute (at the local pressure P), respectively. We can write

$$\left(\frac{x_i}{Py_i}\right)_{eq} = \frac{\nu_{iM}}{\gamma_i^\theta f_{i,P}^\theta} \tag{57}$$

and by substituting from Eq. (57) into Eqs. (54) and (55), we obtain for K(z) and $V_g(z)$

$$K(z) = \frac{\nu_{iM} Z_M RT d_S}{\gamma_i^\theta f_{i,P}^\theta M_S} \tag{58}$$

$$V_g(z) = \frac{\nu_{iM} Z_M^0 R T^0}{\gamma_i^\theta f_{i,P}^\theta M_S} \tag{59}$$

Local K and V_g versus K and V_g at Zero Pressure

Equations (58) and (59) provide for expressing the relationships between K or V_g at the actual (local) pressure and at zero pressure. Just as at P = 0 we have $\nu_{iM} = 1$, $Z_M = 1$, $Z_M^0 = 1$, $d_S = d_{S,0}$, $\gamma_i^\theta = \gamma_{i,0}^\theta$, and $f_{i,P}^\theta = f_{i,0}^\theta$, so at zero pressure do we have, for K and V_g,

$$K_0 = \frac{R T d_{S,0}}{\gamma_{i,0}^\theta f_{i,0}^\theta M_S} \tag{60}$$

$$V_{g,0} = \frac{R T^0}{\gamma_{i,0}^\theta f_{i,0}^\theta M_S} \tag{61}$$

where the subscript 0 indicates that values at zero pressure are concerned. It follows from Eqs. (58) through (61) that

$$\frac{K(z)}{K_0} = \nu_{iM} Z_M \frac{f_{i,0}^\theta}{f_{i,P}^\theta} \frac{\gamma_{i,0}^\theta}{\gamma_i^\theta} \frac{d_S}{d_{S,0}} \tag{62}$$

$$\frac{V_g(z)}{V_{g,0}} = \nu_{iM} Z_M^0 \frac{f_{i,0}^\theta}{f_{i,P}^\theta} \frac{\gamma_{i,0}^\theta}{\gamma_i^\theta} \tag{63}$$

where the individual factors on the right side of Eq. (62) concern, respectively, the nonideality of the carrier gas-solute mixture, the nonideal compressibility of pure carrier gas, the dependence of the fugacity of pure liquid solute on pressure, the dependence of the solute activity coefficient on pressure, and the compressibility of the stationary liquid phase. In order to proceed, the individual pressure-dependent quantities on the right sides of Eqs. (62) and (63) must be expressed as explicit functions of pressure.

Individual Pressure-Dependent Thermodynamic Quantities as Functions of Pressure

It is generally true for ν_{iM} that

$$\ln \nu_{iM} = \frac{1}{RT} \int_0^P \left(\overline{V}_{iM} - \frac{RT}{P} \right) dP \tag{64}$$

where $\overline{V}_{iM}$ is the partial molar volume of solute i in the carrier gas-solute mixture. By using the virial equation of state, truncated after the second term, the following approximate relationship can be derived for $\overline{V}_{iM}$ at very low solute concentrations in the gaseous mixture:

$$\overline{V}_{iM} = \frac{RT}{P} + 2B_{iM} - B_M \tag{65}$$

where B_{iM} and B_M are the second cross-virial coefficient for the carrier gas-solute mixture and the second virial coefficient for pure carrier gas, both at the column temperature T, respectively. By combining Eqs. (64) and (65), integrating, and rearranging, we obtain for ν_{iM}

$$\nu_{iM} = e^{(2B_{iM} - B_M)P/RT} \tag{66}$$

For Z_M it holds approximately that

$$Z_M = 1 + \frac{B_M P}{RT} \tag{67}$$

As $B_M P/RT << 1$ under usual GLC conditions, we can apply the rules for calculating with small numbers and write the approximate relationship $1 + B_M P/RT = e^{(B_M P/RT)}$, thus obtaining for Z_M

$$Z_M = e^{B_M P/RT} \tag{68}$$

The pressure dependence of the fugacity of pure liquid solute can best be derived from Poynting's relationship

$$f_{i,P}^{\theta} = f_i^{\theta} \exp \int_{P_i^{\theta}}^{P} \frac{V_i^L}{RT} \, dP = f_{i,0}^{\theta} \exp \int_0^P \frac{V_i^L}{RT} \, dP \tag{69}$$

where P_i^{θ} is the saturation vapor pressure of pure liquid solute and V_i^L is the molar volume of liquid solute. Assuming that V_i^L is practically constant within the pressure range 0 through P, it can be written $f_{i,P}^{\theta} = f_{i,0}^{\theta} e^{(V_i^L P/RT)}$, and we have for $f_{i,0}^{\theta}/f_{i,P}^{\theta}$

$$\frac{f^{\theta}_{i,0}}{f^{\theta}_{i,P}} = e^{(-V^L_i P/RT)} \tag{70}$$

Pressure changes in the solute activity coefficient, at a constant composition and a constant temperature of the liquid mixture, are described by the equation

$$\frac{\partial \ln \gamma^{\theta}_i}{\partial P} = \frac{\overline{V}^L_{iS} - V^L_i}{RT} \tag{71}$$

so that

$$\ln \gamma^{\theta}_i = \int_0^P \frac{\overline{V}^L_{iS} - V^L_i}{RT} dP \tag{72}$$

and we have for $\gamma^{\theta}_{i,0}/\gamma^{\theta}_i$

$$\frac{\gamma^{\theta}_{i,0}}{\gamma^{\theta}_i} = e^{[-(\overline{V}^L_{iS} - V^L_i)P/RT]} \tag{73}$$

The compressibility of the stationary liquid phase is defined by the coefficient of isothermal compressibility,

$$b_S = -\frac{1}{V_S} \frac{\partial V_S}{\partial P} \tag{74}$$

The right side of Eq. (74) can be rearranged as

$$-\frac{1}{V_S} \frac{\partial V_S}{\partial P} = V_S \frac{\partial (1/V_S)}{\partial P} = \frac{V_S}{m_S} \frac{\partial (m_S/V_S)}{\partial P} = \frac{1}{d_S} \frac{\partial d_S}{\partial P} = \frac{\partial \ln d_S}{\partial P}$$

so that

$$\ln d_S = b_S \int_0^P dP$$

and we can write for $d_S/d_{S,0}$

$$\frac{d_S}{d_{S,0}} = e^{b_S P} \tag{75}$$

Conventionally Measured Chromatographic K and V_g versus K and V_g at Zero Pressure

Substitution from Eqs. (66), (68), (70), (73), and (75) into Eqs. (62) and (63) immediately results in relationships for K(z) and $V_g(z)$ as explicit functions of P, with K_0 and $V_{g,0}$ as parameters:

$$K(z) = K_0 \, e^{[(2B_{iM} - \overline{V}^L_{iS})/RT + b_S]P} \tag{76}$$

$$V_g(z) = V_{g,0} \, e^{[(2B_{iM} - \overline{V}^L_{iS})/RT - B_M/RT + B^0_M/RT^0]P} \tag{77}$$

where B^0_M is the second virial coefficient of the carrier gas at a temperature of $T^0 = 273.15$ K. By Eqs. (76) and (77) we have arrived at the last steps of our derivation, the implementation of which will give us relationships for the conventionally determined K and V_g as explicit functions of P, with K_0 and $V_{g,0}$ as parameters. The expression $(2B_{iM} - \overline{V}^L_{iS})/RT$ in Eqs. (76) and (77) will be denoted by β^+, and in order to make the integration easier, we shall write

$$K(z) = K_0[1 + (\beta^+ \; b_S)P] \tag{78}$$

$$V_g(z) = V_{g,0}\left[1 + \left(\beta^+ - \frac{B_M}{RT} + \frac{B^0_M}{RT^0}\right) P\right] \tag{79}$$

instead of $K(z) = K_0 e^{(\beta^+ + b_S)P}$ and $V_g(z) = V_{g,0} \times e^{[\beta^+ - (B_M/RT) + (B^0_M/RT^0)]P}$. This change is justified provided $(\beta^+ + b_S)P \ll 1$ and $[\beta^+ - (B_M/RT) + (B^0_M/RT^0)]P \ll 1$. Fulfilling this condition is necessary not only for formal reasons (i.e., to facilitate mathematical operations) but mainly because it constitutes the basic premise for the whole model as such to be valid with sufficient accuracy. By using Eqs. (68), (78), (79), and (15) (i.e., we assume that there is an ideal pressure profile in the column), we can put the expression $K(z)P(z)/Z_M$ in Eq. (52) as

$$\frac{K(z)P(z)}{Z_M} = K_0\left\{1 + (\beta \; b_S)\left[\frac{z}{L}(P_i^2 - P_o^2) + P_o^2\right]^{1/2}\right\}\left[\frac{z}{L}(P_i^2 - P_o^2) + P_o^2\right]^{1/2} \tag{80}$$

where $\beta = (2B_{iM} - B_M - \bar{V}^L_{iS})/RT$, and for the mean value $\langle K(z)P(z)/Z_M \rangle$ we obtain according to Eq. (52)

$$\left\langle \frac{K(z)P(z)}{Z_M} \right\rangle = \frac{K_0}{L} \int_0^L \left[\frac{z}{L}(P_i^2 - P_o^2) + P_o^2 \right]^{1/2} dz + \frac{K_0(\beta + b_S)}{L} \int_0^L \left[\frac{z}{L}(P_i^2 - P_o^2) + P^2 \right] dz \tag{81}$$

After carrying out the integration we have

$$\left\langle \frac{K(z)P(z)}{Z_M} \right\rangle = K_0 \left[\frac{2}{3} \frac{P_i^3 - P_o^3}{P_i^2 - P_o^2} + (\beta + b_S) \frac{P_i^2 + P_o^2}{2} \right] \tag{82}$$

where the expression $(P_i^2 + P_o^2)/2$ can be broken up to give

$$\frac{P_i^2 + P_o^2}{2} = \left(\frac{2}{3} \frac{P_i^3 - P_o^3}{P_i^2 - P_o^2} \right) \left(\frac{3}{4} \frac{P_i^4 - P_o^4}{P_i^3 - P_o^3} \right) \tag{83}$$

so that by factoring $2(P_i^3 - P_o^3)/3(P_i^2 - P_o^2)$ out of the last term on the right of Eq. (82), the latter can be rewritten as

$$\left\langle \frac{K(z)P(z)}{Z_M} \right\rangle = K_0 \left\{ \frac{2}{3} \frac{P_i^3 - P_o^3}{P_i^2 - P_o^2} \left[1 + (\beta + b_S) \frac{3}{4} \frac{P_i^4 - P_o^4}{P_i^3 - P_o^3} \right] \right\} \tag{84}$$

The expression $2(P_i^3 - P_o^3)/3(P_i^2 - P_o^2)$ is apparently the mean column pressure $\langle P \rangle$ defined by Eq. (21), and the expression $3(P_i^4 - P_o^4)/4(P_i^3 - P_o^3)$ will be denoted by P". Upon carrying the $\langle P \rangle$ in Eq. (84) over to the left and taking account of Eq. (52), we obtain a relationship for the conventionally measured chromatographic partition coefficient K as an explicit function of P", with K_0 as a parameter:

$$\frac{K}{Z_{Mf}} = \frac{1}{Z_{Mf}} \frac{t_R - t_M}{V_S} \frac{F_{Pf} P_f T}{\langle P \rangle T_f} = K_0[1 + (\beta + b_S)P''] \tag{85}$$

By using Eq. (67), the coefficient Z_{Mf} can be expressed as

$$Z_{Mf} = 1 + \frac{B_{Mf}P_f}{RT_f} \tag{86}$$

and Eq. (85) can be rewritten as

$$K = \frac{t_R - t_M}{V_S} \frac{F_{Pf}P_fT}{\langle P\rangle T_f} = K_0 \left\{ [1 + (\beta + b_S)P''] \left[1 + \frac{B_{Mf}P_f}{RT_f} \right] \right\} \tag{87}$$

where B_{Mf} is the carrier gas second virial coefficient at the flowmeter temperature T_f. Upon converting Eq. (87) back to the more usual exponential form, we get

$$K = \frac{t_R - t_M}{V_S} \frac{F_{Pf}P_fT}{\langle P\rangle T_f} = K_0 e^{(\beta+b_S)P''+B_{Mf}P_f/RT_f} \tag{88}$$

For V_g we arrive, by an analogous procedure [see Eqs. (53) and (79) through (88)] at the relationship

$$V_g = \frac{t_R - t_M}{W_S} \frac{F_{Pf}P_fT^0}{\langle P\rangle T_f} = V_{g,0} e^{(\beta P''+B_{Mf}P_f/RT_f} \tag{89}$$

Let us recall that β, b_S, and P'' in Eqs. (88) and (89) are defined as $\beta = (2B_{iM} - B_M - \bar{V}_{iS}^L)/RT$, $b_S = -(1/V_S)(\partial V_S/\partial P)_T$, and $P'' = 3(P_i^4 - P_o^4)/4(P_i^3 - P_o^3)$.

Let us now recapitulate the more important approximation assumptions that have been applied in the derivation of Eqs. (88) and (89):

1. Ideal pressure profile in the column
2. Carrier gas viscosity independent of pressure
3. Insolubility of the carrier gas in the stationary liquid phase
4. Nonvolatility of the stationary liquid phase
5. Calculation with merely the second virial coefficients
6. Application of the rule $e^a \approx 1 + a$

Martire and Locke [30] have shown the deviations of real pressure profiles in the column from ideal to be quite negligible even at an appreciable nonideality of the carrier gas (Z_M appreciably different from unity), so that approximation (1) can hardly have any substantial effect on the resultant situation. However, all the other approximations may be substantially more consequential, and they actually limit the applicability of Eqs. (88) and (89) to column pressures and pressure drops of up to several tents of megapascals (i.e., several atmospheres) with commonly employed carrier gases at usual temperatures. It has been found [34], by measuring V_g data at absolute column pressures of up to about 10 MPa and analyzing the dependences of the $V_{g,0}$ values on pressure, that the carrier gas solubility in the liquid phase may contribute perceptibly to the dependence of V_g on pressure even under common GLC conditions.

Very rigorous (although necessarily also approximate) models, of gas chromatography, involving more exact mathematics and taking into account the real pressure profile in the column, the pressure-dependent viscosity of the carrier gas, solubility of the carrier gas in the stationary liquid phase, and nonequilibrium effects, have been presented by Cruickshank et al. [14,29]. These models certainly give a truer picture of the real situation, and they can be applied within a broader range of pressures than the model represented by Eqs. (88) and (89), but this is at the price of substantially more complicated relationships. If the factors Z_M, Z_M^0. and Z_{Mf} in Eqs. (56) and (57) are ignored and the mean values of the products K(z)P(z) and $V_g(z)P(z)$ are replaced by the respective products of the individual mean values of K(z), $V_g(z)$, and P(z) (i.e., if it is supposed that $K = \langle K(z) \rangle$ and $V_g = \langle V_g(z) \rangle$) the relationships $K = K_0 e^{(\bar{\beta}^+ + b_S)\langle P \rangle}$ and $V_g = V_{g,0} e^{\beta^+ \langle P \rangle}$ are obtained, which conform (apart from the negligible term b_S) to the relationships published by Everett and Stoddard [26]. If, in addition, $\bar{V}_{iS}^L$ were replaced by V_{iS}^L in the term β^+, the relationships of Desty et al. [27] would be obtained.

Pressure Dependence of Relative Retention Data and Retention Indices

Upon comparing Eqs. (32), (88), and (89), we can immediately write

$$r_{ij} = r_{ij,0}\, e^{(\beta_i - \beta_j)P''} \tag{90}$$

where $r_{ij,0}$ is the relative retention of the studied (i) and reference (j) solutes at zero pressure. The retention index of solute i at zero pressure, $I_{i,0}$, is given according to Eqs. (32), (33), and (90) by

$$I_{i,0} = 100\left\{z + n\frac{\log(K_i/K_z) - [(\beta_i - \beta_z)/2.303]P''}{\log(K_{z+n}/K_z) - [(\beta_{z+n} - \beta_z)/2.303]P''}\right\} \tag{91}$$

After some formal rearrangements in Eq. (91), we obtain the approximate relationship

$$I_i = I_{i,0} - \frac{(\beta_{z+n} - \beta_z)(I_{i,0} - 100z) - 100n(\beta_i - \beta_z)}{2.303 \log (K_{z+n}/K_z)} P'' \tag{92}$$

from which it follows that

$$\frac{dI_i}{dP''} = \frac{100n(\beta_i - \beta_z) - (\beta_{z+n} - \beta_z)(I_{i,0} - 100z)}{2.303 \log (K_{z+n}/K_z)} \tag{93}$$

Equation (92) can be derived as follows. Upon introducing the simplified notation $\log (K_i/K_z) = K_{iz}$, $\log (K_{z+n}/K_z) = K_{zz}$, $(\beta_i - \beta_z)/2.303 = \beta_{iz}$, and $(\beta_{z+n} - \beta_z)/2.303 = \beta_{zz}$, we can write

$$\frac{I_i - 100z}{100n} = \frac{K_{iz}}{K_{zz}} \tag{94}$$

$$\frac{I_{i,0} - 100z}{100n} = \frac{K_{iz} - \beta_{iz}P''}{K_{zz} - \beta_{zz}P''} = f(P'') \tag{95}$$

By expanding the function f(P") to a McLaurin series,

$$f(P'') = f(0) + \frac{1}{1!} \frac{df(P'')}{dP''}(0)P'' + \frac{1}{2!} \frac{d^2f(P'')}{dP''^2}(0)P''^2 + \cdots$$

and truncating the expansion after the second term, we obtain

$$f(P'') = \frac{K_{iz}}{K_{zz}} + \frac{K_{iz}\beta_{zz} - K_{zz}\beta_{iz}}{(K_{zz} - \beta_{zz}P'')^2} P'' \tag{96}$$

and by substitution from Eqs. (94) and (95) and rearrangement, we have

$$I_i = I_{i,0} - 100n \frac{K_{iz}\beta_{zz} - K_{zz}\beta_{iz}}{(K_{zz} - \beta_{zz}P'')^2} P'' \tag{97}$$

which is one of approximate expressions of I_i as a function of P".

Let us now formally break down Eq. (96) as

$$f(P'') = \frac{K_{iz}}{K_{zz}} + \frac{K_{iz}\beta_{zz}P''}{K_{zz} - \beta_{zz}P''} \quad \frac{1}{K_{zz} - \beta_{zz}P''} - \frac{K_{zz}\beta_{iz}P''}{(K_{zz} - \beta_{zz}P'')^2} \tag{98}$$

replace $K_{iz}\beta_{zz}P''$ by $(K_{iz} - \beta_{iz}P')\beta_{zz}P''$ in the numerator of the second term, and neglect $\beta_{zz}P''$ in both the fraction $1/(K_{zz} - \beta_{zz}P'')$ and the denominator of the last term. These changes are justified, since $\beta_{iz}P'' << K_{iz}$ and $\beta_{zz}P'' << K_{zz}$. In this way we arrive at

$$f(P'') = \frac{K_{iz}}{K_{zz}} + \frac{\beta_{zz}P''}{K_{zz}} \frac{K_{iz} - \beta_{iz}P''}{K_{zz} - \beta_{zz}P''} - \frac{\beta_{iz}P''}{K_{zz}} \tag{99}$$

and after substitution from Eqs. (94) and (95) and rearrangement, we have

$$I_i = I_{i,0} - \frac{\beta_{zz}(I_{i,0} - 100z) - 100n\beta_{iz}}{K_{zz}} P'' \tag{100}$$

It is evident from Eq. (100) that if I_i, K_{zz}, β_{iz}, β_{zz}, and P'' are known, it is possible to determine $I_{i,0}$. By once again introducing into Eq. (100) the symbols for the initial notation, we obtain Eq. (92). An equation similar to Eq. (93) has been published (without showing its derivation) by Vigdergauz and Semkin [Ref. 38, Eq. (4)], however, the denominator (factor b_2) on the right side of their equation is incorrectly specified.

IV. SORPTION EQUILIBRIUM AND THE PARTITION COEFFICIENT

A. Problem of Sorption Equilibrium in a Migrating Chromatographic Zone

It is known from chemical thermodynamics that a system of several components and several phases is at equilibrium only if the chemical potentials of all the components in all the phases are the same. Such a situation can easily be imagined in a closed and isolated system, but a migrating chromatographic zone constitutes a typical open and nonstationary system, which is usually thermostated. In spite of this, there may exist in the elution chromatographic zone a region that is very near equilibrium along the entire path of migration of the zone. As has already been outlined on pages 104–105 (see Fig. 1), this is a very narrow segment, situated in the concentration maximum of the zone (migrating under the conditions of quasi-equilibrium linear chromatography). Whereas in the leading half of a migrating zone the transport of solute molecules from the mobile phase to the stationary phase prevails (i.e., sorption of solute takes place), in the trailing half, solute molecules are desorbed from the stationary phase to the mobile. Hence it can be assumed that only in the zone maximum, in which neither sorption nor desorption takes place, is there sorption equilibrium [i.e., the chemical potentials of the solute in both (all) phases are equal to each other]. From this view of the problem it is clear that retention data must be calculated to represent the course of migration of the zone concentration maximum (in quasi-equilibrium linear chromatography) when relationships between chromatographic retention data and thermodynamic properties of the solute in the chromatographic system are to be formulated.

If all the premises of quasi-equilibrium linear chromatography are fulfilled and the input zone is very narrow and symmetrical—and provided there are no significant extracolumn spaces in the chromatographic system—the concentration profile of a developed elution zone should be symmetrical gaussian. However, even in this ideal case, the corresponding peak, recorded at the column outlet by a perfectly linear and rapid detection system, will display a degree of asymmetry and the recorded peak maximum time will not coincide with the time at which the concentration maximum of the zone passes through the detector sensor [39,40]. This is because the zone is subject to spreading and, consequently, dilution while it is passing through the column outlet (detector). This spreading, although symmetrical with respect to the zone center, necessarily creates tailing, and the decrease in the zone maximum solute concentration results in the peak maximum being recorded somewhat sooner than the concentration maximum of the zone. However, the true retention time (representing sorption equilibrium) is given by the instant at which the zone maximum (center) passes through, rather than the time at which the peak maximum is recorded. Since the concentration maximum of a symmetrical zone coincides with the position of a section dividing the solute mass within the zone into two equal parts, it is probably the peak median (i.e., the time at which half of the peak area has been recorded by the detector) that best represents the true retention time [41] in this case.

If there are considerable deviations from equilibrium (still with strictly linear chromatography) in the migrating elution zone, the concentration maximum of the zone migrates more quickly than it would under quasi-equilibrium conditions. This means that the time of elution of the concentration maximum is reduced and the zone becomes asymmetric (tailing) owing to nonequilibrium. In such a case, the retention time of the concentration maximum represents a state of stationary nonequilibrium and true equilibrium is approximately represented by the retention time of the mass center of gravity of the zone.

However, in this case the time course of the detector response is also not a true picture of the corresponding concentration profile. The maximum of the recorded peak will again appear somewhat before the concentration maximum passes through the detector sensor, and it can be assumed that the retention time of the peak center of gravity will be shifted in a similar way with respect to the retention time of the center of gravity of the zone. However, if the cause of zone asymmetry is merely a nonsymmetrical input sample profile (e.g., due to exponential purging of sample from the sample inlet port), and the zone migrates down the column under near equilibrium conditions, the true retention time is again approximated by the time at which the center of gravity of the chromatographic zone reaches the column outlet, but this time must be measured from the instant at which the center of gravity of the input zone entered the column inlet. In other words, in this case the true retention time is the difference between the first statistical moments of the eluted and input zones, as measured at $z = L$ and $z = 0$. However, in almost every practical case, the input profile is unknown, and the only possibility is to take the zone maximum retention time measured from the instant of sample injection, as in the case of a symmetrical zone, as an acceptable approximation.

It follows from the above that determination of the true retention time may be rather problematic. Fortunately, in GLC under usual conditions, the difference between the retention parameters (maximum and center of gravity) of an elution zone and of the corresponding recorded peak are practically negligible, so that it is possible to calculate with peak parameters to a good approximation.

If the zone asymmetry is due to a nonlinear sorption isotherm, the retention time to the center of gravity of the zone is useless, since there is no invariant parameter (except those obtained by extrapolation to infinite solute dilution in the sorbent) with an elution zone developed under conditions of nonlinear chromatography.

The causes of distortion of chromatographic peaks and the problems associated with their interpretation have recently been critically reviewed by Condor [42]. See also Chap. 2.

B. Sources on Nonlinearity in Gas-Liquid Chromatography

Essentially two kinds of nonlinearity may occur in gas-liquid chromatography. One kind stems from the adsorption mechanisms that are operative, and the other is given by the dependence of the bulk liquid/gas partition coefficient on the solute concentration in the stationary liquid phase.

At extremely small solute charges, it can be supposed that both the individual adsorption processes and the dissolution of solute in the bulk stationary liquid phase will obey Henry's law. In this event, the total retention of solute will be given by the sum of the individual contributions, as shown by Eq. (46), and the whole process will be linear. However, the mutual proportions of the individual contributions due to adsorption and bulk dissolution depend on a number of factors, particularly for the loading of the support or capillary column wall with the stationary liquid, properties of the support (capillary column wall), chemical nature of the stationary liquid and solute, and column temperature. Yet as long as all the partial sorption processes take place within the Henry's law region, the retention of a given solute should be independent of the amount of solute introduced into the gas chromatograph.

As the solute charge is gradually increased, the first occurrence is that the linearity limits of the adsorption contributions to solute retention are exceeded because systems with two-dimensional sorbents are much more prone to solute concentration overloading than systems with three-dimensional sorbents. With adsorption, appreciable deviations from linearity begin to occur at a certain probability that solute molecules find occupied active sites on the adsorbent surface as well as free sites. With bulk dissolution, nonlinearity sets in at a certain probability that solute-solute intermolecular interactions take place along with the solute-solvent interactions, provided the two kinds of interaction differ from each other. The loading of the support or capillary column wall with liquid phase is a very important factor in determining which of the two foregoing mechanisms will be responsible for the increase in nonlinearity. As a rule, the situation is such that when nonlinearity caused by concomitant adsorption begins to manifest itself, the bulk liquid dissolution isotherm is still quite linear. Nonlinear adsorption mechanisms can, however, make the entire sorption process nonlinear. In such a case, the individual contributions to retention are no longer additive, and the retention of a given solute, under given conditions, depends on the amount of solute charged into the gas chromatograph.

With larger solute charges, all adsorption-active surfaces may become saturated although the process of bulk dissolution is still fairly linear. This gives rise to a situation in which the concentrations of solute at all the individual adsorbing phase interfaces, along the entire path of zone migration (column length), are sufficiently high to produce flat adsorption isotherms ($dc_{iS}/dc_{iM} = 0$), whereas in the bulk liquid phase the solute concentration is low enough for the dissolution iostherm to be linear. Under these conditions, the retention of the concentration maximum of the zone is given practically only by the bulk dissolution mechanism, the adsorption mechanisms being virtually ineffective. However, saturation of the phase interface surfaces with the solute manifests itself by a tail on the rear of the eluted zone. If the adsorption capacity of the surfaces is small, this tailing may be relatively insignificant, and with larger solute charges and lower detector sensitivity settings the tail may become imperceptible. At still larger charges of solute, the bulk dissolution isotherm eventually also becomes nonlinear.

At still larger charges of solute, the bulk dissolution isotherm eventually also becomes nonlinear.

With extremely dilute solutions of solute in the stationary liquid phase, it is possible to write (when neglecting the gas-phase nonideality and liquid-phase compressibility effects)

$$\lim_{c_{iS}\to 0}\left(\frac{c_{iS}}{c_{iM}}\right)_{eq} = K = \frac{RT}{\gamma_i^\theta P_i^\theta V_S^L} \neq f(c_{iS}) \tag{101}$$

where γ_i^θ and V_S^L (molar volume of the stationary liquid phase) are limiting values at an infinite dilution of the solute. The limiting activity coefficient is given, according to the Flory-Huggins theory [43], by

$$\ln \gamma_i^\theta = 1 + \ln \frac{1}{r} - \frac{1}{r} + \chi_{iS} \tag{102}$$

where r is the ratio of the molar (or partial molar) volumes of the stationary liquid phase and the liquid solute (i.e., $r = V_S^L/V_i^L$) and χ_{iS} is the Flory interaction parameter, which is defined as [44]

$$\chi_{iS} = \frac{V_i^L(\delta_S - \delta_i)^2}{RT} \tag{103}$$

The quantities δ_S and δ_i are the solubility parameters of the stationary liquid phase and of the solute, respectively. The solubility parameter of a given compound is defined as the square root of the density of the molecular cohesion energy of the compound and can be estimated by $\delta = [(\Delta H_v - RT)/V^L]^{1/2}$, where ΔH_v is the molar enthalpy of vaporization of the compound.

At finite solute concentrations in the liquid phase, the following relation is to be used instead of Eq. (101):

$$\left(\frac{c_{iS}}{c_{iM}}\right)_{eq} = K(c_{iS}) = \frac{RT}{\gamma_i^\theta(c_{iS})P_i^\theta V_{S(i)}^L} \tag{104}$$

where $\gamma_i^\theta(c_{iS})$ and $V_{S(i)}^L$ are the activity coefficient of the solute as a function of the solute concentration in the liquid phase and the molar volume of the mixture of stationary liquid phase and solute, respectively. Strictly speaking, relation (104) applies for a static system or for a chromatographic arrangement in which the value of c_{iS} is kept constant along the entire column, such as with step-and-pulse chromatography. For $\gamma_i^\theta(c_{iS})$, the Flory-Huggins theory gives the relation

$$\ln \gamma_i^{\theta}(c_{iS}) = 1 + \ln\left[\frac{1}{r} + \varphi_i\left(1 - \frac{1}{r}\right)\right] - \frac{1}{r} - \varphi_i\left(1 - \frac{1}{r}\right) + (1 - \varphi_i)^2 \chi_{iS} \tag{105}$$

where φ_i is the volume fraction of the solute in the mixture of the stationary liquid phase and the solute. Provided the molar volumes V_S^L and V_i^L are additive,

$$V_{S(i)}^L = V_S^L - (V_S^L - V_i^L)x_i = V_S^L - \frac{\varphi_i V_S^L(V_S^L - V_i^L)}{V_i^L + \varphi_i(V_S^L - V_i^L)} \tag{106}$$

where x_i is the mole fraction of the solute in the mixture of the latter with the stationary liquid phase.

It follows from the preceding discussion that four different regions can be expected to occur in the dependence of the retention of solute on the size of the solute charge. In the first region, which corresponds to extremely small amounts of solute, the retention of solute is given by the sum of linear contributions of all the adsorption mechanisms and also the bulk dissolution mechanism. Hence, the overall sorption isotherm is linear within this region (i.e., the retention of the zone maximum of a given solute, under given conditions, is independent—and relatively largest, compared with the other regions—of the size of the solute charge, and the solute zone is symmetrical). At larger solute charges, the contributions due to adsorption begin to be nonlinear, and consequently the overall sorption isotherm is also nonlinear within this region. Retention of the solute zone maximum decreases upon increasing the solute charge, and the solute zone is tailed. This situation refers to the second region, which extends from the inception of nonlinearity to a stage at which all the active adsorption surfaces become saturated. The third region can be called pseudolinear. The adsorption activity of the surfaces at all the phase interfaces is saturated, and the retention of the concentration maximum of the solute zone is given by the bulk dissolution mechanism only, with the latter still linear. Hence, retention of the solute zone maximum is virtually independent of the amount of solute charged, but the zone is tailed. With even larger solute charges the bulk dissolution process also becomes nonlinear, and the overall sorption isotherm is definitely nonlinear in this fourth region. Retention of the solute zone maximum may then either decrease or increase upon increasing the solute charge, the zone being skewed by tailing, leading, or both. However, with larger solute charges, still other causes of zone asymmetry (pressure and temperature changes within the zone and viscosity effect) must be considered [45].

It should be pointed out that the foregoing concept is merely a model. Such a model may give a true picture of many a real system, but it cannot be generalized. There are, for instance, systems in which the adsorption mechanisms practically do not occur or systems in which the onset of bulk dissolution nonlinearity falls in the curved part of some of the adsorption isotherms.

The range of linearity of the bulk dissolution process alone can be estimated by calculating [see Eqs. (101) through (106)] the partition coefficient of a model solute at different concentrations in a model solvent under typical gas-liquid chromatographic conditions. Such a calculation was done for three systems with one solute and three solvents of very different molar volumes and the same polarity. The properties of the model solute and model solvents were as follows: P_i^θ = 50.66 kPa (0.5 atm), V_i^L = 100 ml/mol, δ_i = 15 $(J/ml)^{1/2}$, V_S^L = 1000, 10 000 and 100 000 ml/mol, δ_S = 20 $(J/ml)^{1/2}$, and T = 333 K. For the sake of simplicity, the densities of the solute, solvents, and mixtures of the solute with the solvents were supposed to be the same, [i.e., the solute concentrations c_{iS} (in grams per milliliter) were taken as identical to the volume or mass fractions of the solute, and the molar volumes V_i^L and V_S^L were considered additive]. The values chosen for the solubility parameters correspond to a slightly polar solvent and a nonpolar solute. The data obtained are summarized in Table 1. The results are rather surprising:

1. The bulk dissolution isotherms are practically linear to solute concentrations in the solvent-solute mixtures of about 10^{-2} g/ml, at which point the deviations from linearity reach about 3%. At solute concentrations of 10^{-3} g/ml, the deviations from linearity amount to about 0.3%.
2. The concentration range of linearity is only slightly dependent on the molar volume (molar mass) of the solvent. In other words, it is the solute concentration (in mass/volume units) rather than mole fraction in the solvent-solute mixture that determines the inception of nonlinearity.

These findings are at great variance with Conder and Young's argument [46] on the problem of linearity of the bulk dissolution isotherm. Naturally, the conclusions reached in this section are correct to the extent to which the Flory-Huggins theory is valid. However, in the framework of our discussion on the linearity of the bulk dissolution isotherm, the validity of this theory may be presumed to be quite sufficient.

The foregoing findings make it possible to formulate a simple and rather general rule for estimating the maximum permissible size of solute charge with which the deviation from linearity does not exceed a certain determined limit. Let us consider a gas chromatographic system with a column of length L containing a volume V_S of the stationary

Table 1. Partition Coefficients of a Model Solute on Three Model Stationary Liquid Phases at Different Solute Concentrations (V_i^L = 100 ml/mol)[a]

	V_S^L = 1000 ml/mol			V_S^L = 10,000 ml/mol			V_S^L = 100,000 ml/mol		
c_{iS} (g/ml)	x_i	$K(c_{iS})$	E (%)	x_i	$K(c_{iS})$	E (%)	x_i	$K(c_{iS})$	E (%)
0	0	90.030	—	0	82.281	—	0	81.544	—
10^{-9}	10^{-8}	90.030	0	10^{-7}	82.281	0	10^{-6}	81.544	0
10^{-7}	10^{-6}	90.030	0	10^{-5}	82.281	0	10^{-4}	81.544	0
10^{-5}	10^{-4}	90.032	2.70×10^{-3}	9.99×10^{-4}	82.283	2.80×10^{-3}	9.90×10^{-3}	81.546	2.81×10^{-3}
10^{-3}	9.91×10^{-3}	90.274	2.71×10^{-1}	9.10×10^{-2}	82.511	2.80×10^{-1}	5.00×10^{-1}	81.773	2.81×10^{-1}
10^{-2}	9.17×10^{-2}	92.491	2.73×10^{0}	5.03×10^{-1}	84.606	2.82×10^{0}	9.10×10^{-1}	83.856	2.83×10^{0}
10^{-1}	5.26×10^{-1}	116.939	2.99×10^{1}	9.17×10^{-1}	107.841	3.11×10^{1}	9.91×10^{-1}	106.971	3.12×10^{1}

[a] $E(\%) = 100[K(c_{iS}) - K(c_{iS} = 0)]/K(c_{iS} = 0)$.

liquid phase. After the introduction of a sample charge, an initial solute zone is formed at the column inlet, occupying a segment of length ΔL with a volume ΔV_S of the stationary liquid phase. If after its vaporization (if charged in the form of a liquid) and mixing with the carrier gas in the sample inlet port the sample charge takes up a volume v, and provided this volume is purged in the form of a plug onto the column, or if a gaseous solute sample of volume v is injected directly onto the column, it is possible to write $\Delta L = L(v/V_R^0)$, $\Delta V_S = (V_S/L)\,\Delta L$, and $\Delta V_S = V_S(v/V_R^0) = V_S[v/(V_M^0 + KV_S)]$. The solute charge, of mass m_i, distributes itself in the initial zone between the stationary and mobile phases, the solute mass present in the stationary liquid phase m_{iS} being given by $m_{iS} = m_iK[V_S/(V_M^0 + KV_S)]$, so that the solute concentration in the stationary liquid phase within the initial zone $c_{iS}^0 = m_i(K/v)$. If we determine that the deviation from linearity in the initial zone should not exceed, for instance, 0.3%, the maximum permissible c_{iS}^0 and m_i will, according to the data in Table 1, be given by $c_{iS,max}^0 \leqslant 10^{-3}$ g/ml and $m_{i,max} \leqslant (v/K) \times 10^{-3}$, respectively. If we choose, for instance, v = 1 ml and K = 100, we have $m_{i,max} \leqslant 10$ μg, which agrees well with experience. Let us recall that the relations for $c_{iS,max}^0$ and $m_{i,max}$ that are stated above apply only to cases in which a liquid or gaseous sample is injected into the flash-vaporizer sample inlet port of the gas chromatograph, or for the on-column injection of gaseous samples. It does not matter whether a packed or capillary column is employed or whether split or splitless sample injection is used. The relations do not hold for the on-column injection of liquid samples.

C. Relations Among the Partition Coefficient and the Differential Standard Thermodynamic Functions of Sorption of the Solute in a Hypothetical Isobaric GLC Column

Since the beginning of the modern conception of chromatography, chromatographers have tried to use thermodynamics to formulate and solve different problems associated with chromatographic retention. Since chromatographic retention data are related in a defined manner to the thermodynamic properties of any given chromatographic system, these endeavors are quite justified. Chromatography does indeed offer many interesting possibilities in this respect, but all of them have certain limitations, which if ignored lead to incorrect results. Moreover, classic thermodynamics constitutes a rigid logical system of theorems, the applicability of which is strictly qualified by the properties of the physicochemical system under investigation, and the terms involved in the nomenclature for the description of thermodynamic functions have strictly determined meanings. These facts are sometimes disregarded by chromatographers; in an effort to present their results in terms of thermodynamics, they incorrectly apply thermodynamic theorems, creating something like "chromatographic thermodynamics" and publishing meaningless data. It seems that the need to specify standard states whenever retention data are related to standard thermodynamic functions is the main source of difficulty, although papers have been published [47,48] in which these problems are precisely described. The purpose of this and the following sections is to show what thermodynamic functions can be obtained from GLC retention data and how these data have to be treated in order to comply with thermodynamic formalism.

Standard Gibbs Functions of Sorption

The chemical potentials of solute i in the stationary and mobile phases at the temperature T and pressure (mean column pressure <P>) of a GLC system, μ_{iS} and μ_{iM}, are defined as:

$$\mu_{iS} = \mu_{iS}^{o} + RT \ln a_{iS} \tag{107}$$

$$\mu_{iM} = \mu_{iM}^{o} + RT \ln A_{iM} \tag{108}$$

where μ_{iS}^{o} and μ_{iM}^{o} are the standard chemical potentials, a_{iS} and a_{iM} are the activities of the solute in the stationary and mobile phases, respectively, and R is the molar perfect gas constant. At equilibrium [i.e., in the concentration maximum of the elution zone (with quasi-equilibrium linear chromatography)], $\mu_{iS} = \mu_{iM}$, and it is possible to

$$\mu_{iS}^{o} - \mu_{iM}^{o} = \Delta G_{sp}^{o} = - RT \ln \left(\frac{a_{iS}}{a_{iM}} \right)_{eq} \tag{109}$$

where ΔG_{sp}^{o} is the differential standard molar Gibbs function of sorption of the solute, and the subscript eq is again used to remind us that the ratio of equilibrium activities is concerned. This ratio represents a thermodynamic distribution constant, the numerical value of which depends on the choice of the standard states for the solute in both phases. Let us recall that the solute activity is defined as the ratio of the actual and standard solute fugacities, or

$$a_{iS} = \frac{f_{iS}}{f_{iS}^{o}} \quad \text{and} \quad a_{iM} = \frac{f_{iM}}{f_{iM}^{o}}$$

In general, standard states can be chosen quite arbitrarily, except the standard temperature, which must be the same as the actual temperature of the system. However, the choice of standard states must conform with the goal being pursued; actually, the choice of standard states can be regarded as a strategy whose task is to ensure that the standard thermodynamic quantities obtained will reflect the subject being investigated. The choice itself involves (with a given manner of expressing the solute concentrations in the stationary and mobile phases) the specification of standard concentrations and standard physical conditions of solute in both phases and a convention for normalizing the solute activity coefficient in the condensed phase. By the term "normalization of activity coefficient" is meant the definition of conditions under which the activity coefficient equals unity. There are essentially two modes of normalizing the activity coefficient: the solute activity coefficient can equal unity either at infinite solute dilution in the solvent or in a state of pure solute. These two modes of normalization will be called conventions I and II, respectively. Activity coefficients normalized according to convention I characterize deviations from Henry's law; those normalized according to convention II characterize deviations from Raoult's law. Activity coefficients defined for a given solute, solvent, and solute concentration according to either convention are different (except a perfect solution, in which both kinds of activity coefficient, at any solute concentration, equal unity). In our discussion, the different kinds of activity coefficient and the related quantities will be distinguished by different superscripts.

The principle of normalization of the activity coefficient is illustrated in Fig. 4. The solid line represents an actual course of the solute fugacity f_{iS} in the liquid phase as a function of the solute mole fraction x_i in this phase, whereas the dashed lines illustrate extrapolations of the limiting linear courses of the functions $f_{iS} = f_{iS}(x_i)$ at $x_i \to 0$ convention I) and at $x_i \to 1$ (convention II). The symbols h_{iS} and f_i^{θ} denote the solute Henry's law constant in the solute-solvent solution and the fugacity of pure liquid solute, and x_i^o is a standard mole fraction of solute in the liquid phase. It follows from Fig. 4 that, whereas with convention I the solute activity coefficient (Henry's law activity coefficient γ_i^*) is defined as $\gamma_i^* = f_{iS}/h_{iS}x_i$, according to convention II we have for the Raoult's law activity coefficient, $\gamma_i^{\theta} = f_{iS}/f_i^{\theta}x_i$. Hence, at a given solute concentration, $\gamma_i^*/\gamma_i^{\theta} = f_i^{\theta}/h_{iS}$. It is also apparent from Fig. 4 that according to conventions I and II the standard fugacities at $x_i = 1$ are $f_{iS}^o = h_{iS}$ and $f_{iS}^o = f_i^{\theta}$, respectively. Thus the physical meaning of the quantity h_{iS} can be understood as the fugacity of pure liquid solute in a hypothetical state of infinite dilution in the solvent, or, alternatively, as the fugacity of pure liquid solute, with the solute-solute intermolecular interactions completely replaced by solute-solvent intermolecular interactions.

The situation in the coexisting gaseous phase is illustrated in Fig. 5, where the solid line again represents an actual course of the solute

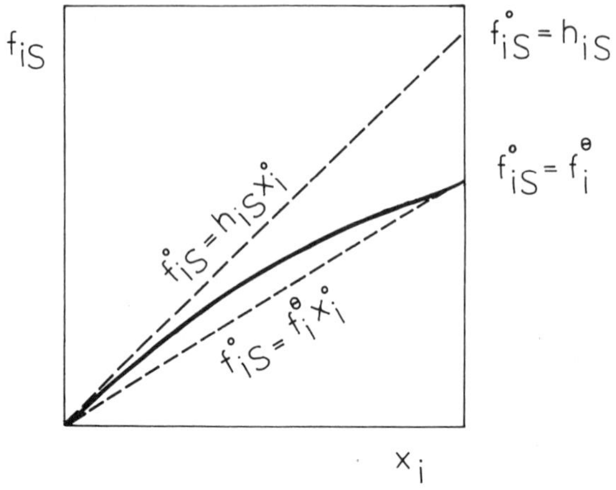

Figure 4. Representation of two conventions for the specification of standard states of the solute in the chemically defined stationary liquid phase. f_{iS}, actual solute fugacity in the liquid solution; x_i, solute mole fraction in the liquid solution; f_{iS}^o, standard solute fugacity; f_i^θ, fugacity of pure liquid solute; x_i^o, standard solute mole fraction; h_{iS}, Henry's law constant of the solute.

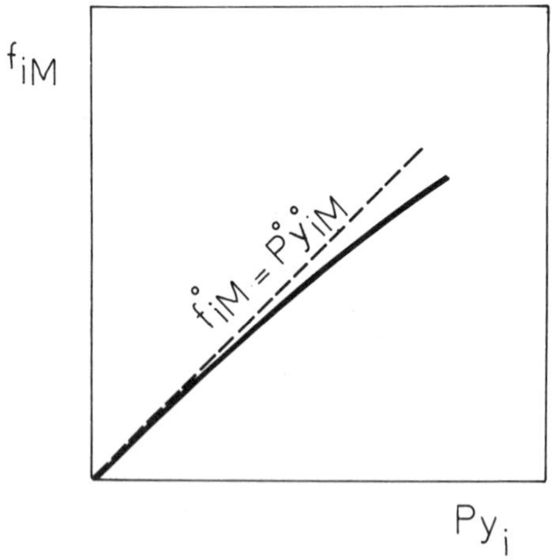

Figure 5. Representation of a convention for the specification of standard states of the solute in the gaseous phase. f_{iM}, actual solute fugacity in the gaseous phase; y_i, solute mole fraction in the gaseous phase; P, actual pressure; f_{iM}^o, standard solute fugacity; y_i^o, standard solute mole fraction; P^o, standard pressure.

fugacity in the gaseous phase as a function of Py_i (at a fixed y_i) in this phase and the dashed line illustrates the extrapolation of the limiting linear course of the function $f_{iM} = f_{iM}(P)$ at $P \to 0$. The quantities f_{iM}^o, y_i^o, and P^o denote standard solute fugacity in the gaseous phase, standard solute mole fraction in the gaseous phase and, standard pressure, respectively. The meaning of f_{iM}^o can be understood as the fugacity of a hypothetical ideal gaseous solute at a given P^o and y_i^o. Since the choice of standard states is arbitrary, there are obviously many formal variants in defining relationships among GLC retention data and standard thermodynamic functions of sorption. We shall quote only a few of them.

Let us recall at this point that all the pressure-dependent quantities occurring in the relationships discussed in this section refer to the mean column pressure $\langle P \rangle$ and column temperature T, unless otherwise specified. For reasons of brevity the respective symbols will not be designated with angular brackets (i.e., the same kind of notation is used here for mean pressure quantities as was used to denote local pressure quantities in Sec. III.B). Further, in contradistinction to the conventionally calculated K according to Eq. (57) [i.e., the K calculated without taking into account the carrier gas nonideal compressibility coefficients Z_M and Z_{Mf} (and/or Z_M^0)], fully corrected (in the framework of the hypothetical mean pressure column model) K and V_g data are considered throughout this section, just as with the local K(z) and $V_g(z)$ [see Eq. (60)].

GLC Systems with Chemically Defined Phases. With these systems, solute concentrations in the phases can be expressed in mole fractions. Let us choose for the solute in the liquid phase a standard state of pure solute in a hypothetical state of infinite dilution in the solvent at the temperature and pressure of the system. This implies that convention I is applied to normalize the activity coefficient, and we can write $f_{iS} = \gamma_i^* h_{iS} x_i$ and $f_{iS}^o = h_{iS}$ for the actual and standard ($x_i^o = 1$, $\gamma_i^* = 1$) solute fugacities in the liquid phase, respectively. For the solute in the gaseous phase we choose a standard state of pure solute in a hypothetical state of perfect gas at a reference pressure P^o and at the tempterature of the system. Hence, we have $f_{iM} = \nu_{iM} P y_i$ and $f_{iM}^o = P^o$ for the actual and standard ($y_i^o = 1$, $\upsilon_{iM} = 1$, P^o) solute fugacities in the gaseous phase, respectively, P being the actual pressure (mean column pressure) in the system. Since $\gamma_i^* = 1$ under the conditions of linear chromatography, we can write, according to Eq. (109) with the standard states specified above,

$$\Delta G_{sp}^* = -RT \ln \left[\frac{P^o}{\nu_{iM}} \left(\frac{x_i}{Py_i} \right)_{eq} \right] \tag{110}$$

where ΔG^*_{sp} is the change in the Gibbs function associated with the transition of 1 mole of solute from the standard state in the gaseous phase to that in the liquid phase.

At equilibrium, $f_{iM} = f_{iS}$, and if convention II is used to express f_{iS} while again defining f_{iM} by the relationship $f_{iM} = \nu_{iM}Py_i$, we can write $\nu_{iM}Py_i = \gamma_i^\theta f_i^\theta x_i$. Hence Eq. (110) can be rewritten as

$$\Delta G^*_{sp} = -\ RT\ \ln \frac{P^o}{\gamma_i^\theta f_i^\theta} \tag{111}$$

where γ_i^θ is the Raoult's law activity coefficient of the solute at its infinite dilution in the solvent. Since $\Delta \overline{G}^E_{iS} = RT \ln \gamma_i^\theta$ and $\Delta G^\theta_{cd} = -RT$ in $(P^{\hat{o}}/f_i^\theta)$, where $\Delta \overline{G}^E_{iS}$ is the partial molar excess Gibbs function of the solute at its infinite dilution in the solvent and ΔG^θ_{cd} is the standard molar Gibbs function of condensation of pure solute, the quantity ΔG^*_{sp} can further be expressed as

$$\Delta G^*_{sp} = \Delta G^\theta_{cd} + \Delta \overline{G}^E_{iS} \tag{112}$$

More specifically, ΔG^θ_{cd} is the Gibbs function change associated with the transition of 1 mole of solute from the standard state in the gaseous phase to a state of normal pure liquid solute at the temperature and pressure of the system.

By applying with our hypothetical mean pressure column the same procedure as was used to derive Eq. (60), we obtain for K

$$K = \left(\frac{x_i}{Py_i}\right)_{eq} \frac{Z_M RT d_S}{M_S} \tag{113}$$

and by combining Eqs. (110) and (113), the following relationship between ΔG^*_{sp} and K is obtained:

$$\Delta G^*_{sp} = -RT\ \ln \frac{KM_S P^o}{\nu_{iM} Z_M RT d_S} \tag{114}$$

A comparison of Eqs. (114) and (111) yields

$$K = \frac{\nu_{iM} Z_M RT d_S}{\gamma_i^\theta f_i^\theta M_S} \tag{115}$$

The relationships defined by Eqs. (114) and (115) can also be expressed in terms of V_g. With regard to Eq. (35) we can immediately write

$$\Delta G^*_{sp} = -RT \ln \frac{V_g M_S P^o}{\nu_{iM} Z^0_M R T^0} \tag{116}$$

$$V_g = \frac{\nu_{iM} Z^0_M R T^0}{\gamma^\theta_i f^\theta_i M_S} \tag{117}$$

where Z^o_M is the coefficient of nonideal carrier gas compressibility at a temperature T^o = 273.15 K and at the mean column pressure.

GLC Systems with a Chemically Nondefined Stationary Phase. If the molar mass of the stationary liquid phase is unknown, it is naturally impossible to use conventions that involve expressing the solute concentrations in such a phase in mole fractions to define standard states. In such cases we must resort to expressing the solute concentration in units in which the amount of the stationary phase is defined by mass or volume. When employing mass or volume fraction scales, conventions analogous to conventions I and II can be used to normalize the activity coefficients, but with scales based on solute amount/solvent volume or solute amount/solvent mass concentration, only a convention of type I is suitable.

Let us consider a case in which the solute concentration in the stationary liquid phase is expressed in units of amount of solute (mole number) per 1 gram of pure solvent, q_i, and the solute activity coefficient, $\gamma^{\square}_i$, is referred to infinite solute dilution (i.e., $\gamma^{\square}_i = 1$ at $q_i \to 0$). Such a mode of normalization is illustrated in Fig. 6, where the solid line illustrates the course of the actual solute fugacity, $f_{iS} = \gamma^{\square}_i k_{iS} q_i$, and the dashed line shows an extrapolation of the limiting linear course at $q_i \to 0$. The quantities f^o_{iS} and q^o_i again stand for the standard fugacity and a standard solute concentration, k_{iS} being a proportionality constant (similar by its nature to the Henry's law constant). At very low solute concentrations, $x_i/q_i \approx M_S$, $\gamma^{\square}_i \approx \gamma^*_i \approx 1$, and $k_{iS} \approx h_{iS} M_S$. When choosing a standard state of solute at a given q^o_i in a hypothetical state of infinite dilution in the solvent, at the temperature and pressure of the system for the solute in the liquid phase, the standard fugacity (q^o_i, $\gamma^{\square}_i = 1$) is $f^o_{iS} = k_{iS} q^o_i$, and if a standard state of perfect gas, pure solute at a reference pressure P^o and at the system temperature is again chosen for the solute in the gaseous phase, we arrive at

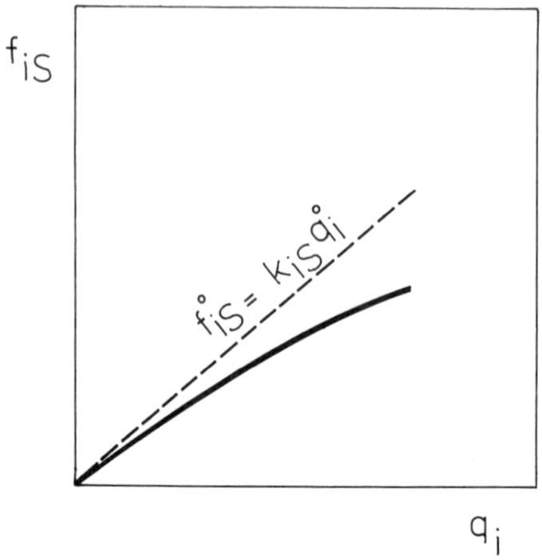

Figure 6. Representation of a convention for the specification of standard states of the solute in the chemically nondefined stationary liquid phase. f_{iS}, actual solute fugacity in the liquid solution; q_i, solute concentration in the liquid solution; f^{o}_{iS}, standard solute fugacity; q^{o}_i, standard solute concentration; k_{iS}, proportionality constant.

$$\Delta G^{\square}_{sp} = -RT \ln \left[\frac{P^{o}}{\upsilon_{iM} q^{o}_i} \left(\frac{q_i}{Py_i} \right)_{eq} \right] \tag{118}$$

And provided that K is still defined as $(c_{iS}/c_{iM})_{eq}$,

$$K = \left(\frac{q_i}{Py_i} \right)_{eq} Z_M RT d_S \tag{119}$$

and $\Delta G^{\square}_{sp}$ is related to K by

$$\Delta G^{\square}_{sp} = -RT \ln \frac{KP^{o}}{\upsilon_{iM} q^{o}_i Z_M RT d_S} \tag{120}$$

Often the relationship $\Delta G^{o} = -RT \ln K$ (or only $\Delta G = -RT \ln K$) is quoted without any further specification and numerical values of ΔG^{o} calculated from this relationship are presented. Let us try to formulate the premises under which such a relationship would be sufficiently defined. The concentration units are given by the definition of the partition coefficient [i.e., $K = (c_{iS}/c_{iM})_{eq}$]. If use is made of the same relation to define the actual solute fugacities in the stationary and gaseous phases (i.e., $f_{iS} = \gamma_{iS}^{\Delta} Q_{iS} c_{iS}$ and $f_{iM} = \gamma_{iM}^{\Delta} Q_{iM} c_{iM}$, where the Qs are proportionality constants), the activity coefficients being normalized as $\gamma_i^{\Delta} = 1$ as $c_i \rightarrow 0$, and if a standard state of solute at unit c_i in a hypothetical state of infinite dilution in either phase at the temperature and pressure of the system is used for the solute in both phases, the standard solute fugacities ($c_i^o = 1$ and $\gamma_i^{\Delta} = 1$) are $f_{iS}^o = Q_{iS}$ and $f_{iM}^o = Q_{iM}$. Provided that both γ_{iS}^{Δ} and γ_{iM}^{Δ} approach unity under the actual GLC conditions (which must always be the case in linear chromatography), the foregoing specification leads to

$$\Delta G_{sp}^{\Delta} = -RT \ln K \tag{121}$$

Let us point out that numerical values of ΔG_{sp}^{*}, $\Delta G_{sp}^{\square}$, and ΔG_{sp}^{Δ}, calculated for a given system, are different. Actually, it is suitable to write $\Delta G_{sp}^o = -RT \ln (C^oK)$, where C^o is a parameter defined by the choice of the standard states. Thus, with ΔG_{sp}^{*}, $\Delta G_{sp}^{\square}$, and ΔG_{sp}^{Δ} defined by Eqs. (114), (120), and (121), for instance, the parameters C^o are $M_S P^o/\upsilon_{iM} Z_M RT d_S$, $P^o/\upsilon_{iM} q_i^o Z_M RT d_S$, and 1, respectively. Hence it follows that numerical values of ΔG_{sp}^o without the specification of the standard states are completely useless.

Role of Temperature and Pressure

By combining Eq. (109) with the general definitions for the dependence of the chemical potential on temperature and pressure, we obtain

$$\left[\frac{\partial}{\partial T}\left(\frac{\Delta G_{sp}^o}{T}\right)\right]_{P,comp} = -\frac{\Delta H_{sp}^o}{T^2} \tag{122}$$

$$\left(\frac{\partial \Delta G_{sp}^o}{\partial P}\right)_{T,comp} = \Delta V_{sp}^o \tag{123}$$

$$\left(\frac{\partial \Delta G_{sp}^o}{\partial T}\right)_{P,comp} = -\Delta S_{sp}^o \tag{124}$$

where ΔH_{sp}^o, ΔV_{sp}^o, and ΔS_{sp}^o are the differential standard molar enthalpies, volumes, and entrophies of the solute in the system, respectively, the standard states for these derived quantities being driven by the choice of the standard states for ΔG_{sp}^o. The subscript "comp" refers to the liquid phase and is relevant at solute concentrations exceeding the Henry's law region. The temperature and pressure dependences of the partition coefficient can easily be derived from the temperature and pressure dependencies of the quantities occurring on the right side of Eq. (115). Thus, we obtain

$$\left(\frac{\partial \ln K}{\partial T}\right)_{P,\mathrm{comp}} = \left[\frac{\partial \ln (\nu_{iM} Z_M)}{\partial T}\right]_{P,\mathrm{comp}} + \frac{\bar{H}_{iS} - H_i^g}{RT^2} + \frac{1}{T} - \alpha_S \tag{125}$$

$$\left(\frac{\partial \ln K}{\partial P}\right)_{T,\mathrm{comp}} = \left[\frac{\partial \ln (\nu_{iM} Z_M)}{\partial P}\right]_{T,\mathrm{comp}} - \frac{\bar{V}_{iS}^L}{RT} + \beta_S \tag{126}$$

where H_i^g is the molar enthalpy of the vapors of pure solute at the temperature of the system and at a very low pressure (perfect gas behavior). With V_g [see Eq. (117) and note that Z_M^o is invariant with T], we have

$$\left(\frac{\partial \ln V_g}{\partial T}\right)_{P,\mathrm{comp}} = \left(\frac{\partial \ln \nu_{iM}}{\partial T}\right)_{P,\mathrm{comp}} + \frac{\bar{H}_{iS} - H_i^g}{RT^2} \tag{127}$$

$$\left(\frac{\partial \ln V_g}{\partial P}\right)_{T,\mathrm{comp}} = \left[\frac{\partial \ln (\upsilon_{iM} Z_M^o)}{\partial P}\right]_{T,\mathrm{comp}} - \frac{\bar{V}_{iS}^L}{RT} \tag{128}$$

The values of the coefficients υ_{iM}, Z_M, and Z_M^0 can be supposed to approach unity under usual GLC conditions, so that the derivatives on the right sides of Eqs. (125) through (128) can be neglected in the first approximation.

It follows from the foregoing discussion that only differential *standard* thermodynamic functions of sorption can be related to the chromatographic partition coefficient regardless of how the latter has been defined. With a certain choice of standard states, it may happen that the standard Gibbs function of sorption will be identical to some other kind of differential Gibbs function or will involve such a kind of function, or both; the situation described by Eq. (112) can be used as an example. Naturally, the same also applies to the differential standard enthalpies, volumes, and entropies of sorption. However, it is necessary that every such situation be evaluated individually in order to arrive at correct conclusions.

When applying the definitions formulated by Eqs. (122) through (124) to ΔG_{sp}^* defined by Eq. (114) or by Eq. (115), or both, employing Eqs. (125) and (126) or Eqs. (127) and (128), respectively, and assuming ν_{iM} and Z_M to practically equal unity, we can write

$$\left[\frac{\partial}{\partial T}\left(\frac{\Delta G^*_{sp}}{T}\right)\right]_{P,comp} = -\frac{\Delta H^*_{sp}}{T^2} = -\frac{\bar{H}_{iS} - H_i^g}{T^2} \quad (129)$$

$$\left(\frac{\partial \Delta G^*_{sp}}{\partial P}\right)_{T,comp} = \Delta V^*_{sp} = \bar{V}^L_{iS} \quad (130)$$

Hence, these differential standard enthalpies and volumes of sorption (i.e., ΔH^*_{sp} and ΔV^*_{sp}) are practically equal to the actual differential enthalpies and volumes of sorption. Since $\Delta \bar{H}^E_{iS} = \bar{H}_{iS} - \bar{H}^{id}_{iS}$, $\bar{H}^{id}_{iS} = H^L_i$, and $H^L_i - H^g_i = \Delta H^\theta_{cd}$ and $\Delta \bar{V}^E_{iS} = \bar{V}^L_{iS} - \bar{V}^{id}_{iS}$ and $\bar{V}^{id}_{iS} = V^L_i$, where $\Delta \bar{H}^E_{iS}$ and $\Delta \bar{V}^E_{iS}$ are the partial molar excess enthalpy and partial molar excess volume of the solute in the solute-solvent mixture, $\bar{H}^{id}_{iS}$ and $\bar{V}^{id}_{iS}$ are the partial molar enthalpy and partial molar volume of the solute in an ideal solution, H^L_i and V^L_i are the molar enthalpy and molar volume of the pure liquid solute, and ΔH^θ_{cd} is the differential standard molar enthalpy of condensation of pure solute [see the comments on Eq. (112)], it holds that

$$\Delta H^*_{sp} = \bar{H}_{iS} - H^g_i = \Delta \bar{H}^E_{iS} + \Delta H^\theta_{cd} \quad (131)$$

$$\Delta V^*_{sp} = \bar{V}^L_{iS} - \frac{RT}{P^0} = \Delta \bar{V}^E_{iS} + V^L_i \quad (132)$$

For the quantity ΔS^*_{sp},

$$\left(\frac{\partial \Delta G^*_{sp}}{\partial T}\right)_{P,comp} = -\Delta S^*_{sp} = \frac{\Delta G^*_{sp} - \Delta H^*_{sp}}{T} \quad (133)$$

Let us recall that $\bar{G}^{id}_{iS} \neq G^L_i$ and $\bar{S}^{id}_{iS} \neq S^L_i$, where $\bar{G}^{id}_{iS}$ and $\bar{S}^{id}_{iS}$ are the partial molar Gibbs function and partial molar entrophy of the solute in an ideal solution and G^L_i and S^L_i are the molar Gibbs function and molar entropy of the pure liquid solute, repectively. However, with our choice of standard states, $\Delta G^*_{sp} = \Delta \bar{G}^E_{iS} + \Delta G^\theta_{cd}$, so that even in this case we have

$$\Delta S^*_{sp} = \frac{\Delta \bar{H}^E_{iS} - \Delta \bar{G}^E_{iS} + \Delta H^\theta_{cd} - \Delta G^\theta_{cd}}{T} = \Delta \bar{S}^E_{iS} + \Delta S^\theta_{cd} \quad (134)$$

Finally, let us try to compare the relationships for K as a function of P, obtained according to the concept of an isobaric mean pressure column and by the more refined theory dealt with in Sec. III.B. Referring to Eqs. (72) and (74), we can write $\ln(\nu_{iM}Z_M) = 2B_{iM}P/RT$. Hence, Eq. (126) can be rewritten as

$$\left(\frac{\partial \ln K}{\partial P}\right)_{T,\text{comp}} = \frac{2B_{iM} - \bar{V}_{iS}^{L}}{RT} + \beta_S \tag{135}$$

Upon integrating Eq. (135) within the limits 0 and $\langle P \rangle$ and rearranging, we obtain

$$\ln K = \ln K_0 + \left(\frac{2B_{iM} - \bar{V}_{iS}^{L}}{RT} + \beta_S\right)\langle P \rangle \tag{136}$$

Whereas the K in Eq. (94) represents a conventionally calculated partition coefficient [see Eq. (57)], in Eq. (136) the K is supposed to have been calculated by making use of the coefficients of nonideal carrier gas compressibility. Hence, when comparing Eq. (136) with Eq. (94) and taking account of all the differences in the two respective models, we see that the results are mutually consistent.

REFERENCES

1. J. C. Giddings, *Dynamics of Chromatography*, Marcel Dekker, New York, pp. 3, 95 (1965).
2. J. C. Giddings, *Dynamics of Chromatography*, Marcel Dekker, New York, p. 8 (1965).
3. M. Kubín, Beitrag zur Theorie der Chromatographie, *Coll. Czech. Chem. Commun.*, *30*:1104 (1965).
4. E. Kučera, Contribution to the theory of chromatography. Linear nonequilibrium elution chromatography, *J. Chromatogr.*, *19*:237 (1965).
5. J. R. Conder and J. H. Purnell, Gas chromatography at finite concentrations, Part 1, Effect of gas imperfection on calculation of the activity coefficient in solution from experimental data, *Trans. Faraday Soc.*, *64*:1505 (1968).
6. J. R. Conder and J. H. Purnell, Gas chromatography at finite concentrations, Part 2, A generalized retention theory, *Trans. Faraday Soc.*, *64*:3100 (1968).
7. J. R. Conder and J. H. Purnell, Gas chromatography at finite concentrations, Part 3, Theory of frontal and elution techniques of thermodynamic measurement, *Trans. Faraday Soc.*, *65*:824 (1969).

8. J. R. Conder and J. H. Purnell, Gas chromatography at finite concentrations, Part 4, Experimental evaluation of methods for thermodynamic study of solutions, *Trans. Faraday Soc.*, *65*:839 (1969).
9. J. R. Conder and C. L. Young, *Physico-Chemical Measurement by Gas Chromatography*, John Wiley & Sons, Chichester, p. 353 (1979).
10. J. N. Wilson, A theory of chromatography, *J. Amer. Chem. Soc.*, *62*:1583 (1940).
11. L. Lapidus and N. R. Amundsen, Mathematics of adsorption in beds, VI, Effects of longitudinal diffusion in ion-exchange and chromatographic columns, *J. Phys. Chem.*, *56*:984 (1952).
12. J. J. van Deemter, F. J. Zuiderweg, and A. Klinkenberg, Longitudinal diffusion and resistance to mass transfer as causes of nonideality in chromatography, *Chem. Eng. Sci.*, *5*:271 (1956).
13. S. Wičar, J. Novák, and N. R. Rakshieva, Non-ideality of the column and retention time in gas chromatography, *Anal. Chem.*, *43*:1945 (1971).
14. A. J. B. Cruickshank, B. W. Gainey, C. P. Hicks, T. M. Letcher, R. W. Moody, and C. L. Young, Gas-liquid chromatographic determination of cross-term second virial coefficients using glycerol, *Trans. Faraday Soc.*, *65*:1014 (1969).
15. A. T. James and A. J. P. Martin, Gas-liquid partition chromatography. A technique for the analysis of volatile materials, *Analyst*, *77*:915 (1952).
16. P. C. Carman, *Flow of Gases through Porous Media*, Macmillan, New York, p. 11 (1956).
17. J. C. Giddings, *Dynamics of Chromatography*, Marcel Dekker, New York, p. 208 (1965).
18. E. Kováts, Gas-chromatografische Charakterisierung organischer Verbindungen, Teil 1, Retentionsindices aliphatischer Halogenide, Alkohole, Aldehyde und Ketone, *Helv. Chim. Acta*, *41*:1915 (1958).
19. J. C. Giddings, The random downstream migration of molecules in chromatography, *J. Chem. Educ.*, *35*:588 (1958).
20. J. C. Giddings, Non-equilibrium kinetics and chromatography, *J. Chem. Phys.*, *31*:1462 (1959).
21. A. L. LeRosen, The characterization of silicic acid-celite mixtures for chromatography, *J. Amer. Chem. Soc.*, *67*:1683 (1945).
22. J. C. Giddings, G. H. Stewart, and A. L. Ruoff, Zone migration in paper chromatography, *J. Chromatogr.*, *3*:239 (1960).
23. R. L. Martin, Adsorption on the liquid phase in gas chromatography, *Anal. Chem.*, *33*:347 (1961).
24. J. R. Conder and C. L. Young, *Physico-Chemical Measurement by Gas Chromatography*, John Wiley & Sons, Chichester, p. 459 (1979).

25. A. Kwantes and G. W. A. Rijnders, The determination of activity coefficients at infinite dilution by gas-liquid chromatography, *Gas Chromatography 1958* (D. H. Desty, ed.), Butterworths, London, p. 125 (1958).
26. D. H. Everett and C. T. H. Stoddard, The thermodynamics of hydrocarbon solutions from G.L.C. measurements, Part 1, Solutions in dinonyl phtalate, *Trans. Faraday Soc.*, *57*:746 (1961).
27. D. H. Desty, A. Goldup, G. R. Luckhurst, and W. T. Swanton, The effect of carrier gas and column pressure on solute retention, *Gas Chromatography 1962* (M. van Swaay, ed.), Butterworths, London, p. 67 (1962).
28. D. H. Everett, Effects of gas imperfection on G.L.C. measurements: A refined method for determining activity coefficients and second virial coefficients, *Trans. Faraday Soc.*, *61*:1637 (1965).
29. A. J. B. Cruickshank, M. L. Windsor, and C. L. Young, The use of gas-liquid chromatography to determine activity coefficients and second virial coefficients of mixtures, *Proc. R. Soc. A*, *295*:259 (1966).
30. D. E. Martire and D. C. Locke, Comparability factor for nonideal carrier gases in gas chromatography, *Anal. Chem.*, *37*:144 (1965).
31. P. D. Schettler, M. Eikelberger, and J. C. Giddings, Nonideal gas corrections for retention and plate height in gas chromatography, *Anal. Chem.*, *39*:146 (1967).
32. S. Wičar and J. Novák, Retention volume in high-pressure gas chromatography, I, Thermodynamics of the specific retention volume, *J. Chromatogr.*, *95*:1 (1974).
33. S. Wičar and J. Novák, Retention volume in high-pressure gas chromatography, II, Comparison of experimental data with the prediction of a pseudo-binary model, *J. Chromatogr.*, *95*:13 (1974).
34. S. Wičar, J. Novák, J. Drozd, and J. Janak, Retention volume in high-pressure gas chromatography, III, Squalane-tetrachloromethane, isooctane, toluene-nitrogen, hydrogen, carbon dioxide systems, *J. Chromatogr.*, *142*:167 (1977).
35. E. Hála, J. Pick, V. Fried, and O. Vilím, *Vapour-Liquid Equilibrium*, Pergamon Press, Oxford (1967).
36. K. Denbigh, *The Principles of Chemical Equilibrium*, Cambridge University Press, Cambridge (1957).
37. W. J. Moore, *Physical Chemistry*, 5th ed., Longman, London (1972).
38. M. Vigdergauz and V. Semkin, Some non-analytical applications of high-pressure gas chromatography, *J. Chromatogr.*, *58*:95 (1971).
39. E. Glueckauf, Theory of chromatography, Part 9, The "theoretical plate" concept in column separations, *Trans. Faraday Soc.*, *51*:34 (1955).

40. J. Janča, Relation between the asymmetry of the elution curve and the efficiency of the separation system in size-exclusion chromatography, *J. Liq. Chromatogr.*, *5*:1605 (1982).
41. J. Å. Jönsson, The median of the chromatographic peak as the best measure of retention time, *Chromatographia*, *14*:653 (1981).
42. J. R. Conder, Peak distortion in chromatography, *J. HRC&CC*, *5*:341 (1982).
43. J. M. Prausnitz, *Molecular Thermodynamics of Fluid-Phase Equilibria*, Prentice-Hall, Englewood Cliffs, New Jersey, p. 293 (1969).
44. J. H. Hildebrand, J. M. Prausnitz, and P. L. Scott, *Regular and Related Solutions*, Van Nostrand Reinhold, New York, p. 207 (1970).
45. J. R. Conder and C. L. Young, *Physico-Chemical Measurement by Gas Chromatography*, John Wiley & Sons, Chichester, pp. 44-49 (1927).
46. J. R. Conder, and C. L. Young, *Physico-Chemical Measurement by Gas Chromatography*, John Wiley & Sons, Chichester, p. 43 (1927).
47. M. R. James, J. C. Giddings, and R. A. Keller, Thermodynamic pitfalls in gas chromatography, *J. Gas Chromatogr.*, *3*:57 (1965).
48. E. F. Meyer, On thermodynamics of solution by gas-liquid chromatography, *J. Chem. Educ.*, *50*:191 (1973).

4
Retention in Gas-Solid Chromatography

Władysław Rudziński
Institute of Chemistry
Maria Curie-Skłodowska University
Lublin, Poland

I. LOW-PRESSURE ADSORPTION ON HOMOGENEOUS SOLID SURFACES AND THE VIRIAL DESCRIPTION FORMALISM

We start by writing the general equation for the absolute retention volume V_N (see Chap. 2, pages 83-84),

$$V_N = \frac{\partial}{\partial \rho_0} [N(\rho_0)] \tag{1}$$

where $N(\rho_0)$ is the number of adsorbate molecules in the surface phase when the molecular density in the free gas phase is equal to ρ_0.

In typical analytic work, the injected samples are usually small and create small concentrations in the free gas phase ρ_0. Therefore, in the first theoretical work on gas-solid chromatography (GSC), attention was focused on the virial description, which is a very powerful method for describing the adsorption at small adsorbate pressures.

This method, begun by the theoretical work of Ono [1], is based on the standard expansions of various thermodynamic quantities into powers of the absolute activity a. By combining these standard expansions, one arrives at other expansions. For instance, in the virial expansion of the average number of molecules in the adsorbed phase N into powers of the density in the free gas phase ρ_0 [2],

$$N = B_{2s}\rho_0 + [B_{3s} + 2B_{2s}B_{20}]\,\rho_0^{\,2} + 0(\rho_0)^3 \tag{2}$$

where B_{2s} and B_{3s} are the so-called second and third gas-solid virial coefficients. They are defined as follows:

$$B_{2s} = \int_{V_s} [e^{(v(\underline{r}_1)/-kT)} - 1] d\underline{r}_1 \tag{3a}$$

$$B_{3s} = \int_{V_s}\int [e^{((v(\underline{r}_1)+v(\underline{r}_2))/-kT)}][e^{\{(\omega(\underline{r}_1,\underline{r}_2)/-kT)\}} - 1] d\underline{r}_1 d\underline{r}_2$$

$$- \int_{V_s}\int [e^{(u(\underline{r}_1,\underline{r}_2)/-kT)} - 1] d\underline{r}_1 d\underline{r}_2 \tag{3b}$$

In Eq. (3), $v(\underline{r})$ is the adsorption potential of a single adsorbate molecule whose center is at the point $\underline{r}$, $\omega(\underline{r}_1, \underline{r}_2)$ is the pair interaction potential of two molecules on the surface, $u(\underline{r}_1, \underline{r}_2)$ is the form of ω in the absence of the solid phase; k is the Boltzmann constant, and T is temperature. Further, V_s is the volume of the whole adsorption space (i.e., the volume of the surface plus the free bulk phase).

The second term on the right-hand side of Eq. (3b) can be identified with the so-called second bulk virial coefficient B_{20}. Assuming that the equilibrium bulk phase is ideal, one can set $B_{20} = 0$, and Eq. (1) takes the following explicit form [3]

$$V_N = B_{2s} + 2B_{3s}\rho_0 + \cdots \tag{4a}$$

$$V_N = B_{2s} + 2B_{3s}\frac{p}{kT} + \cdots \tag{4b}$$

The region of such small adsorbate pressures (densities) in the equilibrium bulk phase, in which the terms of order $O(\rho_0)$, $O(p/kT)$ in expansion (4) becomes negligible, is usually called the Henry's law region, and the second gas-solid virial coefficient B_{2s} is therefore often called the Henry's law constant.

Initially, the attention of scientists working on the virial description was focused on the adsorption of simple molecules on homogeneous solid surfaces, generally noble gases adsorbed on highly graphitized carbons. In such systems, the solid-gas interaction potential well may be represented by the function,

$$v(z) = \frac{n}{m-n}\left(\frac{n}{m}\right)^{m/(n-m)} \varepsilon^*\left[\left(\frac{z_0}{z}\right)^m - \left(\frac{z_0}{z}\right)^n\right] \tag{5}$$

where m and n are integers and m > n. Further, ε^* is the minimum value of v(z), occurring at the distance $z = z^*$:

$$z^* = \left(\frac{n + 3}{m + 3}\right)^{1/(n-m)} z_0 \tag{6}$$

This potential function is obtained by integrating a two-particle potential function describing the interaction of a gas molecule with a material point in a semi-infinite space of adsorbent. The adsorbent phase is assumed to be a structureless medium of uniform density. When the aforementioned two-particle potential function is the Lennard-Jones 6-12 function, then the integration over the semi-infinite phase of adsorbent yields m = 9 and n = 3 in Eq. (5). An exhaustive discussion of this problem can be found in the excellent review by Everett [4].

Barker and Everett [5] have also investigated the high-temperature behavior of the second gas-solid virial coefficient B_{2S} and used it to determine the surface area of adsorbents. They wrote B_{2S} in the following form:

$$B_{s2} = \alpha_S \phi\left(\frac{\varepsilon^*}{kT}\right) \tag{7}$$

where

$$\phi\left(\frac{\varepsilon^*}{kT}\right) = \int_0 [e^{(v(z,\ \varepsilon^*)/-kT)} - 1]\, dz \tag{8}$$

Hansen and co-workers [6] have performed the integration in Eq. (8) analytically and have shown that $\phi(\varepsilon^*/kT)$ can be expressed as

$$\ln\left[\phi\left(\frac{\varepsilon^*}{kT}\right)\right] = \ln\left[\left(\frac{z^*}{z_0}\right)\left(\frac{2\pi}{27T_0}\right)\right]^{1/2} + \frac{T_0}{T} + \frac{175}{216}\frac{T}{T_0} + \frac{109480}{93312}\left(\frac{T}{T_0}\right)^2 \tag{9}$$

where

$$T_0 = -\frac{v(z^*)}{k} \tag{10}$$

While analyzing this expression, Hansen and co-workers came to the conclusion that in the temperature range $3 \leq \varepsilon^*/kT \leq 8$, the function $\phi(\varepsilon^*/kT)$ may well be approximated by the formula

$$\phi\left(\frac{\varepsilon^*}{kT}\right) = e^{0.89\varepsilon^*/kT} \tag{11}$$

which has since been applied by various authors.

As we have already mentioned, the simple potential function (5) can be applied only in the case of adsorption of simple molecules on homogeneous uniform surfaces. Engewald and co-workers [7,8] have done impressive work in calculating the potential energy curves for big adsorbate molecules of complicated chemical structure being adsorbed on graphitized carbons. In this way, they were able to relate the retention values, extrapolated to zero solute, to the chemical structure of various organic compounds. Their work has great importance for certin analytic problems in GSC.

However, graphitized carbons have only limited application in analytical GSC. This technique is mainly used to separate simple gases, since the separation of larger organic molecules is generally done by means of gas-liquid or liquid-solid chromatography (GLC or LSC). Therefore, we turn back to the problem of adsorption of simple molecules on solid surfaces.

Here various silica gels, aluminum oxides, diatomaceous earths, and glasses are probably the most often used column packings. However, Kopečni and Milonjić and co-workers [9-11] have reported that other materials, magnetite for instance, may also be used successfully in various analyses. All these materials are known to possess highly irregular, structurally disordered, and chemically nonuniform surfaces. The variation in the crystallography and stoichiometry of these solid surfaces causes some variation in ε^* and z_0 across a solid surface. This effect has been known in the adsorption literature for a long time and has been called the energetic heterogeneity of the different solid surfaces.

The importance of surface heterogeneity is now widely realized. It is generally believed that the surface heterogeneity manifests itself mainly as the variation in ε^*. Therefore, let χ^* denote the differential distribution of surface points among various values of ε^* normalized to unity:

$$\int_{\Omega^*} \chi^*(\varepsilon^*)\, d\,\varepsilon^* = 1 \tag{12}$$

where Ω^* is the physical domain of the variable ε^*. The second gas-solid virial coefficient for a heterogeneous solid surface $B_{2s}{}^h$ can, with this notation, be written in the form

$$B_{2s}{}^h = \alpha_S \int e^{0.89\,\varepsilon^*/kT}\, \mathfrak{X}^*(\varepsilon^*)\, d\varepsilon^* \tag{13}$$

Steele [12] assumed that Ω^* is the interval $(-\infty, +\infty)$, and that the dispersion of ε^* can be described by the following gaussian function:

$$\chi^*(\varepsilon^*) = \frac{1}{W\sqrt{2\pi}}\, e^{\{-1/2[(\varepsilon^*-\varepsilon_0^*)/W]^2\}} \tag{14}$$

where W is the variance of this gaussian distribution, centered around the point $\varepsilon^* = \varepsilon_0^*$. Steele has applied this function to investigate the surface heterogeneity of adsorbents consisting of carbon black from the temperature dependence of B_{2s}^h. To this purpose he has chosen a highly graphitized carbon black as a standard homogeneous adsorbent and assumed that $\chi^*(\varepsilon^*)$ in this case is simply the Dirac delta distribution $\delta(\varepsilon^* - \varepsilon_0^*)$. Let B_{2s}^0 denote the second gas-solid virial coefficient for this homogeneous standard adsorbent. Then the ratio B_{2s}^h/B_{2s}^0 takes the form

$$\frac{B_{2s}^h}{B_{2s}^0} = \frac{1}{W\sqrt{2\pi}} \int_{-\infty}^{+\infty} e^{[0.89(\varepsilon^*-\varepsilon^*_0)/kT]-1/2[(\varepsilon^*-\varepsilon_0^*)/W]^2}\, d\varepsilon^* \tag{15}$$

After integration, one obtains

$$\frac{B_{2s}^h}{B_{2s}^0} = e^{1/2(0.89W/kT)^2} \tag{16}$$

From Eq. (16) it follows that the logarithm of B_{2s}^h/B_{2s}^0 should be a linear function of $(1/T)^2$. This prediction was confirmed by Steele's analysis of low-coverage adsorption of neon and carbon on various carbon blacks. Other distribution functions have been applied to analyze the experimental coefficients B_{2s}^h and are referred to in the paper by Waksmundzki et al. [13].

Now let us consider the third gas-solid virial coefficient B_{3s}. Its theoretical evaluation demands that the function $\omega(\underline{r}_1, \underline{r}_2)$ be accepted for describing the pair interactions in the vicinity of a solid surface. This problem was long a matter of controversial discussion in the adsorption literature, and it would be far too extensive to refer to it in more detail in this short chapter. We shall therefore confine ourselves to the most essential results.

Let $u(|\underline{r}_1 - \underline{r}_2|)$ be the Lennard-Jones potential function:

$$u(r) = 4u^*\left[\left(\frac{r_0}{r}\right)^{12} - \left(\frac{r_0}{r}\right)^{6}\right] \tag{17}$$

where $r = |\underline{r}_1 - \underline{r}_2|$, u^* is the depth of the potential energy minimum, and r_0 is the intermolecular distance at which $u = 0$. The quantum mechanical consideration of Sinanoglu and Pitzer leads to the following form of the perturbed pair-potential [14]:

$$\omega(r) = u(r) + \frac{\Delta}{r^3}(1 - 3\cos^2\eta) \tag{18}$$

where η is the angle between the vector $(\underline{r}_1 - \underline{r}_2)$ and an axis normal to the solid surface. The constant Δ is related to the second-order energy for the interaction of a single molecule with the surface. In other words, the presence of a solid surface leads to an added repulsion between molecules when they lie at longer distances from one another on the solid surface (horizontal position). On the other hand, when the molecules are directly above and below one another, the presence of a solid surface leads to an additional attraction.

Since the probability of finding two molecules in a horizontal position with respect to a solid surface is the greater of the two, Barker and Everett proposed to evaluate B_{3s} using the following potential function [5]:

$$\omega(r) = 4u^*\left[\left(\frac{r_0}{r}\right)^{12} - \xi\left(\frac{r_0}{r}\right)^{6}\right] \tag{19}$$

where the parameter $2/3 \le \xi \le 1$ reduces the depth of ω in comparison to that at the unperturbed potential u. Thus, the correction of Barker and Everett introduces essentially the same effect as the correction term arising from the quantum mechanical consideration of Sinanoglu and Pitzer. For the planar configuration of molecules, Eq. (18) takes the form

$$\omega(r) = 4u^*\left[\left(\frac{r_0}{r}\right)^{12} - \left(\frac{r_0}{r}\right)^{6} + \psi\left(\frac{r_0}{r}\right)^{3}\right] \tag{20}$$

where

$$\psi = \frac{\Delta}{4u^* r_0^3} \tag{21}$$

Further quantum mechanical works have shown that the additional repulsion term becomes proportional to r^{-6} at longer distances, with the result that the correction of Barker and Everett has the correct asymptotic form for large values or r. The advantage of the Barker-Everett correction method lies in the fact that Eq. (19) can be rewritten in the form of the Lennard-Jones potential function

$$\omega(r) = 4u_m^*\left[\left(\frac{r_m}{r}\right)^{12} - \left(\frac{r_m}{r}\right)^{6}\right] \tag{22}$$

in which the modified parameters u_m and r_m have a simple relation to the unperturbed ones:

$$u_m^* = \xi^2 u^* \qquad r_m = r_0 \xi^{-1/6} \tag{23}$$

However, despite the relatively extensive work that has been done to explain the perturbations in the interactions between adsorbed molecules caused by the presence of the surface, our knowledge of these effects is still uncertain. Wolfe and Sams [15] have shown that conclusions concerning the role of these third-order effects, (i.e., two admolecules plus a surface) depend considerably on the choice of the bulk parameters u and r_0 in the function u(r), in appropriate numerical analysis of experimental data for B_{3s}. Everett [16] draws attention to the fact that the quantum mechanical theories covering this third-order effect indicate that the perturbations should increase with the size of an adsorbed molecule. Meanwhile, Everett's analysis shows a variation in the opposite direction. This analysis has brought him to the conclusion that "the many-body effect in gas adsorption is much smaller than it was previously supposed."

An analysis of fourth-order effects, (i.e., three admolecules plus a surface) has brought Rudziński and Leboda [17] to the conclusion that, contrary to the third-order interaction, the fourth-order interaction introduces an additional attraction between admolecules in a horizontal position in a monolayer. This conclusion was drawn from an analysis of the temperature dependence of the fourth gas-solid virial coefficient B_{4s} for argon adsorbed on the highly graphitized carbon P-33 [18].

Everett believes [16] that the various conclusions about the role of many-body interactions are "of uncertain validity when the surface is energetically heterogeneous." Let us therefore consider in more detail the effect of surface heterogeneity on B_{3s} and higher virial coefficients.

II. EFFECTS OF SURFACE HETEROGENEITY ON THE TEMPERATURE DEPENDENCE OF RETENTION DATA IN GSC

The evaluation of the third gas-solid virial coefficient for a heterogeneous surface involves the consideration of important new physical factors. These relate to the topography of heterogeneous surfaces.

So far, two extreme kinds of the surface topography have been considered. The first, considered by Hill [19] as long ago as 1949, was the "random" topographical model of the heterogeneous surface. It is assumed in this model that various points on the adsorbent surface are distributed randomly among various values of ε^*. In other words, there is no spatial correlation between the surface points exhibiting the same value of ε^*. The average value $\bar{\varepsilon}^*$ for a surface area whose dimensions correspond to the range of intermolecular interactions is the same as for the overall surface area α_S. Therefore, the microscopic density of the adsorbate is the same throughout the entire adsorbent surface, and the adsorption system must be considered a single thermodynamic entity.

The other extreme model of surface topography is the "patchwise" model introduced by Ross and Olivier [20]. It is assumed in this model that the entire surface may be divided into large, homogeneous patches. These patches are large enough that the boundary effects can be neglected and the thermodynamic limits assumed in appropriate statistical thermodynamic considerations. Thus, the patchwise model of the heterogeneous surfaces states that the adsorption system may be considered a collection of independent subsystems that are only in thermal and material contact.

Steele [12] and Pierotti and Thomas [21,22] have developed appropriate general formulas for B_{3s}^{h} for these two kinds of surface topography. For the patchwise model we have

$$B_{3s}^{p} = \iint_{V_s} (G_2 - 1) f_{12}\, d\underline{r}_1 d\underline{r}_2 \tag{24}$$

where

$$G_n = \int_{\Omega^*} \chi^*(\varepsilon^*) e^{nv(\underline{r}_1,\, \varepsilon^*)/-kT}\, d\varepsilon^* \tag{25}$$

$$f_{12} = e^{\omega(\underline{r}_1, r_2)/-kT} - 1 \tag{26}$$

and where the superscript h = p refers to the patchwise surface topography. For random topography, B_{3s}^{h} takes the form

$$B^r_{3s} = \iint_{V_s} [G_1(f_{12} + 1) - G_2 - f_{12}]\, d\underline{r}_1 d\underline{r}_2 \tag{27}$$

An assumption that $\chi^*(\varepsilon^*)$ is the gaussian distribution from Eq. (14) leads to the following explicit expression for B^p_{3s} and B^r_{3s}:

$$-\frac{B^p_{3s}}{2B^2_{2s}} = \frac{B_{2d} - \Lambda}{\alpha_s}\, e^{(W/kT)^2} \tag{28}$$

$$-\frac{B^r_{3s}}{2B^2_{2s}} = \frac{B_{2d} - \Lambda}{\alpha_s} - \frac{1}{2}\,[1 - e^{(W/kT)^2}] \tag{29}$$

where B_{2d} may well be identified with the so-called second two-dimensional virial coefficient,

$$B_{2d} = -\,\pi \int_0 r(e^{\omega(r)/-kT} - 1)\, dr \tag{30}$$

and Λ is the correction term for nonplanarity of the adsorbed molecules, which at lower temperatures can be neglected.

Having the experimentally determined values of B^h_{2s} and B^h_{3s} and the heterogeneity parameter W estimated from the temperature dependence of B^h_{2s} [see Eq. (16)], one can evaluate the experimental value of the term $(B_{2d} - \Lambda)/\alpha_s$ for various temperatures. This value can also be evaluated theoretically and then compared with the values estimated experimentally assuming either the patchwise [Eq. (28)] or the random [Eq. (29)] surface topography. The agreement between the experimentally estimated and theoretically calculated sets of values could be a criterion for a proper choice of the model of surface topography for the adsorption system under investigation.

In so doing, Pierotti and Thomas came to the conclusion that the surface of the heterogeneous carbon black ("Black Pearls No. 71") investigated experimentally by Halsey and co-workers [18] has a random surface topography. Hinman and Halsey [23], on the other hand, analyzing the adsorption of N_2 and Ar on a strongly heterogeneous MgO surface, believe that the model of a patchwise topography should more properly describe the nature of the MgO surface.

Let us remark here, that Eqs. (28) and (29) represent a somewhat relative method of investigating the nature of surface heterogeneity of real solids. This means that a standard homogeneous surface is needed. Such a surface can hardly, or perhaps even cannot at all, be prepared for most solid surfaces. Therefore, Rudziński et al. [13,24] have recently proposed a more independent method for investigating surface heterogeneity from the temperature dependence of the second and third gas-solid virial coefficients. The effects of surface heterogeneity on the fourth and higher virial coefficients have been investigated by Rudziński [25] and by Ripa and Zgrablich [26].

So far we have considered the model of mobile adsorption. At the same time, a strong dispersion of ε^* means the existence of various energy barriers between certain local minima in the gas-solid potential function ε. Therefore, in the case of typical heterogeneous surfaces, the use of models of localized adsorption should be more appropriate in theoretical considerations. This problem has not been sufficiently recognized in the theoretical investigation of the effects of heterogeneity in GSC.

Although the virial approach has been traditionally used to describe mobile adsorption, it can also be used successfully to describe localized adsorption. If adsorption is localized, then the harmonic approximation for $v(\underline{r})$ can be accepted:

$$v(\underline{r}) = -\varepsilon + \frac{1}{2} k_z(z - z^*)^2 + v(\underline{\tau}) \tag{31}$$

where k_z is the force constant for vibrations normal to the surface and $v(\underline{\tau})$ is the energy change for displacement by the vector $\underline{\tau}$ from its position at the minimum. For a homogeneous surface, B_{2s} can then be written in the form

$$B_{2s} = N_0 Z_1^{(s)} \tag{32}$$

where N_0 is the total number of adsorption sites (local minima) on the adsorbent surface, which is equal to (α_S/α_s), where α_s is the surface area associated with a single adsorption site. Further, $Z_1^{(s)}$ is the single-molecule-site configurational integral

$$Z_1^{(s)} = \int_0 dz \int_{\alpha_s} d\underline{\tau}\, e^{\varepsilon - \frac{1}{2}k_z(z-z^*)^2 - v(\underline{\tau})/kT} \tag{33}$$

Incorporating, in addition, an appropriate harmonic approximation for $v(\underline{\tau})$, we arrive at the following equation for $Z_1^{(s)}$:

$$Z_1^{(s)} = q_v\, e^{\varepsilon/kT} \tag{34}$$

where q_v is the molecular vibrational partition function of the adsorbed molecules.

It is to be expected that q_V will generally be a function of ε. This problem was investigated theoretically by Halsey [27] and Appel [28], who suggested expressing q_V in the form

$$q_V = q_{xy} e^{(E_0/kT) - g\varepsilon} \tag{35}$$

where q_{xy} is the factor associated with vibrations parallel to an adsorbent surface, E_0 is the first quantum mechanical vibrational energy level for vibrations normal to the surface, and g is a constant that depends on the nature of the gas-solid interactions. The g values for some adsorption systems have been estimated by Rudziński and Patrykiejew [29]. Some recent experimental and theoretical investigations of adsorption of organic molecules on silica gel surfaces, performed by Rudziński et al. [30], seem to suggest that the term $g\varepsilon$ in Eq. (35) can be safely neglected only at low temperatures. However, since we will now consider the region of low and moderate temperatures, we will neglect, for simplicity, the dependence of q_V on ε.

With the above approximations, the second gas-solid virial coefficient B^h_{2s} for localized adsorption can be written in the form

$$B^h_{2s} = N_0 q_V \int_\Omega \chi(\varepsilon) e^{\varepsilon/kT} d\varepsilon \tag{36}$$

where

$$\chi(\varepsilon) = \frac{\partial N_0}{\partial \varepsilon} \tag{37}$$

is the differential distribution of the adsorption sites among various values of ε and Ω is the physical domain of ε. For mathematical convenience, Ω is, in many theoretical works on the adsorption of gases on heterogeneous solid surfaces, assumed to be the semi-infinite interval $(0, +\infty)$.

Let B(T) denote B^h_{2s} divided by $N_0 q_V$. Then B(T) is simply a Laplace transform of $\chi(\varepsilon)$:

$$B(T) = \int_0^\infty \chi(\varepsilon) e^{\varepsilon/kT} d\varepsilon \tag{38}$$

Having B^h_{2s} measured at various temperatures, one may determine the function $\chi(\varepsilon)$ using appropriate methods of evaluating the inverse Laplace transform. This method has been used by Waksmundzki and co-workers [31] to quantitatively analyze the surface heterogeneity in various adsorption systems.

This method is especially recommended for analyzing adsorption data measured by means of gas chromatography. In that situation, the experimental determination of B_{2s}^{h} at various temperatures is extremely easy. That is, the first two gas-solid virial coefficients are obtained by the following extrapolation procedure [cf. Eq. (4)]:

$$B_{2s}^{h} = \lim_{\rho_0 \to 0} (V_N) \tag{39}$$

$$B_{3s}^{h} = \lim_{\rho_0 \to 0} \left(\frac{V_N - B_{2s}}{\rho_0} \right) \tag{40}$$

Everett [32] was the first to show the importance of heterogeneity effects in GSC. The experimental problems connected with physico-chemical measurements by means of gas chromatography have been described in the excellent monographs by Conder and Young [33] and Laub and Pecsok [34]. However, in the case of strongly heterogeneous surfaces and measurements done at relatively low temperatures, the extrapolations described in Eqs. (39) and (40) may sometimes lead to erroneous results. This is because a Henry's law region of adsorption may not be achieved in GSC measurements, even with the best microsyringes and the smallest possible sample sizes of adsorbate. In the next section we will explain the reason for this.

III. EFFECTS OF SURFACE HETEROGENEITY ON THE PRESSURE DEPENDENCE OF RETENTION DATA IN GSC

We will consider here the Langmuir model of adsorption, assuming a monolayer localized adsorption, with one molecule per one adsorption site, and neglecting the lateral interactions between the adsorbed molecules. The fractional surface coverage θ of the adsorption sites exhibiting the adsorption energy ε is, according to the Langmuir model, given by the equation

$$\frac{N}{N_0} = \theta = \left[1 + \frac{1}{q_v \rho_0} e^{(-\varepsilon/kT)} \right]^{-1} \tag{41}$$

The overall (total) surface coverage of a heterogeneous solid surface θ_t is thus given by

$$\theta_t = \int_{\Omega} \theta(\rho_0, \varepsilon)\chi(\varepsilon)\, d\varepsilon \tag{42}$$

Integration by parts yields the expression [35]

$$\theta_t = \theta\,\mathfrak{X}\Big|_{\varepsilon_1}^{\varepsilon_m} - \int_{\varepsilon_1}^{\varepsilon_m} \frac{\partial\theta}{\partial\varepsilon}\,\mathfrak{X}(\varepsilon)\,d\varepsilon \tag{43}$$

where ε_1 and ε_m are the minimum and the maximum values of the adsorption energy ε within the interval Ω and

$$\mathfrak{X}(\varepsilon) = \int \chi(\varepsilon)\,d\varepsilon \tag{44}$$

The derivative $\partial\theta/\partial\varepsilon$ is a bell-shaped function of the variable ε:

$$\frac{\partial\theta}{\partial\varepsilon} = \frac{(1/kT)e^{(\varepsilon_c-\varepsilon)/kT}}{\left[1 + e^{(\varepsilon_c-\varepsilon)/kT}\right]^2} \tag{45}$$

centered about $\varepsilon = \varepsilon_c$:

$$\varepsilon_c = -kT \ln q_v - kT \ln \rho_0 \tag{46}$$

At the hypothetical temperature $T = 0$, the derivative $\partial\theta/\partial\varepsilon$ degenerates into the Dirac delta distribution $\delta(\varepsilon - \varepsilon_c)$. Thus, when the temperature is low enough, the function $\partial\theta/\partial\varepsilon$ becomes very narrow and behaves like a "sampling function" in evaluations of the second integral on the right side of Eq. (43). Let I_2 denote this integral. Then I_2 may be evaluated effectively by expanding $\mathfrak{X}(\varepsilon)$ into its Taylor series around $\varepsilon = \varepsilon_c$. Doing so, and introducing the new variable t,

$$t = \frac{\varepsilon - \varepsilon_c}{kT} \tag{47}$$

we obtain

$$I_2 = \sum_{n\geq 0} \frac{(kT)^n}{n!}\,C_n \left(\frac{\partial^n \mathfrak{X}}{\partial\varepsilon^n}\right)_{\varepsilon=\varepsilon_c} \tag{48}$$

where C_n is the value of the integral J_n:

$$J_n = \int \frac{t^n e^{-t}}{(1 + e^{-t})^2}\,dt \tag{49}$$

taken between the limits $(\varepsilon_1 - \varepsilon_c)/kT$ and $(\varepsilon_m - \varepsilon_c)/kT$. The first two integrals, J_0 and J_1, can be evaluated analytically:

$$J_0 = \frac{1}{1 + e^{-t}} \tag{50}$$

$$J_1 = -\frac{te^{-t}}{1 + e^{-t}} - \ln(1 + e^{-t}) \tag{51}$$

At lower temperatures (i.e., when $T \to 0$), all the terms except the first one in expansion (48) can be neglected, and from Eq. (43) we obtain

$$\theta_t = \mathfrak{X}(\varepsilon_m)[1 + e^{(\varepsilon_c - \varepsilon_m)/kT}]^{-1} - \mathfrak{X}(\varepsilon_1)[1 + e^{(\varepsilon_c - \varepsilon_1)/kT}]^{-1}$$
$$- \mathfrak{X}(\varepsilon_c)\{[1 + e^{(\varepsilon_c - \varepsilon_m)/kT}]^{-1} - [1 + e^{(\varepsilon_c - \varepsilon_1)/kT}]^{-1}\} \tag{52}$$

Let us note, however, that Eq. (52) has a limited range of applicability. This is because in the case of the finite integration limits $(\varepsilon_1, \varepsilon_m)$, $\chi(\varepsilon)$ should be written in the form

$$\chi(\varepsilon) = \begin{vmatrix} 0 & \varepsilon \le \varepsilon_1 \\ \chi & \varepsilon_1 < \varepsilon < \varepsilon_m \\ 0 & \varepsilon \ge \varepsilon_m \end{vmatrix} \tag{53}$$

Consequently, the function $\mathfrak{X}(\varepsilon)$ has to be written in the form

$$\mathfrak{X}(\varepsilon) = \begin{vmatrix} 0 & \varepsilon \le \varepsilon_1 \\ \mathfrak{X}(\varepsilon) & \varepsilon_1 < \varepsilon < \varepsilon_m \\ 1 & \varepsilon \ge \varepsilon_m \end{vmatrix} \tag{54}$$

Thus, the total adsorption isotherm θ_t should be evaluated as follows:

1. For very small adsorbate pressures in the free gas phase (i.e., when $\varepsilon_c > \varepsilon_m$),

$$\theta_t = [1 + e^{(\varepsilon_c - \varepsilon_1)/kT}]^{-1} \tag{55}$$

2. For moderate adsorbate pressures, when $\varepsilon_c \le \varepsilon_m$, the value of θ_t has to be evaluated according to Eq. (52).

Let us remark that the function θ_t defined in this way is continuous at the point $\varepsilon_c = \varepsilon_m$. Further, from Eq. (55) it follows that, at sufficiently low adsorbate pressures in the free gas phase, the adsorption of the adsorbate on a heterogeneous solid surface will be like that on a homogeneous surface characterized by the adsorption energy $\varepsilon = \varepsilon_1$. In the limit $\rho_0 \to 0$, Eq. (55) reduces to Henry's law:

$$\lim_{\rho_0 \to 0} [1 + e^{(\varepsilon_c - \varepsilon_1)/kT}]^{-1} = \rho_0 q_v e^{\varepsilon_1/kT} \tag{56}$$

in accordance with some basic thermodynamic predictions.

Now let us note that the density ρ_1 at which the Henry's law region is achieved in adsorption on a heterogeneous surface is given by the condition that $\varepsilon_c \geq \varepsilon_m$, or

$$\ln \rho_1 \leq \frac{-\varepsilon_m}{kT} - \ln q_v \tag{57}$$

Thus, when the temperature decreases, the Henry's law region is shifted down to ever lower adsorbate densities (pressures) in the free phase. It may therefore happen that these adsorbate densities will sometimes be difficult to achieve by conventional GSC techniques.

Let us also note, however, that it is not only low temperatures that may lead to this troublesome experimental situation in determining the second gas-solid virial coefficient B_{2s} by means of GSC. One will also face the same situation at moderate temperatures when the solid surface is strongly heterogeneous.

Under such conditions, the spread of $\chi(\varepsilon)$ will be very large and the function (normalized to unity) will reach small values at any point ε. Since the function $\chi(\varepsilon)$ is then very flat, the values of its derivatives at any point ε are also very small. In this situation, all the terms in expansion (48) but the first can practically be neglected. When the spread of $\chi(\varepsilon)$ is very large, the value of ε_m will also be very large. Thus, according to condition (57), very small adsorbate densities must be used to achieve the Henry's law region in the adsorption of solute on the surface of a column packing. We therefore face essentially the same experimental situation as in adsorption at low temperatures.

Let us now summarize our conclusions about the experimental determination of the second gas-solid virial coefficient B_{2s} by means of GSC.

When the surface of a column packing is very heterogeneous and the retention volume is measured at low or moderate temperatures, very small concentrations of the solute in the mobile gas phase must be created to achieve the Henry's law region of adsorption. Otherwise, the extrapolation in Eq. (39) may yield quite undefined values. In the fortunate case in which the concentration of the solute in the gas phase is sufficiently low, and the temperature is low, extrapolation yields a value of B_{2s} like that for a hypothetical homogeneous surface characterized by an adsorption energy equal to ε_1 (c.f. Eq. (55)). It seems that this has not yet been recognized in physicochemical measurements made by means of GSC.

Meanwhile, there is already experimental evidence supporting our conclusion. This occurs in the published work of Gawdzik et al. [36] on the adsorption of benzene, *n*-hexane, and cyclohexane on strongly heterogeneous glass beads.

These authors have applied a very effective method for approximating the experimental function $V_n(\rho_0)$ at low adsorbate densities in the free (mobile) phase. This method seems to be a very important point in the experimental determination of the second gas-solid virial coefficient by means of GSC. Although proposing this new approximation formula, Gawdzik et al. followed the earlier theoretical work of Rudziński [37], who proposed a new exponential virial expansion for the adsorption isotherm $N(\rho_0)$:

$$\frac{N}{N_0} = \theta(\rho_0) = \exp\left(\sum_{m=2}^{\infty} C'_{ms}\, \rho_0^{m-1}\right) - 1 \qquad (58)$$

Rudziński has shown that this new exponential virial expansion is much more effective when one approximates $\theta(p, T)$ at small adsorbate pressures. The result is that with the same number of new "exponential" virial coefficients C'_{ms}, Eq. (58) approximates the function $\theta(p, T)$ better than it does the traditional polynomial exapansion [c.f. Eq. (2)]

$$\theta(\rho_0) = \sum_{m=1}^{\infty} B'_{ms}\rho_0^{m-1} \qquad B'_{ms} = \frac{B_{ms}}{N_0} \qquad (59)$$

One may easily establish the connection between the normal "polynomial" and the "exponential" virial coefficients by expanding the right-hand side of Eq. (58) into powers of the density ρ_0 and then comparing the coefficients at equal powers of ρ_0. Doing this, one arrives at the interrelation

$$C'_{ms} = \sum_{\{j\}} [-1]^{\left(1 + \sum_{n=2}^{m} j_n\right)} \left[\sum_{n=2}^{m} j_n - 1\right]! \prod_{n=2}^{m} \frac{(B'_{ms})^{j_n}}{(j_n)!} \qquad (60a)$$

where the sum is taken over all sets that satisfy the condition

$$\sum_{n=2}^{m} (n - 1)j_n = m - 1 \tag{60b}$$

In particular,

$$C'_{2s} = B'_{2s} \qquad C'_{3s} = B'_{3s} - \frac{1}{2}(B'_{2s})^2 \tag{61}$$

Let us note that, with the simplifying assumption

$$C'_{ms} = 0 \qquad \text{for} \quad m > 2 \tag{62}$$

one arrives at the well-known Jovanovic adsorption isotherm,

$$\theta(\rho, T) = e^{\rho_0 q_v e^{\varepsilon/kT}} - 1 \tag{63}$$

which approximates the isotherm equation for the ideal adsorption on a loalized monolayer almost as well as the full expansion (59) taken to infinity (i.e., the Langmuir equation). Further remarks on the theoretical origin of the Jovanovic isotherm equation may be found in the work of Rudziński and Wojciechowski [38].

Thus, according to Eqs. (1) and (58), the retention volume $V_N(p, T)$ should be effectively approximated by the equation

$$V_N = N_0 \left[\sum_{m=2}^{\infty} (m - 1) C'_{ms} \rho_0^{m-2} \right] \exp\left(\sum_{m=2}^{\infty} C'_{ms} \rho_0^{m-1} \right) \tag{64}$$

Neglecting the terms with $m > 2$, one arrives at the formula

$$V_N = N_0 e^{D'_0 + D'_1 \rho_0} \tag{65}$$

where

$$D'_0 = \ln C'_{2s} = \ln B'_{2s} \tag{66a}$$

$$D'_1 = C'_{2s} = B'_{2s} \tag{66b}$$

which should well approximate the experimental function $V_N(p, T)$ at small adsorbate pressures. Approximation (65) has been used successfully by Gawdzik et al. [36] to determine the value of B^h_{2s} for the adsorption of benzene, *n*-hexane, and cyclohexane on porous glass beads. At the same time, Gawdzik et al. determined the full form of the energy distribution function $\chi(\varepsilon)$ in these three adsorption systems, using a method that will be discussed shortly. The difference $\varepsilon_m - \varepsilon_1$ estimated for these adsorption systems was about 3000 cal/mol, indicating a strong surface heterogeneity of the porous glass beads. Meanwhile, the plots of ln B_{2s}^h versus 1/T were almost ideally linear, such as those for a highly homogeneous surface. There is only one possible explanation for this contradiction, that the slopes of ln B_{2s} plotted according to 1/kT yield the values of the minimum adsorption energy ε_1 in these adsorption systems.

In this way we arrive at the conclusion that a quantitative determination of the surface heterogeneity from the temperature dependence of B_{2s}^h may sometimes be risky in the case of strongly heterogeneous solid surfaces, which is the reason that after publishing a few papers [13,24,31,39-41] on the temperature dependence of B_{2s} and of higher gas-solid virial coefficients, Rudziński and coworkers initiated a new series of works in which the pressure dependence of V_N was exploited for quantitative determination of surface heterogeneity.

IV. QUANTITATIVE ESTIMATION OF SURFACE HETEROGENEITY BY MEANS OF GSC METHODS

In the first two papers published on this problem [42,43], Rudziński et al. accepted the so-called condensation approximation (CA) approach to determine $\chi(\varepsilon)$ from the pressure dependence of V_N. In this approach it is assumed that adsorption on a heterogeneous solid surface proceeds in an ideally stepwise fashion:

$$\theta = \theta_c = \begin{vmatrix} 0 & \varepsilon < \varepsilon_c'(p, T) \\ 1 & \varepsilon \geq \varepsilon_c'(p, T) \end{vmatrix} \tag{67}$$

where ε_c' is the value of ε at which a sudden increase ("condensation step") appears in the pressure p and temperature T on the "condensation isotherm" $\theta_c(p, T)$. The function $\varepsilon_c'(p, T)$ can be found from Cerofolini's variational principle, which leads to the following simple condition for ε_c' [43]:

$$[\theta(\varepsilon, p, T)]_{\varepsilon=\varepsilon_c'} = \frac{1}{2} \tag{68}$$

From this, it can easily be determined, that ε_c' is the same function as the function ε_c defined in Eq. (46).

The approximate "condensation" isotherm θ_c is then used to replace the true kernel in Eq. (42). The integration can then be performed very easily:

$$\theta_t = \int_{\varepsilon_c}^{\varepsilon_m} \chi(\varepsilon)\, d\varepsilon = \mathfrak{X}(\varepsilon_m) - \mathfrak{X}(\varepsilon_c) \tag{69}$$

and an approximate solution for $\chi(\varepsilon)$ is obtained by the following simple differentiation:

$$\chi_{CA}(\varepsilon_c) = - \frac{\partial\theta_t}{\partial\varepsilon_c} = - \frac{\partial\theta_t}{\partial p}\frac{dp}{d\varepsilon_c} = \frac{p}{kT}\frac{\partial\theta_t}{\partial p} \tag{70}$$

where the subscript CA is used to denote the approximate form of the energy distribution function $\chi(\varepsilon)$ obtained by using the CA approach. In the terms of the retention volume V_N, Eq. (70) takes the form

$$\chi_{CA}(\varepsilon_c) = \frac{1}{N_0}\frac{p}{(kT)^2} V_N(p) \tag{71}$$

Thus, in order to obtain the approximate energy distribution χ_{CA}, one only has to replace the experimental pressure p in the Eq. (71) by the function $y(\varepsilon)$:

$$y = q_v e^{-\varepsilon/kT} \tag{72}$$

in which a certain value for q_v must be assumed. This problem will be discussed shortly.

The function χ_{CA} is only a crude approximation for $\chi(\varepsilon)$. A more exact solution is obtained by applying the so-called asymptotically correct condensation approximation (ACCA) approach. In this approach, the true kernel $\theta(\varepsilon, p, T)$ in Eq. (42) is replaced by the following combination of Henry's law and step isotherms:

$$\theta = \theta_{ACCA} = \begin{cases} q_v\, p_0\, e^{\varepsilon/kT} & \varepsilon < \varepsilon_c(p, T) \\ 1 & \varepsilon \geq \varepsilon_c(p, T) \end{cases} \tag{73}$$

The solution for $\chi(\varepsilon)$ in this case, reads [43]

$$\chi_{ACCA}(\varepsilon_c) = -\frac{\partial\theta_t}{\partial\varepsilon_c} - kT\frac{\partial^2\theta_t}{\partial\varepsilon_c^2} \tag{74}$$

Or, in terms of the retention volume V_N,

$$\chi_{ACCA}(\varepsilon_c) = -\frac{1}{N_0}\left(\frac{p}{kT}\right)^2\left(\frac{\partial V_N}{\partial p}\right)_T \tag{75}$$

Although other methods for evaluating $\chi(\varepsilon)$ from the pressure dependence of V_N have also been proposed [44,45], Eq. (75) developed by Rudziński et al., appears to be probably the most popular solution for $\chi(\varepsilon)$ in GSC. It requires only the experimental determination of $V_N(p)$ and a subsequent graphic differentiation of this function. Experimental problems connected with the determination of $V_N(p)$ have been described in the excellent monograph by Kiselev and Yashin [46] and in the paper by Volf et al. [47].

Rudziński's solution (75) has been used by many authors [48-51] to quantitatively analyze the surface heterogeneity of various adsorbents. Gawdzik et al. [36] have introduced into this method the very effective analytic approximation for the experimental function $V_N(p)$,

$$V_N = \exp\left(\sum_{m=0}^{s} D_m p^m\right) \tag{76}$$

the simplified form of which was given in Eq. (65). This expansion was used next by Leboda et al. [49,50] to investigate the surface heterogeneity of various adsorbents and catalysts. It is interesting to note that the plots of D_0 versus 1/T reported by Leboda et al. were also highly linear, like those reported earlier by Gawdzik et al. However, both Gawdzik et al. and Leboda et al. interpreted the slope of the function D_0 versus 1/kT as a "mean adsorption energy," not as a minimum adsorption energy ε_1.

Let us note here that a combined study of both the temperature and pressure dependence of retention should have an important advantage since a weak point in the determination of $\chi(\varepsilon)$ from the pressure V_N lies in the arbitrary choice of both q_V and ε_1. An a priori theoretical estimation of these quantities is generally impossible. In effect, one can evaluate the shape of the energy distribution function $\chi(\varepsilon)$, but it is difficult to fix this function on an energy scale. However, according to the theoretical works of Rudziński and Jagiełło [35], it should be possible to estimate q_V and ε_1 in the case of strongly heterogeneous adsorbents, from the temperature dependence of the retention volume extrapolated to the zero sample solute condition $V_N(0)$:

$$\ln V_N(0) = \ln (N_0 q_v) + \frac{\varepsilon_1}{kT} \tag{77}$$

where the slight temperature dependence of q_v is neglected. The adsorbent capacity N_0 can be found from the normalization to unity of the full energy distribution function χ_{ACCA}, obtained from the pressure dependence of V_N. When accepting approximation (76), the unnormalized energy distribution function is given by the equation

$$\chi_{ACCA}(\varepsilon_c) = -\frac{y^2}{(kT)^2}\left(\sum_{m=1}^{s} mD_m y^{m-1}\right)\exp\left(\sum_{m=0}^{s} D_m y^m\right) \tag{78}$$

where $y(\varepsilon)$ is given by Eq. (72).

Numerous calculations of $\chi(\varepsilon)$ have been reported in various works. In general, the calculated functions $\chi(\varepsilon)$ show a complicated form with two or more maxima. However, it is interesting to note that when these functions $\chi(\varepsilon)$ are smoothed, they exhibit very similar features for all heterogeneous surfaces. This problem was investigated theoretically by Cerofolini, and the results have been stressed in his theory of "equilibrium surfaces" [52]. There, some general rules have been formulated about the formation of real solid surfaces. Cerofolini has pointed out that, in a crude approximation, all heterogeneous, true solid surfaces should be characterized by the same energy distribution function,

$$\chi(\varepsilon) = [\alpha + 2\beta(\varepsilon - \varepsilon_1)]e^{-\alpha(\varepsilon-\varepsilon_1)-\beta(\varepsilon-\varepsilon_1)^2} \tag{79}$$

in which α and β are heterogeneity parameters. Of course this must be reflected in certain universal features of the experimental isotherms $N_t = N_0 \theta_t$. At very low adsorbate pressures, the experimental isotherms will tend to Henry's law, and the problem is then quite trivial. Therefore, let us consider the region of moderate adsorbate pressures (i.e., when $\varepsilon_c < \varepsilon_m$).

The adsorption isotherm θ_t is then given by Eq. (52), which can be considerably simplified if ε_c is not very close to ε_m. Then, the existence of ε_m is no longer essential, and one can safely accept $(+\infty)$ as the upper integration limit in Eq. (42). With this assumption, the coefficients C_n in expansion (48) take the simple form [53]

$$C_n = n!(\pi)^n \left| \begin{array}{ll} -1 & \text{for } n = 0 \\ 0 & \text{for } n = 2m + 1 \\ B_n & \text{for } n = 2m \end{array} \right. \tag{80}$$

where B_n is Bernoulli's number. At the same time, the first term on the right-hand side of Eq. (43) vanishes. Neglecting, in expansion (48), the terms of order higher than $0(kT)^2$, we arrive at the following equation for θ_t [53]:

$$\theta_t = -\mathfrak{X}(\varepsilon_c) - \frac{C_2}{2}(kT)^2 \frac{\partial^2 \mathfrak{X}}{\partial \varepsilon_c^2} \tag{81}$$

The first term on the right-hand side of Eq. (81) is thus identical to the result given by the crude CA approach [cf. Eq. (69)], whereas the second term and the others that are neglected in Eq. (81) are correction terms to the result given by the CA approach. Let us consider the crude CA approach first.

In this case, the equation for θ_t related to the universal energy distribution (79) takes the form

$$\theta_t = e^{\alpha(\varepsilon_c - \varepsilon_1) - \beta(\varepsilon_c - \varepsilon_1)^2} \tag{82}$$

Considering the function $\varepsilon_c(p, T)$ in Eq. (46), one can rewrite the universal isotherm equation in the form

$$\ln \theta_t = A_0 - A_1 \ln \frac{p}{p_s} - A_2 \left(\ln \frac{p}{p_s}\right)^2 \tag{83}$$

where

$$A_0 = \alpha\varepsilon_1 + (2\beta\varepsilon_1 - \alpha)\varepsilon_c^0 - \beta[(\varepsilon_c^0)^2 + \varepsilon_1^2] \tag{84a}$$

$$A_1 = kT[2\beta(\varepsilon_1 - \varepsilon_c^0) - \alpha] \tag{84b}$$

$$A_2 = \beta(kT)^2 \tag{84c}$$

and where

$$\varepsilon_c^0 = -kT \ln \frac{p_s q_v}{kT} \tag{85}$$

with p_s the saturation pressure of the adsorbate. When $\beta = 0$, Eq. (83) reduces to the well-known Freundlich adsorption isotherm. When, on the other hand, the coefficient A_1 is small, Eq. (83) yields the well-known Dubinin-Radushkevich adsorption isotherm. In general, the behavior of the experimentally measured adsorption isotherms will be a hybrid between the Freundlich and Dubinin-Radushkevich type of behaviors. Let us, however, note that even when the coefficients A_1 and A_2 are comparable, the Dubinin-Radushkevich behavior will become more prevalent as the adsorbate pressure decreases. This is because the term $[\ln(p/p_s)]^2$ will grow more rapidly than the term $\ln(p/p_s)$.

From the foregoing consideration, it follows that the universal features of θ_t should be reflected in some universal features of the pressure dependence of V_N. These general features of $V_N(p)$ can be deduced by taking the derivative $(\partial\theta_t/\partial p)_T$ of the function θ_t defined in Eq. (83). However, these general features have not yet been investigated.

So far we have considered the monolayer localized ideal adsorption on real heterogeneous solid surfaces. However, the theoretical expressions developed above can easily be generalized to include the effects of the lateral interactions between the adsorbed molecules. Extensive research along these lines has recently been done by Rudziński et al. [53-56]. We will briefly discuss the most essential results.

To this purpose, we will accept the simplest Bragg-Williams approximation to account for the effects of the interactions between adsorbed molecules. Under these conditions, all the equations developed above, beginning with Eq. (67), remain unchanged, except that the function $\varepsilon_c(p, T)$ includes a new term arising from the interactions between adsorbed molecules. Let u_{12} denote the interaction energy between two molecules adsorbed at two neighboring sites and c be the number of the nearest neighbors adsorption sites of an arbitrarily chosen site. Then, for a patchwise surface topography, the function ε_c must be replaced by its generalized form ε_c^p:

$$\varepsilon_c^p = \varepsilon_c - \frac{1}{2} cu_{12} \tag{86}$$

whereas for a random topography we have

$$\varepsilon_c^r = \varepsilon_c - cu_{12}\theta_t \tag{87}$$

Additionally, expansion (81) is still valid except that for the patchwise surface topography the coefficient C_2 must be replaced by C_2^p:

$$C_2^p = -3.29\, e^{-1.64(u_{12}/kT)^{1.24}} \tag{88}$$

For a random surface topography, the coefficient C_2 in expansion (81) is still given by Eq. (80).

Rudziński and Baszyńska [54], and Morel et al. [55], have shown that replacing ε_c by ε_c^p or ε_c^r in Eq. (83) permits a much better correlation of experimental data at moderate adsorbate pressures. Further, most of the solid surfaces that have been investigated have exhibited a random topographical distribution of adsorption sites.

Recently, Rudziński et al. [56] showed that the second term on the right side of Eq. (81) is a very effective correction for θ_t evaluated by means of the crude CA approach. Thus, they came to the conclusion that Eq. (81) could also be used to give an accurate determination of the adsorption energy distribution from experimental adsorption isotherms.

Originally, Eq. (81) is a second-order linear differential equation for $\chi(\varepsilon)$ that could be solved numerically. Further, Rudziński et al. have found that the problem can be considerably simplified without a remarkable loss in the accuracy of the solution obtained for $\chi(\varepsilon)$.

Specifically, since the second term on the right side of Eq. (81) is a correction term, Rudziński et al. evaluated it by means of the CA approach. Doing so, they arrived at the following solution for $\chi(\varepsilon)$:

$$\chi(\varepsilon) = \chi_{CA}^s(\varepsilon_c^s) - \frac{C_2}{2}(kT)^2 \frac{\partial^2 \chi_{CA}^s}{\partial(\varepsilon_c^s)^2} \tag{89}$$

where the subscript s is used to denote the solution for either the patchwise (s = p) or the random (s = r) surface topography.

Appropriate numerical model investigation has shown that for typical adsorption systems solution (89) does not diverge from the exact solution $\chi(\varepsilon)$ more than by about 5%. Thus, solution (89) is much more accurate than solution (75) proposed earlier by Rudziński and widely used by various investigators in GSC. However, since solution (89) was only published recently, it has not been used in GSC until now.

The results of the extensive investigation recently done by Rudziński et al. on the role of surface topography in adsorption on heterogeneous surfaces can be summarized as follows. A dispersion of adsorption energy $\chi(\varepsilon)$ much more strongly affects the adsorption on surfaces characterized by a patchwise than by a random surface topography. It relates to the behavior of adsorption isotherms, heats of adsorption, and other observed adsorption variables, such as heat capacities. In other words, surface heterogeneity effects manifest themselves more strongly in the case of surfaces characterized by a patchwise topography than for surfaces with a random topography.

ACKNOWLEDGMENT

The author would like to express his warm thanks to Dr. Jolanta Narkiewicz-Michalek for her technical assistance in preparing the manuscript.

LIST OF SYMBOLS

Latin Terms

A_0, A_1, A_2	coefficients defined in Eqs. (83) and (84)
B_n	Bernoulli's number
$B(T)$	function defined in Eq. (38)
B_{2d}	two-dimensional second virial coefficient
B_{20}	second bulk virial coefficient
B_{ns}	nth gas-solid virial coefficient
B_{ns}^{o}	value of B_{ns} when $\varepsilon^* = \varepsilon_0^*$
B_{ns}^{h}	nth gas-solid virial coefficient for a heterogeneous solid surface
c	number of nearest neighbors adsorption sites
C_n	coefficient in expansion (48)
C_N	value of the integral J_n taken between the integration limits $(\varepsilon_1 - \varepsilon_c)/kT$ and $(\varepsilon_m - \varepsilon_c)/kT$
C'_{ms}	mth exponential gas-solid virial coefficient defined in Eq. (58)
D_m	virial coefficient defined in Eq. (76)
D'_n	virial coefficient defined in Eq. (65) and Eq. (66)
f_{12}	Myer function defined in Eq. (26)
g	parameter in the vibrational partition function q_v defined in Eq. (35)
G_n	function defined in Eq. (25)
J_n	integral defined in Eq. (49)
k	Boltzmann constant
k_z	force constant for vibrations normal to a surface
N	number of adsorbate molecules in the surface phase
N_0	number of adsorption sites on a solid surface (surface capacity)

p	adsorbate pressure in the free gas phase
p_s	saturation pressure of adsorbate
q_v	molecular vibrational partition function
r	distance between two interacting adsorbate molecules
r_0	distance parameter in the pair potential function (17)
r_m	effective Barker-Everett distance parameter in the modified pair potential defined in Eqs. (22) and (23)
t	function defined in Eq. (47)
T	absolute temperature
T_o	parameter defined in Eq. (10)
u	pair interaction potential between two adsorbate molecules
u	depth in the pair potential (17)
u_m	effective interaction depth in the modified Barker-Everett potential, defined in Eqs. (22) and (23)
v	gas-surface interaction potential
W	variance in the gaussian distribution (14)
y	function of ε defined in Eq. (72)
z	distance of an adsorbed molecule from the solid surface
z^*	distance from the surface at which the potential function v reaches its minimum value
z_o	distance parameter in the gas-solid potential function (5)
$Z_1^{(s)}$	single molecule-site configuration integral defined in Eq. (33).

Greek Terms

α	heterogeneity parameter in the generalized adsorption energy distribution (79)
α_s	adsorbent area associated with a single adsorption site
α_S	total area of adsorbent surface
β	heterogeneity parameter in the generalized distribution function (79)
Δ	parameter in the Sinanoglu-Pitzer potential function (18)
ε	adsorption energy, defined in Eq. (31)
ε_c	function defined in Eq. (46)
ε^*	the depth in the gas-solid potential function v(z)
$\varepsilon_1, \varepsilon_m$	minimum and maximum value of ε on a given heterogeneous surface
$\varepsilon_c^p, \varepsilon_c^r$	functions defined in Eqs. (86) and (87)
ε_c^0	function defined in Eq. (85)
ε_0^*	most probable value of ε^* on a heterogeneous surface
η	angle between the vector $\lvert \underline{r}_1 - \underline{r}_2 \rvert$ joining the centers of masses of two adsorbed molecules and the axis normal to surface
θ	relative surface coverage, equal to (N/N_0)
Λ	Barker-Everett's correction for the nonplanarity of an adsorbed phase

ξ	Barker-Everett's perturbation parameter defined in Eq. (19)
ρ_0	adsorbate density in the free gas phase (molecules/volume unit)
$\underline{\tau}$	vector for the displacement of an adsorbed molecule in a direction parallel to the adsorbent surface function defined in Eq. (8)
χ	differential distribution of adsorption sites among various values of ε
$\mathfrak{X}$	indefinite integral of χ
χ_{CA}	solution for χ obtained by using the CA approach
χ_{ACCA}	solution for χ obtained by using the ACCA approach
ψ	parameter defined in Eqs. (20) and (21)
ω	interaction potential between two adsorbate molecules in the vicinity of an adsorbent surface
Ω	physical domain of the variable ε
Ω^*	physical domain of the variable ε^*

REFERENCES

1. S. Ono, Statistical mechanics of adsorption, *J. Phys. Soc. Japan, 6*:10 (1951).
2. S. Sokołowski, The virial theory of adsorption—a critical discussion, *J. Colloid Interface Sci., 74*:26 (1980).
3. W. Rudziński, Study on the possibility of determining the third gas-solid virial coefficient in physical adsorption by gas chromatography, *J. Chromatogr., 66*:1 (1972). Estimation of the surface area of adsorbents from the third gas-solid virial coefficient, *J. Chromatogr., 72*:221 (1972).
4. D. H. Everett, The application of high temperature/low coverage gas adsorption measurements to the determination of the surface areas of solids, *Surface area determination, Proc. Int. Symp. 1969* (D. H. Everett, ed.) Butterworth, London, p. 181 (1970).
5. J. A. Barker and D. H. Everett, High temperature adsorption and the determination of the surface area of solids, *Trans. Faraday Soc., 58*:1608 (1962).
6. R. S. Hansen, J. A. Murphy, and T. C. McGee, Gas chromatographic measurement of gas-solid interaction potentials and solid surface areas. *Trans. Faraday Soc., 60*:597 (1964).
7. W. Engewald, J. Graefe, A. V. Kiselev, K. D. Shcherbakova, and T. Welsch, Molekülstruktur und Retentionsverhalten, I, Strukturelle Zuordnung von thermischen Umlagerungsprodukten der Cyclododecatriene-(1.5.9) mittels Gas-Adsorption-Chromatographie, *Chromatographia, 7*:229 (1974).
8. H. J. Hofman, W. Engewald, D. Heidrich, J. Pörschmann, K. Thieroff, and P. Uhlmann, Molekülstruktur und Retentionsverhalten, V. Berechnung der Adsorptionsenergie aromatischer Kohlen-

wasserstoffe an graphitiertem thermischen Russ, *J. Chromatogr.*, *115*:299 (1975).
9. S. K. Milonjić and M. M. Kopečni, Natural magnetite as an adsorbent in gas-solid chromatography, *J. Chromatogr.*, *172*:357 (1979).
10. N. M. Djordjević, M. M. Kopečni, and S. K. Milonjić, Chromatographic and adsorptive properties of mercury sulfide, *Chromatographia*, *13*:226 (1980).
11. N. M. Djordjević, S. K. Miljonjić, and M. M. Kopecni, Gas-solid chromatographic studies of organic compounds on zirconium oxide beads, *Bull. Chem. Soc. Japan*, *54*:3162 (1981).
12. W. A. Steele, Monolayer adsorption with lateral interaction on heterogeneous surfaces, *J. Phys. Chem.*, *67*:2016 (1973).
13. A. Waksmundzki, S. Sokołowski, and W. Rudziński, Evaluation of the energy distribution function in physical adsorption on heterogeneous surfaces: virial analysis of low-coverage physical adsorption, *Z. Phys. Chem. (Leipzig)*, *257*:833 (1976).
14. O. Sinanoglu and K. S. Pitzer, Interactions between molecules adsorbed on a surface, *J. Chem. Phys.*, *32*:1279 (1960).
15. R. Wolfe and J. R. Sams, Three-body effects in physical adsorption, *J. Chem. Phys.*, *44*:2181 (1966).
16. D. H. Everett, Interactions between adsorbed molecules, *Trans. Faraday Soc.*, *73*:177 (1965).
17. W. Rudziński and R. Leboda, Fourth-order interaction in physical adsorption, *Czech. J. Phys. B*, *23*:141 (1973).
18. J. R. Sams, Jr., G. Constabaris, and G. D. Halsey, Third- and fourth-order interactions of argon with a graphitized carbon black, *J. Chem. Phys.*, *36*:1334 (1962).
19. T. L. Hill, Statistical mechanics of adsorption. VI. Localized unimolecular adsorption on a heterogeneous surface, *J. Chem. Phys.*, *17*:762 (1949).
20. S. Ross and J. P. Olivier, *On Physical Adsorption*, Interscience, London (1964).
21. R. A. Pierotti and H. E. Thomas, Virial analysis of low-coverage physical adsorption data on heterogeneous surfaces, *Trans. Faraday Soc.*, *70*:1725 (1974).
22. R. A. Pierotti and H. E. Thomas, Physical adsorption: The interaction of gases with solids, in *Surface and Colloid Science, Vol. IV*, (E. Matijevic, ed.) Wiley, New York (1971).
23. D. C. Hinman and G. D. Halsey, Adsorption of Ar on a nonuniform MgO surface, *J. Chem. Phys.*, *64*:3353 (1976).
24. W. Rudziński and S. Sokołowski, Evaluation of the energy distribution function in physical adsorption as the problem of Laplace transform, I, Evaluation of the energy distribution function from the second gas-solid virial coefficient, *Ann. Univ. Mariae Curie-Skłodowski, Lublin, Poland*, *31/32*:203 (1976/977). Evaluation of the energy distribution function in physical adsorp-

tion as the problem of Laplace transform, II, Spatial distribution of adsorption energy evaluated from the third gas-solid virial coefficient, *Ann. Univ. Mariae Curie-Skłodowska, Lublin, Poland* *31/32*:214 (1976/77).

25. W. Rudziński, Influence of heterogeneity on higher gas-solid virial coefficients, *Phys. Lett.*, *42A*:519 (1973).
26. P. Ripa and G. Zgrablich, Effect of the potential correlation function on physical adsorption on heterogeneous substrates, *J. Phys. Chem.*, *79*:2118 (1975).
27. G. D. Halsey, The role of surface heterogeneity in adsorption, *Advan. Catal.*, *7*:259 (1952).
28. J. Appel, Freundlich's adsorption isotherm, *Surface Sci.*, *39*:237 (1973).
29. W. Rudziński and A. Patrykiejew, On the theoretical origin of Haul and Gottwald empirical equation, *Vacuum*, *27*:545 (1977).
30. W. Rudziński, J. Jagiełło, M. Kopecni, and S. Milonjić (unpublished work).
31. A. Waksmundzki, W. Rudziński, and Z. Suprynowicz, On the possibility of determination of adsorption potential distribution on adsorbent surface by gas chromatography, *J. Gas Chromatog.*, *2*:93 (1966).
32. D. H. Everett, The interaction of gases and vapours with solids, *Gas Chromatography 1964*, (A. Goldup, ed.) Institute of Petroleum, London, p. 219 (1965).
33. J. R. Conder and C. L. Young, *Physicochemical measurement by gas chromatography*, Wiley, New York (1979).
34. R. J. Laub and R. L. Pecsok, *Physicochemical applications of gas chromatography*, Wiley, New York (1978).
35. W. Rudziński and J. Jagiełło, Low-pressure adsorption of single gases on heterogeneous solid surfaces, *Vacuum*, *32*:577 (1980).
36. J. Gawdzik, Z. Suprynowicz, and M. Jaroniec, Determination of the pre-exponential factor of Henry's constant by gas adsorption chromatography, *J. Chromatogr.*, *121*:185 (1976).
37. W. Rudziński, Exponential virial expansion for adsorption isotherm, *Czech. J. Phys. B*, *22*:431 (1972).
38. W. Rudziński and B. W. Wojciechowski, On the Jovanovic model of adsorption, I, An extension for mobile adsorbed layers, *Colloid Polymer Sci.*, *255*:869 (1977).
39. W. Rudziński, Estimation of the adsorption site distribution from the second and third gas-solid virial coefficients, *Phys. Lett.*, *38A*:443 (1972).
40. W. Rudziński, A. Waksmundzki, M. Jaroniec, and S. Sokołowski, Use of the second, third and fourth gas-solid virial coefficients to the problem of estimating adsorptive properties of adsorbents, *Ann. Univ. Mariae Curie-Sklodowska, Lublin, Poland 29/30*:99 (1974/75).

41. A. Waksmundzki, R. Leboda, Z. Suprynowicz, and W. Rudziński, Adsorptive properties of Polsorb C investigated by gas chromatography, Part I, Energy properties and distribution of adsorption sites, *Ann. Soc. Chim. Polonorum, 47*:825 (1973), and W. Rudziński, A. Waksmundzki, R. Leboda, Z. Suprynowicz, and M. Lasoń, Investigations of adsorbent heterogeneity by gas chromatography, II, Evaluation of the energy distribution function, *J. Chromatogr., 92*:25 (1974).
42. W. Rudziński, A. Waksmundzki, R. Leboda, and M. Jaroniec, New possibilities of investigating adsorption phenomena by gas chromatography: Estimation of adsorbent heterogeneity from the pressure dependence of retention data, *Chromatographia, 11*:663 (1974).
43. G. F. Cerofolini, Adsorption and surface heterogeneity, *Surface Sci., 24*:391 (1971).
44. M. Jaroniec, R. Leboda, S. Sokołowski, and A. Waksmundzki, Some remarks on application of gas-adsorption chromatography data for investigations of the adsorptive properties of adsorbents, *Separation Sci., 11*:411 (1976).
45. A. Waksmundzki, S. Sokołowski, J. Rayss, Z. Suprynowicz, and M. Jaroniec, Application of gas-adsorption chromatography data to investigation of the adsorptive properties of adsorbents, *Separation Sci., 11*:29 (1976).
46. A. V. Kiselev and Ya. I. Yashin, *Gazoadsorpcionnaja Chromatographia* (in Russian), Nauka, Moscow (1967). *Gas-Adsorption Chromatography*, Plenum Press, New York (1969).
47. J. Volf, J. Koubek, and J. Pasek, A contribution to the measurement of adsorption isotherms by the pulse chromatograph technique, *J. Chromatogr., 81*:9 (1973).
48. A. Waksmundzki, M. Jaroniec, and Z. Suprynowicz, A modification of the chromatographic Hobson method for studying heterogeneity of adsorbents, *J. Chromatogr., 110*:381 (1975).
49. R. Leboda and S. Sokołowski, Investigation of the influence of chemical modification of adsorbents on their adsorption properties: Adsorption of cyclohexane and cyclohexene on silica gels, *J. Colloid Interface Sci., 61*:365 (1977).
50. R. Leboda, S. Sokołowski, J. Rynkowski, and T. Paryjczak, Chromatographic investigations of mixed catalysts, *J. Chromatogr., 138*:309 (1977).
51. S. C. Boudreau and W. T. Cooper, Determination of Surface Polarity by Heterogeneous Gas-Solid Chromatography, *Anal. Chem. 59*:353 (1987).
52. G. F. Cerofolini, A model which allows for the Freundlich and the Dubinin-Radushkevich adsorption isotherms, *Surface Sci., 51*:333 (1975). Equilibrium Surfaces, *Surface Sci., 61*:678 (1976).

53. W. Rudziński and J. Jagiełło, Low-temperature adsorption of gases on heterogeneous solid surfaces: Surfaces with random topography, *J. Low Temp. Phys.*, *45*:1 (1981). Low-temperature adsorption of gases on heterogeneous solid surfaces: Effects of surface topography, *J. Low Temp. Phys.*, *48*:307 (1982).
54. W. Rudziński and J. Baszyńska, Physical adsorption on heterogeneous surfaces: The influence of topography of adsorption sites on calorimetric effects of adsorption, *Z. Phys. Chem. (Leipzig)*, *262*:533 (1981).
55. D. Morel, H. F. Stoeckli, and W. Rudziński, Physical adsorption of simple gases on the (111) face of sulfur: A model for the adsorption by heterogeneous surfaces, including lateral interactions, *Surface Sci.*, *114*:85 (1982).
56. W. Rudziński, J. Jagiełło, and Y. Grillet, Physical adsorption of gases on heterogeneous solid surfaces: Evaluation of the adsorption energy distribution from adsorption isotherms and heats of adsorption, *J. Colloid Interface Sci.*, *87*:478 (1982).

5
Mixed Retention Mechanisms in Gas-Liquid Chromatography

Jan Åke Jönsson and Lennart Mathiasson
University of Lund
Lund, Sweden

In gas-liquid chromatography (GLC), the main mechanism of retention is partition of the analyte between the mobile gas phase and the stationary liquid phase (SLP). Under the conditions normally used in modern GLC with sensitive detectors, the maximum concentration of analyte in the SLP is so low that Henry's law is valid. The assumptions of linear chromatography then apply, and consequently the retention times and peak shapes will be practically independent of the analyte concentration. This greatly simplifies both the quantitative and qualitative interpretation of the chromatogram.

It has long been known that adsorption on various surfaces in a column can disturb this satisfactory picture. These adsorbing surfaces include the surface of the SLP, the interface between the SLP and the solid support, the surface of the uncovered support, and the column walls. The adsorption of analyte on these surfaces will contribute to the overall retention and change the behavior of the chromatographic system, most often in an undesirable way.

Additionally, adsorption may occur externally to the chromatographic column (i.e., in the injection port, connecting tubing, or elsewhere). This chapter deals mainly with processes occurring in the column proper, even if some of the conclusions reached are also applicable to extracolumn adsorption effects.

In this chapter the adsorption effects in GLC are referred to as "mixed retention mechanisms." These have previously been reviewed by several authors, notably Locke [1], Conder and Young [2], and Berezkin [3].

The extent of adsorption of analyte molecules on a given surface varies with the analyte concentration. Generally, the adsorption will follow an analog of Henry's law at very small concentrations. In this case, no effects on the chromatographic process are seen, other than an increase in the apparent partition coefficient. When the concentration is increased, the amount of analyte adsorbed will be relatively less and nonlinearity will appear. This causes the chromatographic peaks to become skewed and the retention time to decrease. At still higher concentrations of analyte, adsorption effects may level off owing to saturation of the surface.

Thus the disturbing effects of adsorption will be observed only within a limited range of concentrations. The upper and lower limits of this range vary according to the characteristics of each particular surface-adsorbate combination. It may be difficult or even impossible to reach a concentration low enough for all adsorption processes to follow Henry's law. On the other hand, with relatively large samples the adsorption effects are often not observable.

The significance of concentration dependence is often overlooked, which is probably the source of a controversy that has existed in the literature on the importance and even existence of mixed retention mechanisms, especially adsorption on the surface of the SLP [4]. Some observed discrepancies in different studies of solution properties by GLC may well also be explained in this way.

Mixed retention mechanisms are not always an obstacle; sometimes a different selectivity of adsorption processes than of partition can be used to achieve a better separation than is otherwise possible. The technique of using adsorption processes in gas-liquid chromatography has been called gas-liquid-solid chromatography, and it is discussed in Sec. VI.D.

I. ADDITIVITY OF RETENTION INCREMENTS

A. Model

The observed net retention volume V_N in GLC can, under certain assumptions, be considered the sum of several terms (retention contributions), each corresponding to a certain retention mechanism. We shall here consider a general description of a GLC column, pictured in Fig. 1, which shows a solid support material partly covered by a liquid. The bulk liquid and gas phases are designated L and G, respectively. These are the "normal" and intended phases in the column. Additionally, we will consider three adsorbing surfaces: the gas-liquid surface (I), the gas-solid surface (A), and the liquid-solid surface (S). Each of these surfaces can be considered an interfacial region, the dividing planes being chosen so that the bulk phases have homogeneous properties throughout, and the properties

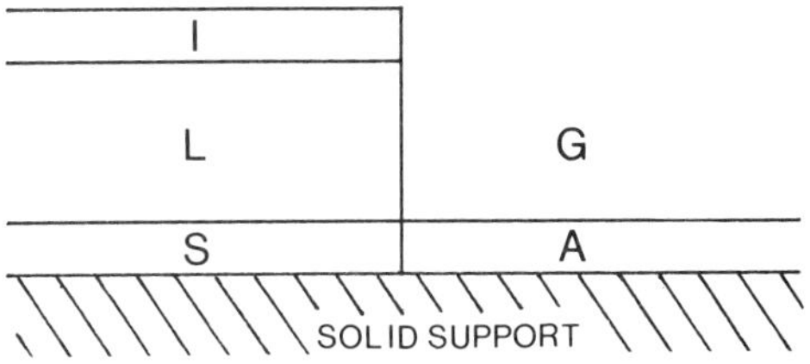

Figure 1. Schematic diagram of the various phases and interface regions in a GLC column.

within the interfacial regions vary from those of one surrounding phase to those of another. These interfacial regions will (slightly improperly) be called surface phases, and for convenience of notation we will temporarily assign each of them a volume. Formally, we can just as well consider them strictly two dimensional and characterized by an area.

This model of the column may be made more detailed by incorporating into it regions of stagnant mobile phase in deep pores, different modificational forms of the liquid, and so forth. With proper definitions of these "phases," much of the following treatment will still be applicable.

To each of the stationary phases thus defined, a sorption mechanism is related. The analyte molecules entering the column will be dissolved in the bulk liquid and will become adsorbed onto the surfaces. For each of these mechanisms, a functional relationship, an isotherm, can be stated:

$$C_i = f_i(C_j, C_k, \ldots) \tag{1}$$

with $i \in \{L, I, A, S\}$ and $j, k, \ldots \in \{G, L, I, A, S\}$. Thus, the equilibrium concentration C_i in one of the stationary phases is a function of the concentration in the gas phase. It can also, as in this general expression, be considered dependent on the concentration in the other stationary phases. For the surface phases, excess concentrations are considered (i.e., the number of moles present in a unit of volume of the interface region in addition to the concentration in the adjoining bulk phases). This definition of excess concentrations necessitates the introduction of a hypothetical dividing surface between the two dividing planes discussed above. The exact position of the dividing surface is determined by the principles of the Gibbs treatment of adsorption thermodynamics, which is discussed in Sec. II.A. The thickness of the surface phase is normally negligible compared with that of the SLP, so that uncertainties in defining the exact bounds of the phases will have very little practical significance.

B. Retention Equations

Chromatographic retention is governed by the relation between the number of analyte molecules that are moving (M) and those that are not moving (NM). The relation can be expressed as

$$C_{NM} = g(C_M) \tag{2}$$

According to the model, $C_M = C_G$ and

$$C_{NM} = \frac{\sum_i V_i C_i}{V_x} \tag{3}$$

where V_i is the volume of the phase i, V_x is the total volume of the nonmoving phases, and i $\in$ {L, I, A, S}. The function g is thus a general "isotherm" function, incorporating all the retention mechanisms.

According to chromatographic retention theory (see, for example, Chap. 2, pages 83-84) the retention volume is given by

$$V_N = \frac{\partial g(C_M)}{\partial C_M} V_x \tag{4}$$

This simple equation describes the retention volume for a zone of constant concentration, nonideality effects such as diffusion and slow kinetics being disregarded.

Thus we can write, considering Eqs. (2) and (3),

$$V_N = \sum_i V_{N,i} = \sum_i V_i \frac{\partial C_i}{\partial C_G} \tag{5}$$

This equation was given by several authors and is the general basis for further discussion about mixed retention mechanisms.

The isotherm expressions in Eq. (1) can often be considerably simplified. Most often the dependence of C_i on concentrations other than C_G is neglected (i.e., the retention mechanisms are considered independent). This assumption changes nothing in Eq. (5). An additional simplification is introduced by assuming that C_i is directly proportional to C_G. This means that we assume that Henry's law is valid for the phase equilibrium in question:

$$C_i = K_i C_G \tag{6}$$

Here, K_i is the corresponding equilibrium constant. With these assumptions applied to all phase equilibria, Eq. (5) becomes

$$V_N = \sum_i V_i K_i \tag{7}$$

Many authors have used various versions of this equation. (For a review, see Ref. 3.) At sufficiently low concentrations of analyte, all retention mechanisms become linear and Eq. (5) degenerates into Eq. (7). However, the conditions of practical gas chromatography are often such that the conditions in Eq. (6) are not valid throughout, and Eq. (5) is needed to describe the behavior of the system.

For the surface phases, the areas are more accessible experimental parameters than the volumes, which motivates some slight redefinitions of the quantities above. If we define C_I^a, C_A^a, and C_S^a as the number of moles per unit area of the adsorbing surface [i.e., $C_I^a = C_I(V_I/A_I)$, and so on], we obtain $K_I^a = C_I^a/C_G = K_I(V_I/A_I)$, and so on. Thus K_A^a, and K_S^a will have the dimension of length. Below, we drop the superscript for simplicity, and we can rewrite Eq. (7) as follows. [Eq. (5) can be rewritten in an analogous way]:

$$V_N = V_L K_L + A_I K_I + A_A K_A + A_S K_S \tag{8}$$

Equations (5), (7), and (8) express the retention volume V_N as the sum of several retention increments, each related to a certain retention mechanism. For this treatment to be valid, the column must be divided into several, physically separated regions, as in Fig. 1. Thereby it is possible to distinguish unequivocally between a molecule that is dissolved in a bulk phase and one that is adsorbed on one of the surfaces. In other words, we relate a certain retention mechanism to a physical domain of the column, which is the usual, but most often implicit, basis for the definition of the concept of retention mechanisms.

II. INDIVIDUAL RETENTION MECHANISMS

Most of the processes that constitute the individual retention mechanisms in GLC as described above are exploited in their own right as main mechanisms in other chromatographic techniques. Thus, adsorption on gas-solid surfaces is the basis for GSC, and adsorption on liquid-solid surfaces is of prime importance in liquid chromatography. These types of adsorption, together with the bulk liquid solution process, have been treated in detail in other chapters of this book, and only short summaries with special reference to the variation of the various retention increments with sample concentration are needed here. Only adsorption on gas/liquid interfaces was not described in other chapters and merits a more detailed treatment here.

A. Adsorption at the Gas/Liquid Interface

Adsorption phenomena occuring at gas/liquid interfaces, especially those of interest in GLC, are considerably less frequently studied than adsorption on solid surfaces. Most studies have been performed on aqueous surfaces, which are not of prime importance in gas chromatography. The matter of gas-liquid adsorption is intimately connected with the theory of surface tension, and the fundamental treatment of Gibbs, discussed below, is also valid for the types of liquids encountered as stationary phases in GLC.

Excess Concentration

The concept of the excess concentration of analyte in the gas/liquid interfacial region, C_I, deserves some comment. When treating adsorption on a surface of a solid, it is customary (although not necessary) to assume the existence of "sites" on the surface onto which the analyte molecules are adsorbed (localized adsorption). At the gas/liquid interface there are no such sites, and C_I denotes merely an accumulation of analyte molecules in the interface region. C_I can be negative (i.e., there can be deficiency of analyte material in the interface as compared with bulk concentration). In Fig. 2, three dividing planes A, B, and S are shown. Starting from below, the bulk liquid phase is homogeneous up to plane B, where its properties start to change gradually. The change is completed at plane A, above which is the homogeneous gas phase. Plane S is the formal division between the liquid and gas phases. The position of S is arbitrary, since the properties of the system change continuously between A and B, but the position of S influences the definition of concentrations. The excess concentration C_I is the number of molecules per volume of the interface region in excess of those that would be present if the bulk phases were homogeneous up to the dividing surface.

Let the distances A - B and S - B be denoted a and b, respectively. Then, if we consider a unit area of the surface region, the total amount of analyte in the interfacial region will be

$$N_I = bC_L + (a - b)C_G + aC_I \tag{9}$$

and we see that the value of b will influence the numerical value of C_I for constant values of N_I, C_L, C_G, and a.

Note that the volume between S and A is considered to belong to the gas phase. Various conventions for the choice of the position of the surface S exist, and they are described in detail by Adamson [5]. At low concentrations of analyte, the difference between these conventions becomes insignificant.

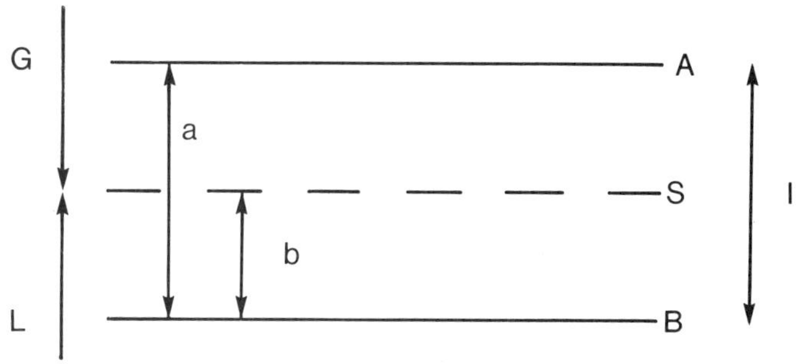

Figure 2. Schematic diagram of the gas/liquid interfacial region.

Gibbs Adsorption Equation

The basic treatment of the thermodynamics of adsorption on liquid interfaces leads to the Gibbs equation. Its derivation is found in most textbooks on physical chemistry and is not reproduced here. (See, for example, Refs. 5 and 6.) The equation is

$$C_I = \frac{-a}{RT}\left(\frac{d\sigma}{da}\right) \tag{10}$$

Here, a denotes the activity of the solute in the bulk of the solution and σ is the surface tension of the solution. Equation (10) is valid only when the surface S is placed so, that the excess concentration of the solvent is zero. The equation does not specify the actual behavior of the concentration C_I; it merely relates it to the behavior of the surface tension of the solution. Since this is a function of C_L and thus of C_G, Eq. (10) formally complies with the form of Eq. (1).

The variation of surface tension with solution composition may be classified into three types after McBain (see Fig. 3) [7]. In type I there is a moderate and regular decrease of σ with increasing solute concentration; in type II there is a corresponding increase; and in type III systems the surface tension is strongly decreased at low concentration values but more or less constant thereafter. In water solutions, type III behavior is typical for solute compounds that can be oriented at the surface (i.e., having one polar and one nonpolar region) [7]. It may be expected that similar conditions are also true for organic liquids.

K_I at Infinite Dilution

At infinite dilution, the activity of the analyte is given by

$$a = \gamma_L^{\infty} x_2 \tag{11}$$

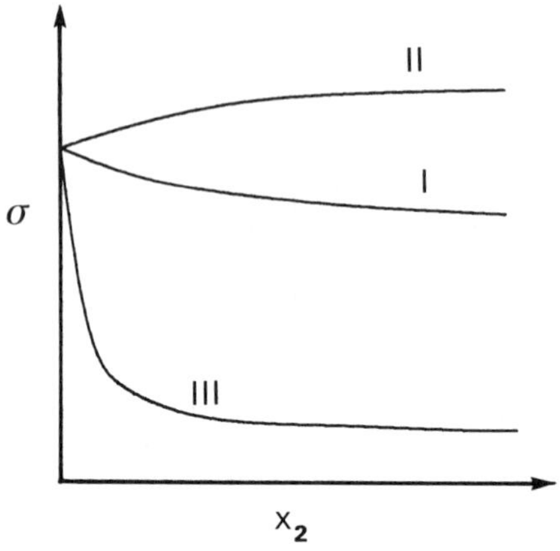

Figure 3. Variation of surface tension σ versus molar fraction x_2 of solute. (Redrawn after McBain [7].)

where x_2 is the mole fraction of solute and γ_L^∞ is the infinite dilution activity coefficient, signifying the deviation from ideality (i.e., the deviation from Raoult's law). We obtain, from Eq. (10)

$$C_I = -\frac{x_2}{RT}\frac{d\sigma}{dx_2} \tag{12}$$

The adsorption coefficient K_I can then be expressed, recognizing that $C_L = x_2/V_1{}^0$ at infinite dilution, where $V_1{}^0$ is the molar volume of the solvent:

$$K_I = \frac{-V_1^0}{RT}\frac{d\sigma}{dx_2}K_L \tag{13}$$

To continue the derivation, we need expressions for the surface tensions of solutions. Such relations were reviewed by Adamson [5]. If we have a binary mixture that forms an ideal solution and an ideal surface phase, it holds that

$$x_1 e^{(\sigma-\sigma_1)A_1^0/RT} + x_2 e^{(\sigma-\sigma_2)A_2^0/RT} = 1 \tag{14}$$

Here, x_1, x_2, A_1^0, A_2^0, σ_1, and σ_2 are the mole fractions, molar areas, and surface tensions for the two components, solvent (SLP), and solute (analyte), respectively, in pure form. From this equation, which describes a curve of McBain's type II, $d\sigma/dx_2$ can be calculated and inserted into Eq. (13), which becomes

$$K_I = \frac{V_1^0 K_L}{A_1^0}\left(e^{(\sigma_1-\sigma_2)A_2^0/RT} - 1\right) \tag{15}$$

By means of this equation, K_I can be calculated from accessible quantities. However, the applicability of this simple equation is restricted because of several important assumptions made during its derivation, especially that of ideality of the solution and surface phase.

To take nonideality into account Eon and Guiochon [8] introduced infinite dilution activity coefficients for both the bulk liquid and the surface phase. They developed the following analog to Eq. (14):

$$x_1 e^{(\sigma-\sigma_1)A_1^0/RT} + x_2 \frac{\gamma_{2,L}}{\gamma_{2,I}} e^{(\sigma-\sigma_2)A_2^0/RT} = 1 \tag{16}$$

where $\gamma_{2,I}$ and $\gamma_{2,L}$ are the activity coefficients of the solute in the indicated phases. The activity coefficients are defined according to the symmetrical convention of Everett [9]: $\gamma_{i,L}$ signifies the deviation from Raoult's law (for both the solvent and the solute), as is common in gas chromatography [2]; $\gamma_{i,I}$ is defined analogously for the surface phase. If $\gamma_{2,L}/\gamma_{2,I}$ is larger than 1, the corresponding $\sigma - x_2$ curve will tend to McBain's type III. Now, from Eq. (16) we obtain

$$K_I = \frac{V_1^0 K_L}{A_1^0}\left(\frac{\gamma_{2,L}^\infty}{\gamma_{2,I}^\infty} e^{(\sigma_1-\sigma_2)A_2^0/RT} - 1\right) \tag{17}$$

where $\gamma_{2,L}^\infty$ and $\gamma_{2,I}^\infty$ signify the activity coefficients at infinite dilution.

Using this equation, it is possible to calculate $\gamma_{2,I}^\infty$ from experimental values for K_I and K_L. Unfortunately, Eon and Guiochon made a calculation error, so their values for $\gamma_{2,I}^\infty$, quoted in several reviews [2,10], are not valid. Castells et al. [11] recently recalculated these values in the correct way. The values found for $\gamma_{2,I}^\infty$ in various systems range from 0.77 to 230 and are found to vary in a way similar to that of the bulk activity coefficient $\gamma_{2,L}^\infty$, in contrast to previous observations [2,8] based on the erroneous values.

Defay and Prigogine [12] derived equations for the surface tension of nonideal solutions. Following their treatment, Riedo and Kováts [13] have derived an equation for K_I that is very similar to Eq. (17). The only difference is that the activity coefficient factor has another form, containing the molar enthalpy of mixing q and a parameter β, signifying the decrease in coordination number of a solute molecule at the surface compared with that in bulk solution. By comparing the equation of Riedo and Kováts with Eq. (17), we find

$$\frac{\gamma_{2,L}^{\infty}}{\gamma_{2,I}^{\infty}} = e^{q\beta/RT} \tag{18}$$

Relations like this, and the possibility of experimental determination of the parameters involved, are important from the viewpoint of physical chemistry. For the estimation of K_I from the properties of the components of the solution—our main object at present—the most easily applied equation is Eq. (15), which, however, can only provide a rough estimation of the magnitude of K_I.

Experimentally Measured Values of K_I

A number of chromatographic studies, aiming at experimental determinations of numerical values for K_I in different systems, have been published. Those were comprehensively reviewed by Conder and Young [2]. Considerable experimental and theoretical difficulties arise, mainly in two areas. First, the separation of the gas-liquid retention increment from retention increments due to other, necessarily concurrent retention mechanisms is not trivial. This problem is discussed in Sec. V. The second difficulty is related to measuring the actual surface area A_I, a matter that is not straightforward. It is thoroughly discussed in Refs. 1 and 2 (see also pages 209-212).

These two problems were considered to different extents with different assumptions and different degrees of care by the various authors who have published K_I data based on gas chromatographic measurements. It is our opinion that the literature on the subject is not entirely reliable and that the data provided can only be trusted to give the correct order of magnitude, at best. Relative comparisons of adsorption effects for different compounds on a particular stationary phase should be safer.

Here, only a few selected examples of measured K_I values will be presented, in order to provide an idea of the relative importance of the gas-liquid adsorption effect in different chromatographic systems. The effect is most prominent in systems in which the solubility of the solute in the stationary phase is poor (i.e., in which the solute and solvent are rather different in their chemical nature). Accepting common chromatographic vocabulary, this type of adsorption should thus be noted with polar samples on nonpolar stationary phases and with nonpolar samples on polar stationary phases, but not when the polarity of the stationary phase and the sample is similar.

With nonpolar solutes and polar stationary phases, the most reliable data seem to be those of Martire et al. [14]. These data were obtained by static measurements of the surface tension of dilute solutions and subsequent application of Eq. (13). The $\sigma - x_2$ curves found are of McBain's type I [7] (see pages 195-196). The values obtained by these authors for benzene and cyclohexane on β,β'-thiodipropionitrile (TDPN) at 30°C were $K_I = 1.61 \times 10^{-6}$ and 0.98×10^{-6} m, respectively. These values agree (in magnitude) with gas chromatographic data for other, similar, systems [2,8]. With polar solutes and nonpolar stationary phases, the experimental problems are greater, owing to the possibility of strong adsorption on the solid support (A and S adsorption). Static surface tension measurements made in a way similar to that described above were performed by Pecsok and Gump [15]. Their data for K_I range from 2.0×10^{-6} m (acetone) to 10×10^{-6} m (diethylamine) on squalane at 30°C. Recent gas chromatographic work [16], in which great care was exercised to obtain reliable measurements, gave $K_I = 55 \times 10^{-6}$ m for diisopropyl ether on *n*-octadecane at 60°C, roughly one order of magnitude higher than the values of Pecsok and Gump. However, inspecting Fig. 2 in Ref. 15, which shows $\sigma - x_2$ curves of McBain's type III, the limiting slope (which is proportional to K_I) can equally well be drawn with a considerably higher slope, corresponding to values of K_I in the range found in Ref. 16. This illustrates that the adsorption effects in such systems are strongly dependent on concentration: at very low concentrations, the adsorption can be quite considerable, but with moderate concentrations it may be negligible.

Gas-Liquid Adsorption Isotherms

From the discussion in the introduction to this chapter, it is clear that a knowledge of the concentration dependence of the retention increment caused by gas-liquid adsorption is desired. Unfortunately, rigorous descriptions of the corresponding adsorption isotherms are not available in general cases. Only after extensive simplification is it possible to handle the problem in a tractable way

First, it is necessary to consider the definition of C_I in more detail. As was stated above, the Gibbs equation [Eq. (10)] is valid only when the excess concentration of the solvent is zero. This restriction is insignificant when infinite dilution of the sample is considered, as in Sec. II.A, but in the general case, the Gibbs equation cannot directly give an absolute value of the C_I as a function of σ. If we assume that the adsorption takes place strictly in a monolayer composed of molecules of both kinds and that the dividing surface S in Fig. 2 is placed between bulk liquid and the monolayer [17], it is possible to proceed. We will also have to assume that the molecular dimensions of the solute and sample are the same ($V^0 = V_1^0 = V_2^0$; $A^0 = A_1^0 = A_2^0$).

Under these restricted conditions, the excess concentration C_I^* in the monolayer can be written [14,17]

$$C_I^* = x_1 C_I \tag{19}$$

Here, C_I is the excess concentration in the Gibbs meaning from Eq. (10).

The retention increment $V_{N,I}$ arising from the adsorption in the monolayer will be (see Sec. I.B)

$$V_{N,I} = A_I \frac{\partial C_I^*}{\partial C_G} \tag{20}$$

Let us further assume that $a = \gamma_{2,L} x_2$ and that the gas phase is ideal. Additionally, we assume that the concentration of sample is low, which will normally be the case in chromatographic systems. Then the variation of the activity coefficients will be small and they can be approximated by their values at infinite dilution: $\gamma_{2,L} \approx \gamma_{2,L}^\infty$ $\gamma_{2,I} \approx \gamma_{2,I}^\infty$ and $\gamma_{1,L} \approx \gamma_{1,I}^\infty = 1$.

By differentiation of Eq. (19) and combination with Eqs. (10 and (20), we get:

$$V_{N,I} = A_I \frac{K_L V^0}{RT} \left[(x_2 - x_1) \frac{\partial \sigma}{\partial x_2} - x_1 x_2 \frac{\partial^2 \sigma}{\partial x_2^2} \right] \tag{21}$$

To obtain the concentration dependence of the surface tension, we derive, from Eq. (16), expressions for the derivatives in Eq. (21), finally leading to

$$V_{N,I} = \frac{A_I K_L V^0}{A^0} \left[\frac{E + 1}{(1 + X_2 E)^2} - 1 \right] \tag{22}$$

where

$$E = \frac{\gamma_{2,L}}{\gamma_{2,I}} e^{(\sigma_1 - \sigma_2)A^0/RT} - 1 \tag{23}$$

and

$$x_2 = K_L V^0 C_G \tag{24}$$

At infinite dilution, Eq. (22) correctly reduces to Eq. (17) and

$$V_{N,I}^{\infty} = \frac{K_L \cdot V^0 \cdot E}{A^0} \cdot A_I = K_I \cdot A_I \tag{25}$$

so that $E = A^0 \cdot K_I/(V^0 K_L)$. Observe that in Sec. II.A we considered different molar dimensions for the solvent and the solute; here these dimensions must be assumed equal.

Equation (22), just as Eq. (17), describes the additional retention caused by an excess accumulation of solute in the upper monolayer of the solution. When x_2 is large enough, $V_{N,I}$ will be negative. This happens when $x_2 > [(E + 1)^{1/2} - 1]/E \approx 1/E^{1/2}$. The region where $V_{N,I}$ is thus positive is narrow for large values of E. The maximum negative value for $V_{N,I}$ is $A_I K_L V^0/A^0$ [i.e., equal to the bulk liquid retention volume contribution from a monolayer of the solution (V^0/A^0 is the thickness of a monolayer)].

It is possible to express the liquid surface contribution $V_{N,I}$ in an alternative way, as the absolute contribution to retention from the monolayer, by adding to Eq. (22) the monolayer dissolution contribution (equivalent to removing the term -1). With that definition, $V_{N,I}$ will always be positive, but the V_L must be correspondingly adjusted not to include the volume of the monolayer. Because the volume of a monolayer is usually very small relative to the total V_L, the two alternative ways of expressing $V_{N,I}$ are practically equivalent. It should be noted that if the term -1 is neglected, Eq. (22) has nearly exactly the form of Eq. (28) below, which is obtained when a Langmuir adsorption isotherm is assumed. Apparently, the adsorption on a gas/liquid interface can in this simple case also be described approximately with the Langmuir equation.

From the experimental values of K_I given in the preceding section, E can be calculated according to Eq. (25) using available values of A^0, V^0, and K_L. Typical values of E range from 40 to 2000. In Fig. 4, $V_{N,I}$ from Eq. (22), divided by $V_{N,I}^{\infty}$, is plotted versus x_2 for different values of E. It can be seen that the relative de-

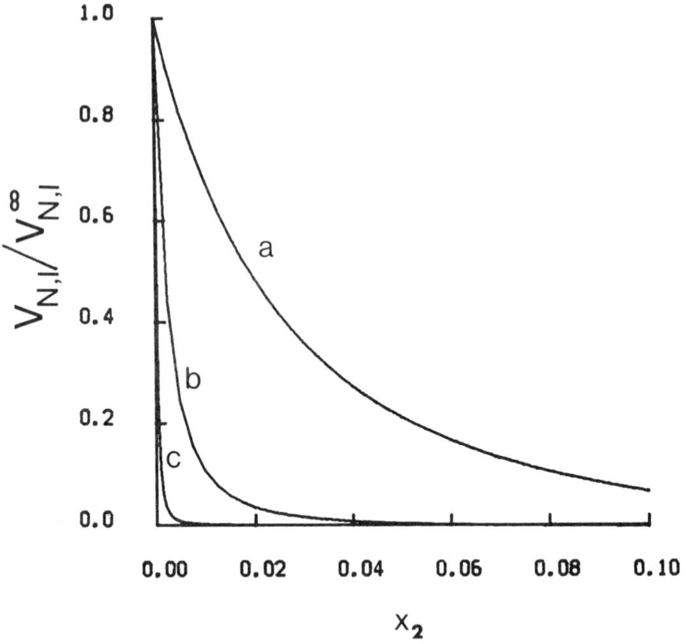

Figure 4. The relative retention increment $V_{N,I}/V_{N,I}^{\infty}$ due to adsorption on the gas/liquid interface as a function of solute mole fraction x_2: (a) E = 20; (b) E = 200; (c) E = 2000.

crease of the retention increment with x_2 occurs nearly entirely at the lowest molar fractions. If E = 2000, $V_{N,I}$ has fallen to 1% of its initial value at $x_2 = 4.4 \times 10^{-3}$ and to zero when $x_2 = 0.022$. With E = 200, the corresponding values are 0.036 and 0.066, respectively. This illustrates once again that the liquid surface adsorption effects can be very dependent on concentration and also that they may be negligible except in the lowest concentration range. It can also be seen that when the adsorption is high (i.e., when E is larger) the relative decrease with concentration is faster and the concentration range in which the effect is noted is more narrow than when E is small.

Equation (22) is accurately valid only under very restricted assumptions. In real chromatographic systems, showing large adsorption effects, the solute and the solvent are usually of widely different types. Therefore, the assumptions about molecular dimensions and about the activity coefficients are hardly met. Consequently, the treatment above is approximate and should only be considered to provide information about the general trends of gas-liquid adsorption isotherms.

Behavior of Real Systems

Some common tendencies of relative adsorption effects can be discerned from both theory and experimental evidence. In systems with nonpolar solutes on polar liquid surfaces, showing $\sigma - x_2$ curves of type I (see Sec. II.A), the values of E are relatively small and $V_{N,I}$ is relatively invariant with concentration. On the other hand, in systems with polar solutes on nonpolar stationary phases, $\sigma - x_2$ curves may show type III behavior. This leads to high values of E in connection with a steep decrease of $V_{N,I}$ within a narrow concentration range, although the adsorption at somewhat higher concentrations may be insignificant.

Examples of liquid-surface adsorption isotherms of the rare "anti-Langmuir" type [i.e., corresponding to a surface tension curve of type II (see Fig. 3)] were given by Karger and co-workers [18, 19] for the adsorption of hydrocarbons on water surfaces. The curvature, in the opposite direction that that predicted by Eq. (22), occurs only at high concentrations; in the concentration range normally used in chromatography, these isotherms are linear, which means that $V_{N,I}$ is practically independent of concentration. The explanation [18] for this unusual behavior is that the hydrocarbons are believed to form gaseous, mobile films on the water surface, violating the assumptions of Eq. (19).

B. Adsorption on Bare Support Surfaces

The adsorption of gases and vapors on solids has been thoroughly studied. Its connection to gas-solid chromatography was reviewed in Chap. 4. Other treatments of this topic are abundant, for example in Refs. 2, 5, 6, 8, 20, 21, and many others. A short description follows of some gas-solid adsorption isotherms suitable to our discussion.

The Langmuir Isotherm

A simple description of adsorption is offered by the Langmuir adsorption isotherm:

$$\theta = \frac{bp}{1 + bp} \tag{26}$$

where θ is the fractional coverage of adsorbate on the surface, p is the pressure of the adsorbate, and b is a constant—the adsorption coefficient—equal to $b_0 e^{\varepsilon_0/RT}$, where ε_0 is the adsorption energy, the enthalpy of adsorption. The derivation of the Langmuir equation can be found in Ref. 5, for example. Equation (26) is based on the assumption that the surface is energetically homogeneous (i.e., that the adsorption energy is the same for all "sites" on the surface). Furthermore, there should be no lateral interactions between adsorbed molecules and the adsorption should be limited to a monolayer.

Let us now define $C_A = \theta/A_2^0$; the subscript 2 refers to the adsorbate. In Sec. I we considered C_A an excess concentration, but owing to the low density of the gaseous phase, the distinction is insignificant here. We will also assume that the carrier gas has no influence on the adsorption process and that the molar area of the adsorbate is independent of surface coverage. Furthermore, we assume that the gas phase is ideal (i.e., $p = C_G RT$). We then get

$$C_A = \frac{1}{A_2^0} \frac{b_1 C_G}{1 + b_1 C_G} \tag{27}$$

with $b_1 = bRT$. The retention increment $V_{N,A}$ from gas-solid adsorption is then

$$V_{N,A} = A_A \frac{\partial C_A}{\partial C_G} = \frac{A_A b_1}{A_2^0 (1 + b_1 C_G)^2} \tag{28}$$

For infinite dilution the denominator becomes unity, and Henry's law is valid:

$$V_{N,A} = A_A K_A \tag{29}$$

with $K_A = b_1/A_2^0$. Equation (28) predicts that the retention increment $V_{N,A}$ decreases when C_G increases and that $V_{N,A}$ will become effectively zero at high enough values for C_G. The concentration C_G^1, where $V_{N,A}$ is 1% of its initial value, is given by

$$C_G^1 = \frac{9}{b_1} \tag{30}$$

Thus the concentration range in which $V_{N,A}$ influences the total retention volume is narrower when b_1 is large (i.e., when K_A is large). This is a picture similar to that in Sec. II.A in connection with Eq. (22). The quantity b_1 is not immediately comparable to E in Eq. (23), but to the quantity EV^0K_L, which can be seen by inserting Eq. (24) into Eq. (22).

Heterogeneity

If the adsorbing surface is energetically heterogeneous (i.e., if the various adsorption sites show different adsorption energies), the situation is more complex. It is, however, of great importance to consider the heterogeneity, since most adsorbents encountered in practice are markedly heterogeneous and, consequently, the Langmuir equation cannot accurately describe the adsorption effects they produce.

If we assume that the distribution of adsorption energies is given by the normalized distribution function $\chi(\varepsilon)$, we get (see Chap. 4, p. 168).

$$\Theta = \int_{\Omega} \theta(\varepsilon)\chi(\varepsilon)\, d\varepsilon \tag{31}$$

where $\theta(\varepsilon)$ is a local Langmuir isotherm [Eq. (26)] with $b = b_0 e^{\varepsilon/RT}$, and Ω is the range of possible adsorption energies. Θ is the convolution product of χ and θ (see Chap. 2, page 41). Equation (26) can be considered a special case of Eq. (31), with $\chi(\varepsilon) = \delta(\varepsilon - \varepsilon_0)$.

The evaluation of Eq. (31) is, in general cases, complicated. However, a few simple isotherm equations have been presented that can be used to introduce surface heterogeneity into the present treatment. The equation that seems to be most adequate for our purpose was suggested by Toth [22]:

$$\Theta_T = \frac{bp}{[1 + (bp)^m]^{1/m}} \qquad 0 < m \leq 1 \tag{32}$$

The energy distribution χ_T, which is related to this isotherm equation, has been evaluated [23] using special mathematical methods. It is an asymmetric bell-shaped function, expanded on the left-hand side. The parameter m can be interpreted as a measure of heterogeneity since it controls the width of the energy distribution function: when m = 1, the width is zero (i.e., χ_T degenerates to a δ function). When m decreases, the width increases.

From the definition $C_A = \Theta / A_2^0$ as above, we get the following expression for $V_{N,A}$ on a heterogeneous surface by differentiation of Eq. (32):

$$V_{N,A} = \frac{A_A b_1}{A_2^0 [1 + (b_1 C_G)^m]^{(1+1/m)}} \tag{33}$$

This equation correctly reduces to Eq. (28) when m = 1. It also reduces (for any m) to Henry's law, Eq. (29), at infinite dilution.

In Fig. 5, some examples are shown to illustrate the influence of m. It can be seen that the adsorption takes place less abruptly when m decreases. In analogy to Eq. (30), we find that

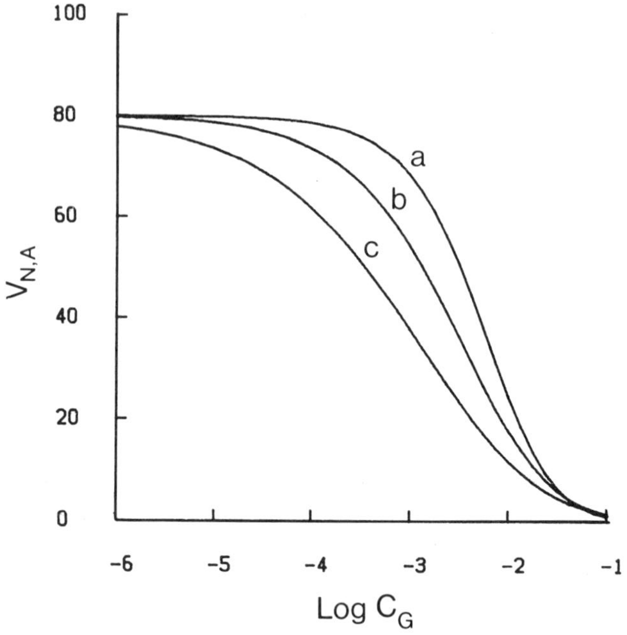

Figure 5. Retention increment $V_{N,A}$ due to adsorption on uncovered support surface as a function of log C_G and the heterogeneity parameter m: (a) m = 1; (b) m = 0.7; (c) m = 0.5.

$$C_G^1 \approx \frac{100^{1/(1+m)}}{b_1} \tag{34}$$

Thus, other parameters being equal, the concentration range in which the adsorption is significant is somewhat wider for heterogeneous surfaces (m < 1) than for homogeneous surfaces (m = 1). For example, if m = 0.5, $100^{1/(1+m)} \approx 21.5$, which should be compared the constant 9 in Eq. (30).

C. Adsorption at the Liquid/Solid Interface

The adsorption on the solid support from the bulk liquid solution is principally the same phenomenon as the main retention mechanism in liquid-solid chromatography (LSC). The liquid solvents considered in our present context are different from those normally encountered in LSC, but this has no fundamental consequence. Chapter 6 gives a thorough treatment of the theory of LSC, to which the reader is referred for detailed information on the matter. Several other sources may also be consulted [5,8,20].

Adsorption on a liquid/solid interface has some characteristics of both adsorption types previously described. The liquid-solid adsorption effects are determined by the properties of both the liquid and the solid surface. Consequently, the matter can be formulated in several alternative ways.

Langmuir and Related Theory

In dilute solutions, as normally in chromatography, the Langmuir equation [Eq. (26)] applies in simple cases [5]. Accordingly, it should be noted that the adsorption is competitive (i.e., the solute and solvent compete for the adsorption sites). The adsorption energy is thus the difference between the adsorption energies for the solute and the solvent, and we may write an expression for $V_{N,S}$ analogous to Eq. (28):

$$V_{N,S} = \frac{A_S b_2}{A^0(1 + b_2 C_G)^2} \tag{35}$$

where $b_2 = b_0' e^{(\varepsilon_2 - \varepsilon_1)/RT}$, and ε_2 and ε_1 are the adsorption energies of solute and solvent, respectively. It must be assumed that $A_1^0 = A_2^0 = A^0$. Heterogeneity of the solid surface can be treated much in the same way as in Sec. II.B.

Gibbs Theory

The case of liquid-solid adsorption may also be treated using the Gibbs equation in a way similar to that for gas-liquid adsorption in Sec. II.A [9]. The assumptions and simplifications are analogous. An equation for K_S similar to Eq. (15) can be written:

$$K_S = \frac{V_1^0 K_L}{A^0}\left[e^{(\sigma_1^s - \sigma_2^s)A_2^0/RT} - 1\right] \tag{36}$$

In the same way, Eqs. (17) and (22) can be rewritten for this case. In these equations, liquid/solid interfacial tensions σ_i^s appear, which are very difficult to measure directly, and few reliable data can be found in the literature. However, Eon [24] pointed out the relation between the liquid/solid interfacial tensions and the solvent strength parameter ε_i^0 of Snyder [20]:

$$\sigma_r^s - \sigma_i^s = \varepsilon_i^0 RT 4.50 \times 10^{-5} \tag{37}$$

(with surface tensions in N m^{-1}). The subscript r refers to a reference solvent, usually pentane.

Relative Magnitude of Liquid-Solid Adsorption Effects

The relations between the adsorption process on the liquid/solid interface and that on the gas/liquid interface on one hand and on the gas/solid interface on the other hand suggest the possibility of connecting these three types of adsorption in order to compare their relative magnitudes in a given system. Such relations have been developed [5,25] but suffer from the fact that they necessarily assume a simple behavior of the solutions and the interfaces. Activity coefficients in both solutions and gas/liquid interfaces are usually far from unity in systems of chromatographic interest, sorbents are usually energetically heterogeneous, and other difficulties exist. This seriously detracts from the usefulness of these relations in the present context.

Everett [25] developed the following approximate relation for perfect systems:

$$\ln \frac{K_S}{K_I} = \frac{\varepsilon_2 - \varepsilon_1}{RT} \tag{38}$$

where ε_i is the adsorption energy of component i at the gas/solid interface. Thus, if the SLP shows a stronger adsorption to the solid than is shown by the solute (i.e., $\varepsilon_1 > \varepsilon_2$), we obtain $K_S < K_I$, which is intuitively reasonable.

From the gas-solid adsorption constants $K_{A,i}$ (at infinite dilution) for solute 1 and solvent 2, K_S can be calculated [5]:

$$K_S = \frac{K_{A,2} P_2^0 V_1^0 K_L}{K_{A,1} P_1^0 A_1^0} \tag{39}$$

Here P_i^0 denotes the saturated vapor pressures of the components. All parameters in Eq. (39) are experimentally readily accessible, but the application is hindered by the necessary assumption of solution ideality.

D. Dissolution in the Bulk Liquid Phase

The main cause of retention in GLC is the dissolution of the analyte (solute) in the bulk of the SLP. The thermodynamics of this process have been treated in many reviews (e.g., Refs. 1, 2, and 10), as well as in Chap. 3. For the following discussion of the relative importance of the various retention mechanisms and their interplay, a knowledge of the course of the solution isotherm $C_L = f_L(C_G)$ is desired. It is well known that this isotherm is linear over a relatively wide concentration range (i.e., it can be written $C_L = K_L C_G$). At high concentrations (in the sense of modern GLC), deviations from linearity occur, usually in the positive direction, manifested as "leading" peaks and increased retention times.

In Sec. IV.B of Chap. 4, the question of deviation from linearity is investigated with the aid of the theory of regular solutions, and an expression for the upper concentration limit for isotherm linearity is given.

The general conclusion is that the deviation from linearity reaches significant values only at relatively high concentrations of solute in the stationary phase. At a concentration of 10^{-2} g/ml, the deviation is approximately 3%. The following approximate relation can be deduced from the figures given in Chap. 3, Table 1, page 140.

$$V_{N,L} = V_L K_L (1 + 3C_L M_2) \tag{40}$$

where M_2 is the molar weight of the solute.

III. SURFACE AREAS IN A GLC COLUMN

As can be seen in Sec. I, each retention contribution $V_{N,i}$ is a product of an intensive factor (a partition coefficient, denoted $\partial C_i/\partial C_G$, or K_i) and an extensive factor (the corresponding phase volume V_i or surface area A_i). Thus it is clear that the relative importance of retention increments depends not only on the intensive factors discussed in Sec. II but also on the relation between surface areas and SLP volumes.

A. The Concept of Wetting

The spatial distribution of the stationary liquid phase on a porous solid support material or on the inner wall of a capillary column is largely determined by the surface tension and the wetting properties of the materials involved.

The conditions for wetting were thoroughly studied by Zisman [26], who defined the critical surface tension of wetting σ_c (γ_c with Zisman's notation) as a property of a solid surface. If the surface tension of a liquid is greater than σ_c, it cannot wet the solid in question. The value of σ_c varies strongly with the chemical structure of the surface, fluorinated polymers such as Teflon showing the smallest values. Compilations of σ_c values can be found in Zisman's review article [26], as well as in standard texts and tables [5,27].

B. Surface Areas with a Wetted Support

If the SLP wets the solid support material well, the entire support surface will be covered, provided that the amount of SLP is sufficient to form at least a complete monolayer. In such cases, the area of uncovered support A_A will be zero and the area of the solid/liquid interface A_S will be equal to the physical surface area A_{supp} of the support or (for capillary columns) the inner column surface. The gas/liquid interface area A_I is also approximately equal to A_{supp}, if the SLP forms a film with a thickness that is negligible in comparison with the dimensions of pores, interstitial volumes, column diameter, and other factors. This simple state of affairs occurs with small loadings of most stationary phases on nondeactivated diatomaceous supports and in nondeactivated capillary columns.

On porous supports, the thickness of the liquid film is generally not proportional to V_L. Instead, the liquid is predominantly accumulated in pores by capillary forces and most of the solid surface is covered by an adsorbed thin film. This model of liquid distribution was suggested by Giddings [28] and further substantiated by other workers. (For a review, see Ref. 2, Chap. 11.)

As is easily perceived, the Giddings model of liquid distribution leads to a marked decrease of A_I when the SLP loading is increased. Unfortunately, the geometry of the three-dimensional structure of a typical solid support material is very irregular and inadequately known. This precludes the derivation of a general equation for the decrease of A_I with liquid loading. Berezkin and co-workers [3,29] gave a semiempirical equation that, at least in some cases, adequately fit experimental data:

$$A_I = A_{supp}(a - b \ln V_L) \tag{41}$$

The constants a and b in this equation can be interpreted in terms of pore diameter distributions [29].

When the amount of SLP is so small that a complete monolayer cannot be formed, A_A is not zero and $A_S = A_{supp} - A_A$. Thus, starting from a liquid loading of zero, A_I quickly increases to a maximum value $\approx A_{supp}$, from which it subsequently decreases as described above. The point of monlayer formation corresponds, in a typical case, to a liquid-support ratio of 0.1% w/w [30].

In Fig. 6, the approximate plots of the surface areas A_I, A_S, and A_A are shown for the case in which the SLP completely wets the solid support. Measurements of A_I can be made either by using the BET method (when the coverage is complete) [5,31] or chromatographically, assuming that all retention increments other than $V_{N,I}$ are either insignificant or can be compensated for. A knowledge of K_I then directly gives A_I from Eq. (8).

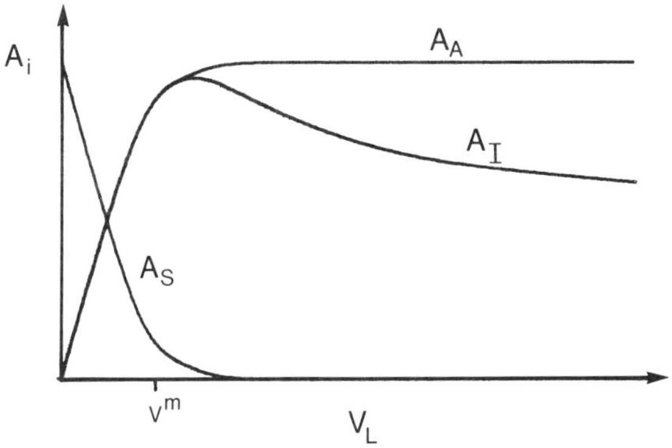

Figure 6. Approximate dependence of A_I, A_A, and A_S on V_L when the SLP wets the support. V^m = the point of monolayer formation.

C. Surface Areas with a Nonwetted Support

If the wetting of the support by the SLP is incomplete, the situation is more complicated than described above. First, A_A will be significant and generally of unknown magnitude even at high liquid loadings. Second, the magnitude of A_I will be highly variable and sensitive to experimental conditions and will thus be difficult to estimate. This type of system is found with nonactive supports, such as the Teflon or diatomaceous type, which have been deactivated by silanization or other techniques.

The critical surface tension σ_c (page 209) for Teflon is 18 dyn/cm, which means that it cannot be completely wetted by the liquids normally used as liquid phases in GLC. For silanized diatomaceous surfaces, σ_c is about 24 dyn/cm [26,32], and such surfaces will be completely wetted only by short-chain hydrocarbons and fluorocarbons. If the solid support is not wetted, there are several alternatives for the physical form of the liquid.

Droplets. On a nonporous support, such as the inner wall of a deactivated glass capillary column, the liquid can be expected to distribute in droplets or small pools, preferably situated on occasional high-energy points (such as mechanical or chemical defects). If the liquid phase is highly dispersed, the total area of the droplets (i.e., A_I) may then be fairly large. With a large amount of liquid, the droplets tend to coalesce and A_I will decrease. Droplets can also be imagined on porous materials, especially if the surface tension of the liquid is relatively large. The droplet model has been discussed in several studies [30,33,34]. We [34] have made a simple mathematical analysis of the problem, showing that at the point of coalescence the total liquid surface area may decrease 1.5-3 times. Thus, a maximum in the $A_I - V_L$ curve will be noted. This maximum occurs at an amount of SLP well above that which corresponds to a hypothetical monolayer.

Capillary Liquid. As was advocated by Serpinet [32,35], a nonwetting liquid may also be situated mainly in the most narrow pores, exposing only a very small gas/liquid interface area [i.e., as in the Giddings theory of liquid distribution [28] (page 210) but without the adsorbed film]. This condition may be expected with low liquid loadings and when the surface tension of the liquid is not much higher than σ_C.

An attempt is made in Fig. 7 to draw approximate curves for A_I, A_S, and A_A versus V_L for the case of a non-wetting SLP on a porous support. In region 1, the liquid fills the most narrow pores as described above and A_I is very small and approximately constant. With a higher loading (roughly a few percentage w/w on common types of supports) droplets are formed in region 2 and A_I increases rapidly. At 3, the coalescence of the drops becomes significant and A_I decreases to a fairly stable value at 4, decreasing only slowly thereafter owing to smoothing of cavities and large pores, as in the case of wetting liquids.

The general shape of the curve in Fig. 7 was experimentally substantiated chromatographically [34,36] over the entire range of practically useful liquid loadings and also separately for low loadings [35].

The measurement of A_I with the BET method, if questionable with wetting liquids, fails totally with nonwetting liquids, since it measures the sum $A_I + A_A$.

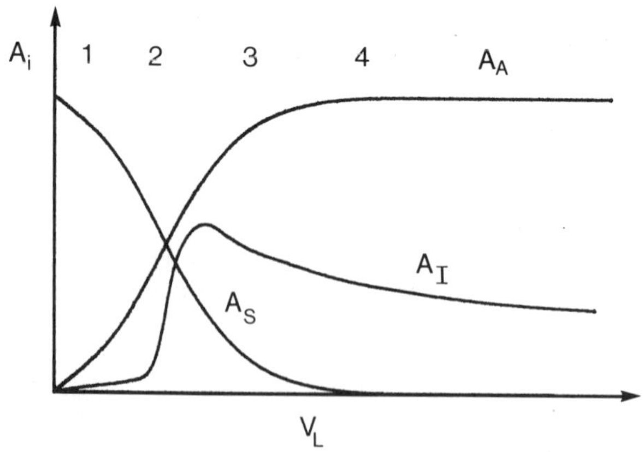

Figure 7. Approximate dependence of A_I, A_A, and A_S on V_L when the SLP does not wet the support.

IV. CONCURRENCE OF SEVERAL RETENTION MECHANISMS

In the preceding sections the most important factors contributing to retention in GLC have been individually examined, and we are now ready to tie them together. The following illustrations, summarizing the combined effects of mixed retention mechanisms, are necessarily approximate and may sometimes be objectionable, but it is hoped that they will provide the reader with a practical overview of the field.

A. V_N as a Function of Solute Concentration

It was briefly mentioned in the introduction to this chapter that V_N often varies with solute concentration in a way that can be schematically depicted as in Fig. 8. Such diagrams were given by several authors [37-41]. In the low concentration region, denoted α, Henry's law is valid for all mechanisms, V_N is constant, and the chromatographic process is linear. In region β, V_N decreases with C_G owing to Langmuir-like adsorption effects and the process is nonlinear. At γ, the curve is again horizontal owing to saturation of the adsorption "sites" and the process is quasi-linear. Finally, at δ, nonlinearity of the dissolution process becomes significant. Note the increased comprehensibility when V_N is plotted versus $\log C_G$. All adsorption mechanisms effective in the system are assumed to be Langmuir-like (i.e., the corresponding retention increment approaches zero when the solute concentration increases). It is true that, at very high adsorbate pressures (more or less close to the saturation vapor pressure), most adsorption isotherms will rise sharply because of multilayer formation (type II of the BET classification; see, for example, Ref. 5, p. 566). However, in the limited concentration range used in gas chromatographic practice, this will not be seen. To get an overview of the concentration dependence of adsorption, it is illustrative to do some indicative calculations for typical cases.

Let us assume that the adsorption takes place preferentially on one surface (I, A, or S). This will be a fair approximation in a majority of practical cases. To model the adsorption contribution to V_N, we use Eq. (33), which is derived from Toth's equation for adsorption on heterogeneous surfaces, an equation directly valid for adsorption on a solid surface (A or S). For the I surface, we saw in Sec. II.A that in simple cases a Langmuir approximation is appropriate, corresponding to m = 1 in Eq. (33). In other cases, when $V_1^0 \neq V_2^0$ or the activity factors are not constant, deviations that have not yet been elucidated occur. However, these effects will be similar to those arising from energetic heterogeneity, and consequently, Eq. (33) will in this case also be a useful approximation.

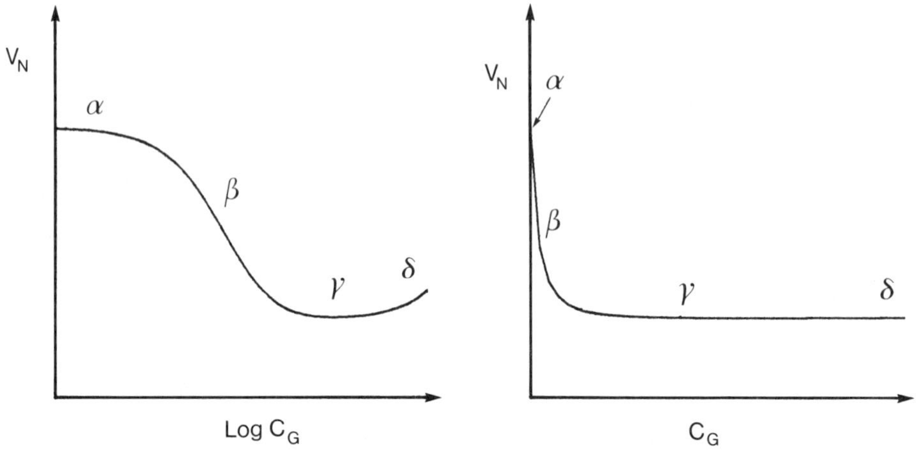

Figure 8. Variation of total V_N with solute concentration C_G.

By combining Eq. (33) with Eq. (40), the following equation for the total retention volume as a function of the gas phase concentration is obtained:

$$V_N = V_{N,L} + V_{N,i} = V_L K_L (1 + 3M_2 K_L C_G) + \frac{A_i b_i}{A^0[1 + (b_i C_G)^m]^{1+1/m}} \tag{42}$$

Here, i is I, S, or A, b_I is equal to EV^0K_L, and other symbols will be apparent from Eqs. (33), (35), and (40).

In the following examples, various values (see Table 1) of the parameters in Eq. (42) are selected, and the following quantities for 1 g of column packing will be calculated (Table 2). The resulting curves are plotted in Figs. 9 and 10.

- $V_{N,L}^{\infty}$ contribution from bulk dissolution at infinite dilution
- $V_{N,i}^{\infty}$ contribution from adsorption at infinite dilution
- v^{∞} $V_{N,i}^{\infty}/V_{N,L}^{\infty}$, relative contribution from adsorption
- C_G^{α} upper concentration limit for the α region, where $V_{N,i} = 0.99V_{N,i}^{\infty}$
- C_G^{β} upper concentration limit for the β region, where $V_{N,i} = 0.01V_{N,L}^{\infty}$
- C_G^{γ} upper concentration limit for the γ region, where $V_{N,L} = 1.01V_{N,L}^{\infty}$

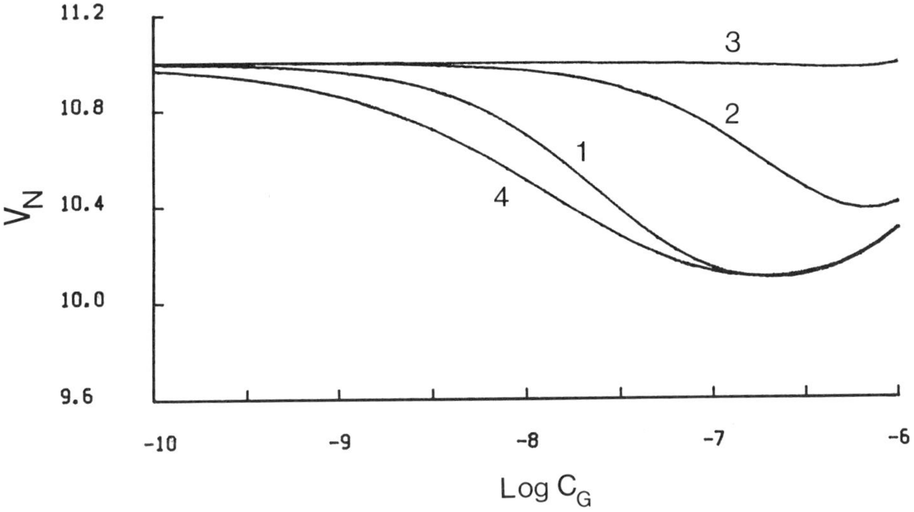

Figure 9. Retention volumes V_N versus log C_G for Examples 1-4 (see text).

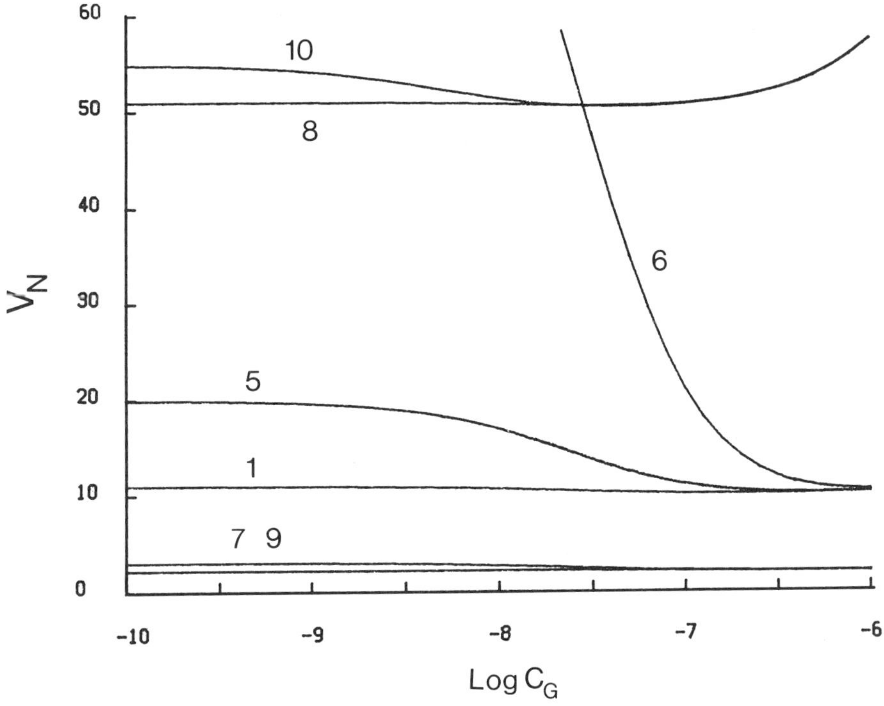

Figure 10. Retention volumes V_N versus log C_G for Examples 1-10 (see text).

Table 1. Parameters for Numerical Examples

Example	A_i/cm^2	K_L	b_i/cm^3mol^{-1}	E	m
1	100	100	2×10^7	2,000	1
2	1000	100	2×10^6	200	1
3	10,000	100	2×10^5	20	1
4	100	100	2×10^7	2,000	0.7
5	1000	100	2×10^7	2,000	1
6	10,000	100	2×10^7	2,000	1
7	100	20	2×10^7	10,000	1
8	100	500	2×10^7	400	1
9	100	20	4×10^6	2,000	1
10	100	500	1×10^8	2,000	1

Example 1. High-energy Adsorption on a Small Surface. As typical starting values we choose V_L = 0.1 ml, K_L = 100, M_2 = 100 g/mol, and $A^0 = 2 \times 10^9$ cm^2/mol (Table 1). A high-energy adsorption is signified by a b_i value in the region of 2×10^7 cm^3/mol (corresponding to E = 2000 for gas-liquid adsorption). With a surface area of 100 cm^2 (0.01 m^2), we obtain the values on the first line of Table 2 and curve 1 in Fig. 9.

We find $v^\infty = 0.1$ (i.e., the maximum adsorption effects amount to 10% of bulk dissolution). We also find that $C_G^\beta < C_G^\gamma$, which means that the γ region is significant and that $V_N^\gamma \approx V_{N,L}^\infty$.

Examples 2 and 3. Lower Adsorption Energy. Here we decrease the value of b_i to 2×10^6 cm^3/mol (Example 2) and to 2×10^5 m^3/mol (Example 3) and increase the surface area A_i correspondingly, so that the value of v^∞ is not affected. Both C_G^α and C_G^β are increased, so that the α region is successively enlarged and the γ region disappears. From the curves in Fig. 9 we clearly see the effects. Note that curve 3 is nearly horizontal, almost corresponding to a linear isotherm, although a significant fraction of the total retention volume is due to adsorption.

Example 4. Heterogeneity. Starting from the values in Example 1, we decrease m to 0.7, thereby introducing surface heterogeneity into the calculations. Only C_G^α is markedly affected by this (i.e., the region of infinite dilution is shortened). This can also be seen from Fig. 9, curve 4.

Table 2. Results of Numerical Calculations

Example	$V^{\infty}_{N,L}$	$V^{\infty}_{N,i}$	v^{∞}	$\log C^{\alpha}_{G}$	$\log C^{\beta}_{G}$	$\log C^{\gamma}_{G}$	γ region
1	10	1	0.1	−9.6	−7.0	−6.5	y
2	10	1	0.1	−8.6	−6.0	−6.5	n
3	10	1	0.1	−7.6	−5.0	−6.5	n
4	10	1	0.1	−10.7	−7.0	−6.5	y
5	10	10	1	−9.6	−6.3	−6.5	n
6	10	100	10	−9.6	−5.8	−6.5	n
7	2	1	0.5	−9.6	−6.5	−5.8	y
8	50	1	0.02	−9.6	−7.7	−7.2	y
9	2	0.2	0.1	−8.9	−6.3	−5.8	y
10	50	5	0.1	−10.3	−7.7	−7.2	y

Examples 5 and 6. Variation of Surface Area. Still starting from the parameters in Example 1, we increase the surface area A_i to 1000 and 10,000 cm^2, respectively. The effect of this is to increase C^{β}_{G} and, of course, v^{∞} (see Fig. 10).

Examples 7-10. Variation of K_L. Now K_L is changed to 20 (Examples 7 and 9) and to 500 (Examples 8 and 10). In Examples 7 and 8, b_i is kept constant and the resulting effects apply to the case of support adsorption with constant parameters. Since the parameter E for gas-liquid adsorption depends on both b_i and K_L, b_i is adjusted to keep E constant in Examples 9 and 10. These examples demonstrate the effects of a variation of K_L on gas-liquid adsorption with constant parameters. In Fig. 10, these examples are illustrated graphically.

From these examples, some general conclusions can be drawn.

1. With a large adsorption coefficient (high-energy adsorption), the effect of adsorption may be negligible within the linear part of the bulk dissolution isotherm (i.e., the γ region exists).
2. With low adsorption coefficients (low-energy adsorption), the adsorption is significant over a much wider concentration interval and there will be no horizontal γ region within the linear part of the dissolution isotherm.

3. The effect of heterogeneity is mainly to decrease C_G^{α} (i.e., the upper limit of validity for Henry's law).
4. For support adsorption, the effect of different values of K_L is, with a constant b_i, straightforward and related only to the value of $V_{N,L}$.
5. For gas-liquid adsorption with a constant E, the influence of different values of K_L also applies to the C_L values, thereby also influencing $V_{N,i}$.

With regard to adsorption on gas/liquid interfaces, it was shown in Sec. II.A that high E values are typical for systems with polar solutes on nonpolar stationary phases. Such systems should be characterized by point 1. On the other hand, systems with nonpolar stationary phases are usually characterized by point 2.

B. V_N as a Function of Liquid Loading at Infinite Dilution of Solute

At infinite dilution, Henry's law is valid for all retention mechanisms and Eq. (8) is applicable. The retention volume in a given solute-solvent-support system is then determined by the relation between the various surface areas and the volume of the stationary liquid phase. We treated in Sec. III the question of surface areas in a GLC column in some detail, and here it will only be necessary to give a few examples of the combined effects of several types of surfaces and the corresponding retention mechanisms. As in Sec. III, we will consider wetted supports and nonwetted supports separately.

Wetted Supports

Figure 6 shows the approximate dependence of the surface areas A_I, A_S, and A_A on the volume of liquid phase V_L for a typical porous, well-wetted support. The course of the total retention volume V_N as a function of V_L will also depend on the relative values of the distribution constants K_i. In the case of a polar stationary phase and a relatively nonpolar solute, usually $K_S < K_I$ (cf. page 208). Depending on the type of solute and on the support material, K_A can have different values in relation to K_S and K_I. If the support is not deactivated, K_A is probably considerably greater than K_I. Taking this into account, the total $V_N - V_L$ curve will consist of the contributions shown in Fig. 11. Curves similar to those in Fig. 11 have been published by several authors [42,43]. In other types of systems, K_S may dominate K_I. The corresponding change that this introduces in Fig. 11 is obvious.

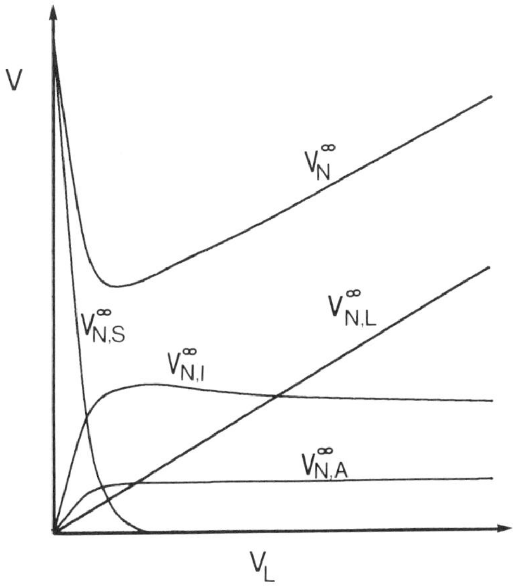

Figure 11. Approximate dependence of the total retention volume V_N^∞ and the contributions $V_{N,L}^\infty$, $V_{N,I}^\infty$, $V_{N,A}^\infty$, and $V_{N,S}^\infty$ on the SLP volume V_L for a wetted porous support material, assuming $K_A > K_I > K_S$.

Nonwetted Supports

Conditions under which the porous support is not wetted by the SLP are more complicated and uncertain than those in the case of complete wetting. In Fig. 7 the approximate situation is depicted. Since a condition for nonwetting is that the support surface be well deactivated, relatively small values of K_S and K_A are expected. If we then assume $K_I > K_A > K_S$, we get the curves shown in Fig. 12. If we also assume that the deactivation is not complete (i.e., that there is a small part of the support showing a much higher K_A value than the rest of the support surface), a contribution shown in the figure as a dotted line will be observed.

The most prominent feature of the $V_N^\infty - V_L$ curve in Fig. 12 is the hump. Depending on the relative magnitude of K_I versus the other K_i values, the hump can be more or less prominent. We [34,36] and others [35,38,44] have observed such curves.

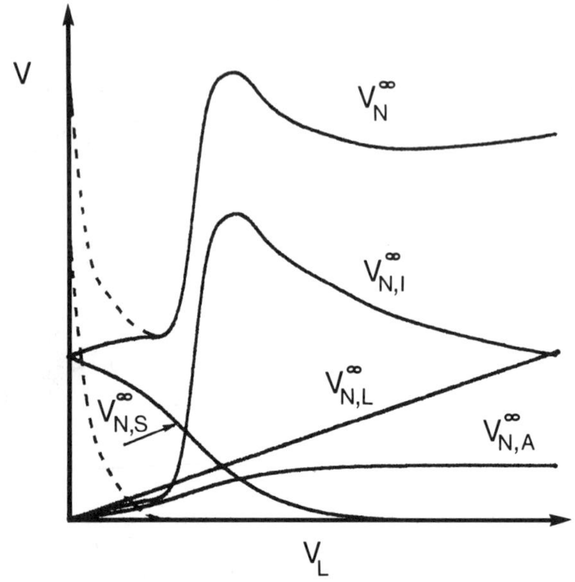

Figure 12. Approximate dependence of the total retention volume V_N^∞ and the contributions $V_{N,I}^\infty$, $V_{N,A}^\infty$, and $V_{N,S}^\infty$ on the SLP volume V_L for a nonwetted porous support material, assuming $K_I > K_A > K_S$.

V. METHODS FOR EXPERIMENTAL SEPARATION OF RETENTION MECHANISMS

The observed net retention volume V_N is, as we have seen in previous sections, the sum of several increments $V_{N,i}$, each of which is due to an individual retention mechanism. In order to investigate these phenomena and to extract useful information from observed retention volumes, procedures for separating the contributions from different retention mechanisms are needed. Several such procedures can be devised, which can be grouped into two categories: procedures for the separation of bulk dissolution (partition) effects from adsorption effects and procedures for the separation of different types of adsorption effects from each other.

A. Separation of the Bulk Dissolution Contribution from the Combined Adsorption Contributions

Variation of Liquid Loading

If it is possible to attain infinite dilution conditions with respect to *all* the retention mechanisms involved, Eq. (8) is valid. By division by V_L we obtain [30]

$$\frac{V_N}{V_L} = K_L + \frac{\text{Ads}}{V_L} \tag{43}$$

where Ads = $A_I K_I + A_A K_A + A_S K_S$, the combined adsorption contribution to V_N at infinite dilution.

Several columns, differing only in the loading of SLP (i.e., in the value of V_L) are needed, providing a set of V_N values. A plot of V_N/V_L versus $1/V_L$ will then give a straight line, providing Ads is a constant. The intercept is equal to K_L and the slope equal to Ads.

From the definition of Ads it is seen that it is constant only if A_I, A_A, and A_S are all independent of V_L, a condition that may in some cases be approximately met. If the SLP wets the supporting material well, A_A is zero and A_S is constant, equal to the area of the solid. If, additionally, the thickness of the liquid phase is not too large, A_I is also approximately constant and equal to the area of the solid (see Sec. III.B).

When the quantity Ads varies with the extent of liquid loading, which is the usual case, no straight line is obtained from Eq. (43). If the variation is moderate, it may still be possible to extrapoalte to $1/V_L = 0$ in a nonlinear way, necessarily introducing an uncertainty. If the variation of Ads is due to a variation of A_I on a wetted support, Eq. (41) may aid in the nonlinear extrapolation.

This method for separating the bulk partition contribution from the total V_N can be regarded as a standard method that has been used by many authors [1-3]. It will perform satisfactorily in those cases in which it is possible to reach the Henry's law region for all topical retention mechanisms. In this region, called α above (see Fig. 8), the retention volume is invariant with concentration.

The retention volume may also be constant at relatively high concentrations when the adsorption sites are saturated and the adsorption effects have leveled out (i.e., in the region called γ above). In such cases, the extrapolation described in this section may be used to extract values for K_L and values of the current adsorption contribution, which is not easily related to K_I, K_A, and K_S.

When the total isotherm is nonlinear (i.e., when peaks are markedly asymmetric and the retention volume varies with sample concentration), the application of Eq. (43) is less straightforward. This applies to regions β and δ. Conder [2,45] devised procedures involving the construction of several curved plots according to Eq. (43), each for a fixed solute concentration. These curves are extrapolated to a common value at $1/V_L = 0$, giving K_L. For a thorough discussion of these procedures, see Chap. 11 in Ref. 2.

Also, in this case, the infinite dilution adsorption contribution is not immediately found and caution must be exercised if studies of adsorption effects are attempted.

Variation of Sample Concentration

If the retention volume is measured over a wide range of sample concentrations, curves similar to those shown in Fig. 8 are found when adsorption contributes appreciably to retention. It is essential that the concentration be varied widely, preferably within the entire range of detector linearity. A flame ionization detector, or another detector with similar sensitivity, is necessary.

From the treatment above, especially in Sec. IV, it can be expected that an equation that takes into account isotherm nonlinearity and surface heterogeneity should fit the decreasing parts (α, β, and γ) in a plot of V_N versus sample size (see Fig. 8). From the parameters of such an equation, the various retention contributions may be calculated.

A suitable equation may be based on Toth's equation for adsorption on heterogeneous surfaces [Eq. (33)]. With this we obtain

$$V_N = A + \frac{C}{[1 + (BC_G)^D]^{1+1/D}} \tag{44}$$

If Eq. (44) is least-squares fit to experimental data with the help of a computer, the parameters A-D thus determined have the following meaning [compare with Eq. (42)]: A, the bulk dissolution contribution $V^{\infty}_{N,L} = K_L V_L$; B, Henry's constant [$b_1$ in Eq. (33)]; C, the combined adsorption contribution Ads at infinite dilution; and D, a heterogeneity parameter [m in Eq. (33)], which is unity if the adsorption follows Langmuir's equation.

It is interesting to note that this approach suggests a possibility for determining surface areas. Let us assume that only one adsorption mechanism (i) is effective. Then C = Ads = $K_i A_i$, and the relation $K_i = B/A_2^U$ [see Eq. (29)] permits the simultaneous calculation of K_i and A_i from B and C. This is an alternative to the more straightforward determination of K_i from Ads and an independent measurement (i.e., by the BET technique) of A_i. No systematic study of this procedure has been performed.

We successfully applied the idea of fitting an equation to experimental $V_N - C_G$ data (see Fig. 13) [16,34,36,40]. The equation used in this work was, however, not exactly Eq. (44), but a very similar one characterized by the exponent 2 instead of 1 + 1/D in the denominator. With that change, the parameters B and D do not have a simple physical meaning.

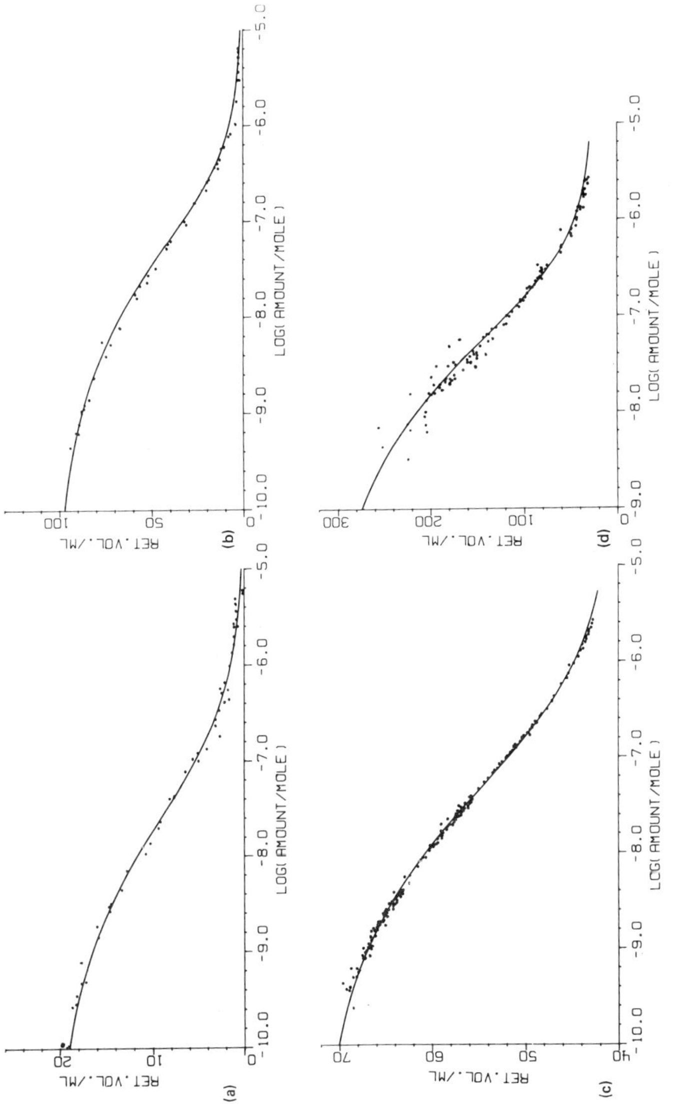

Figure 13. Fit of an equation similar to Eq. (50) to experimental points System (a) = diisopropyl ether on Supasorb AW DMS column; (b) methyl ethyl ketone on Supasorb AW DMS column; (c) diisopropyl ether on 10% octadecane column; (d) = methyl ethyl ketone on 10% octadecane column. (After Jönsson and Mathiasson [40], with permission.)

Equation (44) was applied to some of the data from the experiments described (unpublished work). The values of A and C obtained in this way agreed well with the values in Ref. 34. The values for B were different and gave reasonable values for K_i, calculated as mentioned above. Assuming that all of the adsorption of diisopropyl ether on a 15% n-octadecane column with a silanized support is due to adsorption at the gas/liquid interface, K_I can be calculated from the appropriate B value and A^0 for diisopropyl ether (2.6×10^6 m^2/mol [16]). The value found is 40×10^{-6} m, which compares favorably with the value $55 \pm 15 \times 10^{-6}$ m determined in another way (see page 199) [16].

The application of Eq. (44) and similar equations for the separation of adsorption and dissolution contributions to the retention volume rests on some assumptions partially discussed earlier:

1. The adsorption effects must be Langmuir-like, i.e., decreasing with concentration.
2. The curvature of the dissolution isotherm should not be significant at the concentration at which the adsorption contribution is negligible. It was shown in Sec. IV.A that this condition is most likely to be met when the adsorption coefficient at infinite dilution is high.

The bulk dissolution contribution A determined by fitting an equation such as Eq. (44) to experimental data can and should be verified by plotting A versus V_L. This plot should give a straight line through the origin, the slope of which is equal to K_L.

Recommended Procedure

For reliable separation of the retention volume increment due to dissolution from that due to adsorption, it is advisable to apply a combination of the two approaches described in the preceding paragraphs. The following general strategy is suggested.

The retention volume should first be recorded at a few widely different concentrations. If the peak shape and retention volume are unaffected within a range of sample sizes that includes the smallest sample size measurable, it can be assumed that infinite dilution is reached. This behavior is typical of a low-energy system [i.e., one characterized by a low value of the adsorption coefficient (see pages 201-202 and 217). Note that this does not necessarily mean that the adsorption contribution is small; the area of the adsorbing surface may be correspondingly large. Adsorption of nonpolar substances on a surface of a polar liquid is an example of such systems (page 203). The method of variation of liquid loading (page 220) in this case provides an adequate determination of the bulk dissolution contribution $V_{N,L}$.

If, however, the retention volume is found to markedly depend on sample size, even at very low concentrations, infinite dilution is not attained. Such cases are characterized by relatively high values of the adsorption coefficient, typical for adsorption on solid surfaces or for the adsorption of polar substances on the surfaces of nonpolar liquids. Here Eq. (44) (or similar equations) can be used for extrapolation. A suitable number of corresponding values of V_N and C_G should be obtained in order to allow the construction of curves similar to those in Fig. 13 and permit a reliable fitting of Eq. (44). The value of C_G is found from the height of the peak and the appropriate detector calibration factor. The parameter A in Eq. (44) found by this procedure is equal to $V_{N,L}$, provided that certain assumptions, discussed on page 224, are valid, which is more likely when the nonlinearity is prominent.

Although it is in principle possible to obtain $V_{N,L}$ and thus K_L from one column only by using this procedure, it is strongly recommended, especially with unknown systems, that the determination be repeated with several columns of differing V_L. Then Eq. (43) can be applied to the A values found. In this way the applicability of Eq. (44) to the system in question can be verified and residual adsorption contributions can be eliminated.

B. Separation of Different Adsorption Contributions

When the contribution $V_{N,L}$ from bulk dissolution has been determined as described in the preceding section, the rest of the observed retention volume is due to one or several adsorption effects. It is not an easy task to correctly divide the observed adsorption retention volume into parts, each corresponding to an identified type of surface, and no generally applicable procedure can be given.

General Approach

If it were possible to systematically vary each of the surface areas A_I, A_S, and A_A in a given support-SLP system, the problem might easily be handled by the simultaneous solving of several (at least three equations of the type

$$Ads_j = K_I A_{I,j} + K_A A_{A,j} + K_S A_{S,j} \tag{45}$$

where Ads_j is the adsorption contribution at infinite dilution determined by either Eq. (43) or Eq. (44) on the jth column, with surface areas $A_{I,j}$, $A_{S,j}$, and $A_{A,j}$. Although solid supports with different porosities, different liquid loadings, and other characteristics would to some extent permit a systematic variation of the area of all three surfaces, this approach is not generally available.

Practical Approach

The assignment of adsorption effects to certain types of surfaces must be made indirectly, considering several circumstances (we will restrict this treatment to the conditions of infinite dilution):

1. The relative areas of different surfaces in different systems. (As discussed in Sec. III; for example, A_A will be near zero on wetted supports and the surface areas will vary in characteristic ways with V_L.
2. The relative values of K_I, K_S, and K_A in various systems. (As discussed in Sec. II; for example, K_S is often less than both K_A and K_I. Rough estimations of K_I, K_A, and K_S can be calculated from literature data, independent experiments, and other sources.
3. Thermodynamic arguments. Variation of the temperature can give thermodynamic information valuable for the assignment of an adsorption effect to a particular mechanism.)

Examples

There are very few works in the literature in which the retention contributions from several concurrent adsorption mechanisms are quantitatively determined. In a relatively recent work [42], reasoning according to points 1 and 2 in the preceding list was extensively employed. Four different ranges of V_L, corresponding to different physical conditions of the SLP, were identified, and the different adsorption coefficients were determined separately.

By varying the temperature in a system with nonpolar solutes on a polar SLP that wetted the support, Suprynowicz and co-workers [46] separated the retention volume from adsorption into two increments, each characterized by a certain adsorption energy. By comparison with the adsorption energy on an uncovered support, which had an intermediate value, they determined that the adsorption contribution with the lowest adsorption energy was due to the liquid-solid surface, the adsorption energy of the solute being decreased by competition with the solvent. From the results quoted, reasonable values for K_I can be calculated ($K_I \approx 0.75 \times 10^{-6}$ m for cyclohexane on dinonylphthalate at 30°C).

VI. PRACTICAL CONSEQUENCES OF MIXED RETENTION MECHANISMS

As is well known to every chromatographer, the appearance of adsorption effects in gas-liquid chromatography usually decreases the performance of the chromatographic system. Such adsorption effects can, for example, be observed as skewed peaks, as variable retention times, or as irreproducible peak heights.

This section will treat the most important manifestations of mixed retention mechanisms, with emphasis on the consequences for practical analyses and measurements.

A. Peak Shape

The shape of chromatographic peaks was treated in detail in Chap. 2. For considerations here we note that the effect of adsorption in gas-liquid chromatography is usually Langmuir-like, leading to isotherms that are curved in the direction shown in Fig. 4, curve a, page 82. This leads to peaks that are more or less tailing. If the adsorption effects are not overwhelming, other peak-shaping effects (such as diffusion and nonequilibrium) cannot be neglected in comparison with the tailing. Consequently, the conditions of nonideal, nonlinear chromatography apply. This is the most complicated case of chromatography, and it is not possible to describe the resulting peak shapes in an explicit form. For the present, a qualitative discussion is sufficient.

Peak Shape Variation with Concentration

Let us consider the variation of peak shape with sample concentration in a given chromatographic system exhibiting significant adsorption effects in the general manner shown in Fig. 8.

Figure 14 shows the typical shapes of peaks referring to the concentration regions α, β, and γ (Fig. 8). The peaks are normalized to the same area. The first peak represents the region of infinite dilution α, where the isotherm is effectively linear. This peak is symmetrical, and the retention volume is V_N^∞ according to Eq. (8). The question of the exact peak shape with linear isotherms, as well as of the correct measurement of the retention volume, was discussed in Chap. 2. The most asymmetric peak refers to the β region, the region of medium concentrations, where the variation in retention volume is largest. When the concentration is still larger, approaching the γ region, where the retention volume is invariant with concentration, the "main" part of the peak again becomes relatively symmetrical to the eye and the tail becomes less important. However, it is important to note that the tail always exists, since a certain number of molecules are still retarded, up to the volume V_N^∞.

Thus, every peak is spread out over a range of retention volumes, from V_N^∞ down to the retention volume observed for the peak in question.

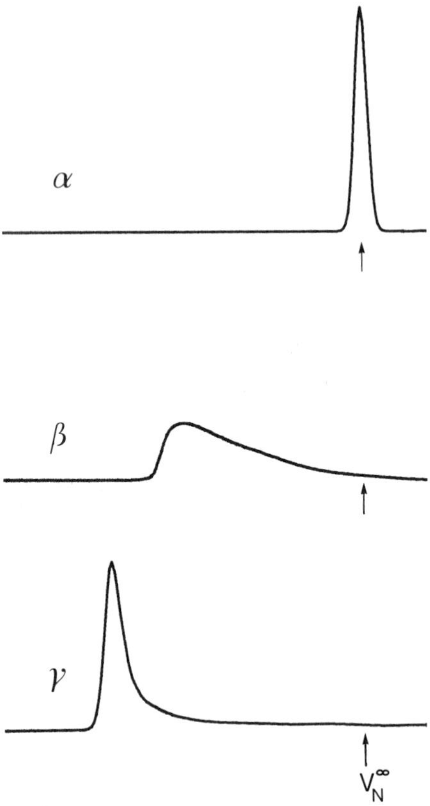

Figure 14. Typical peak shapes in the concentration regions α, β, and γ.

Peak Shape Variations in Different Chromatographic Systems

The general course of peak shape described here is followed by all systems that show Langmuir-like adsorption effects. However, as we saw earlier in this chapter, the relative magnitude of the adsorption effects might be considerably different. Figure 14 shows only the part of the chromatogram that comprises the retention volume variation due to adsorption. The point of injection can be imagined far to the left, off the page, for systems with small adsorption effects and a high efficiency. It can also be imagined within the limits of the picture in systems with large adsorption effects.

We also saw earlier in this chapter that the concentration range over which adsorption effects are noted can be considerably different. In certain systems characterized by high-energy adsorption on relatively small surface areas, the α region is very small and can be difficult to attain. This is typical for solid surface adsorption and the adsorption of polar solutes on nonpolar liquids. Under such conditions, skewed peaks of type β in Fig. 14 will be observed for small samples, together with the accompanying retention volume variation. With successively larger samples, peaks of type γ are observed, and the tail may eventually become insignificant (i.e., the adsorption effects are practically eliminated). This behavior is well known to everyone working in the trace analysis of polar compounds.

In some cases a small number of very strongly adsorbing sites are present, corresponding to a very high (perhaps "infinite") retention of a small amount of sample. This results in a very long and low tail and no α region.

In other types of systems, the α region can be so extended that in practice most peaks are quite symmetrical and only large peaks show tailing of the β type. This indicates low-energy adsorption on relatively large, homogeneous surface areas, such as the adsorption of nonpolar solutes on polar liquids. Also, the types of systems used in gas-liquid-solid chromatography (page 234) have these characteristics.

In summary, it can be said that those peak tailing effects that are most significant for small peaks arise from high-energy adsorption on small surfaces (if the surfaces were large, the adsorption effects would be excessive). On the other hand, if the tailing appears on relatively large peaks only, the adsorption energy is low but the adsorbing surface is correspondingly larger.

B. Qualitative Analysis

From the preceding sections it follows that adsorption mechanisms in GLC impair the precision and accuracy of measured retention data, and consequently, decrease the reliability of qualitative analysis by gas chromatography. This is also true when such concepts as relative retention and retention indices are used.

Relative Retention

A quantitative expression for the influence of mixed retention effects on the relative retention r can be derived starting from Eq. (8), assuming infinite dilution:

$$r = \frac{V_N}{V_{N,s}} = \frac{K_L V_L + \Sigma K_i A_i}{K_{L,s} V_L + \Sigma K_{i,s} A_i} \tag{46}$$

The subscript s denotes the standard (reference) substance, and i stands for A, S, and I. Following Berezkin [3], we expand the right side of Eq. (46) into a McLaurin series around $1/V_L = 0$:

$$r = \frac{K_L}{K_{L,s}} + \frac{\lambda}{V_L} + O\left(\frac{1}{V_L^{\,2}}\right) \tag{47}$$

The quantity λ is given by

$$\lambda = \frac{K_L}{K_{K,s}} \sum \left(\frac{K_i}{K_L} - \frac{K_{i,s}}{K_{L,s}}\right) A_i \tag{48}$$

Thus, the relative retention is not independent of liquid loading when adsorption contributes to the retention of either the analyte or the standard (or both). Considering the discussion above about the relative magnitude of the surface areas A_i in different types of columns, it is clear that λ will vary with V_L in a way that might be difficult to predict.

If conditions of infinite dilution cannot be assumed, the K values in Eqs. (46) through (48) should be substituted for by the corresponding derivatives [see, e.g., Eq. (5)], which will introduce a concentration dependence of λ, further complicating the situation.

Consequently, it is essential to eliminate all adsorption effects if accurate determinations of relative retention values, having general validity, are attempted. This is discussed in Sec. VI.F, where some guidelines for the elimination of adsorption can be found.

Retention Index

To extend and unify the concept of relative retention, Kováts [47] suggested the use of normal alkanes as reference substances. He defined the retention index I as

$$I = 100 \cdot \left[z + \frac{\log\,(V_N/V_{N,z})}{\log\,(V_{N,z+1}/V_{N,z})} \right] \tag{49}$$

$V_{N,z}$ is here the net retention volume of a normal alkane with z carbon atoms. A comprehensive review [48] of the history, use, and characteristics of the Kováts retention index system was recently published.

As can be seen from Eq. (49), the definition of the retention index contains two relative retention values. Accordingly, it will also be influenced by adsorption effects (see the preceding paragraph). This has been clearly demonstrated in several publications [3,37,49-52].

For nonpolar compounds, nonpolar stationary phases are normally used, providing adequate chromatographic conditions also for the alkane reference substances. In such a case disturbances due to adsorption are minimized. For polar compounds, polar stationary phases are usually used. Here the alkanes will be more or less significantly adsorbed on the surface of the liquid phase. (See Sec. II.A.) A more or less significant variation of the retention index with column parameters is then expected. This is an inherent disadvantage of the original definition of the retention index, prescribing n-alkanes as reference compounds.

The problem can be partly bypassed by using a homologous series of polar compounds as reference compounds instead of n-alkanes. Ethers, esters, alcohols, and other substances have been suggested, usually for reason somewhat other than those mentioned here (for a short review, see Ref. 49). This approach, although attractive from the point of view of accuracy and precision, detracts from the original idea of a common, generally applicable retention parameter.

In cases of incomplete support deactivation, adsorption of the polar analyte on the support surface can be notable, also influencing the retention index. Evans and Smith [51] found significant variations with solid support type. Apparently, it is imperative to use the best type of solid support (or capillary column) deactivation method available.

Polar analytes on nonpolar stationary phases represent a combination that is not very useful for analytic work but can be encountered when analyzing complex mixtures. Here, retention indices sometimes vary with sample concentration and column parameters to such an extent [49] that their measurement is meaningless.

C. Quantitative Analysis

The precision and accuracy of quantitative analysis by gas chromatography depend to a great extent on the quantitative evaluation of the chromatographic peaks. Two parameters are significant here: the peak area and the peak height.

Peak Area

The area of a peak is, in principle, directly proportional to the amount of eluted substance, irrespective of the shape of the peak. As has been comprehensively reviewed [53], the area of a chromatographic peak can be measured in several ways, including manual methods (planimetry, cut-and-weight, triangulation, and others) and automatic methods (ball-and-disk integrator, (digital or analog) electronic integrator, computer, and others). At present, the most important measurement technique is the application of electronic integrators. The accuracy and precision of such devices will largely be a function of the construction of a specific instrument (including software, when applicable). In spite of this, some general observations can be made.

Integration is best performed when peaks are reasonably narrow and symmetrical. This does not preclude the integration of skewed peaks, but the choice of suitable integrator parameters will be more critical, and the precision and accuracy will be impaired.

The most crucial task for the integrator is finding the correct baseline. A long low tail (see, for example, Fig. 14, curve γ) can easily be mistaken for a drifting baseline, especially at high sensitivity settings (i.e., for small peaks) at which noise may be disturbing. This will cause systematic errors, usually in the negative direction.

For unresolved peaks, integrator performance is generally worse than for single peaks, the difficulty being least for symmetrical peaks of roughly equal size.

In region α of infinite dilution (see Figs. 8 and 14), adequate conditions for integration prevail. When the concentration is increased into the β region, the peak will be skewed, and the conditions of integration will be successively worse. In the γ region, peak symmetry improves but the tail always exists, and thus some area may inevitably be lost. At high enough concentrations, the area of the tail will be insignificant and the quantitation of the peak will again be adequate. A sensitive procedure to use in checking for the presence of adsorption effects influencing quantitation is to plot A/m versus m (instead of the usual "calibration curve" plotting of A versus m). Here m is the amount of analyte injected and A is the corresponding measured peak area. In the ideal case such a plot is a straight horizontal line. The effects of adsorption on integration accuracy as described above will result in the curve sketched approximately in Fig. 15. The shape of such a curve in a practical experiment will be very dependent on the setting of the integrator parameters, and this approach will aid in finding the optimal settings.

In the common case of high-energy adsorption on small surface areas (Sec. VI.A), the α region may not be attainable, and a plot corresponding to Fig. 15 will show only a decrease in A/m when m is decreased.

With manual integration techniques, the influence of adsorption is similar. Some widely used methods, such as triangulation and the multiplication of peak height by width at half-height assume a gaussian peak, an approximation that apparently is not valid with skewed

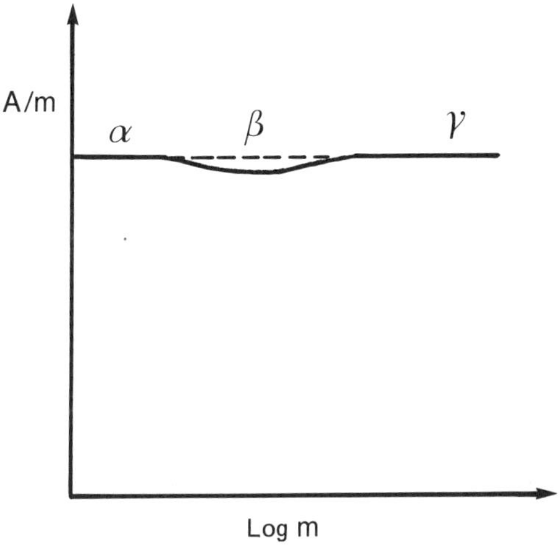

Figure 15. Relative peak area (measured) A/m versus the logarithm of the injected amount m, with adsorption (full line) and without (broken line).

peaks. Planimetry and similar methods will be more accurate, but even here it will be difficult to distinguish between baseline distrubances and low tails.

Peak Height

In the case of symmetrical peaks without adsorption effects, the peak height is proportional to the peak area. The proportionality factor depends, however, on many experimental parameters.

It is apparent that the height of an asymmetric peak is relatively lower than that of a symmetrical peak. The relative peak height h/m is expected to vary with m according to the approximate curve in Fig. 16. Since the peak is narrower in the γ region than in the α region, the relative height is larger in the former. Note that Figs. 15 and 16 show different phenomena. The variation of A/m in Fig. 15 is due to systematic errors in the measurement of a principally constant quantity and can be minimized by optimization of the integrator parameters. The variation of h/m is more fundamental and, in practical cases, much greater than that of A/m.

In systems with significant adsorption effects, the use of peak height for quantitative measurements is not recommendable.

Separation

For a successful quantitation of adjacent peaks, the peaks must be sufficiently resolved. generally, peak aysmmetry increases the peak

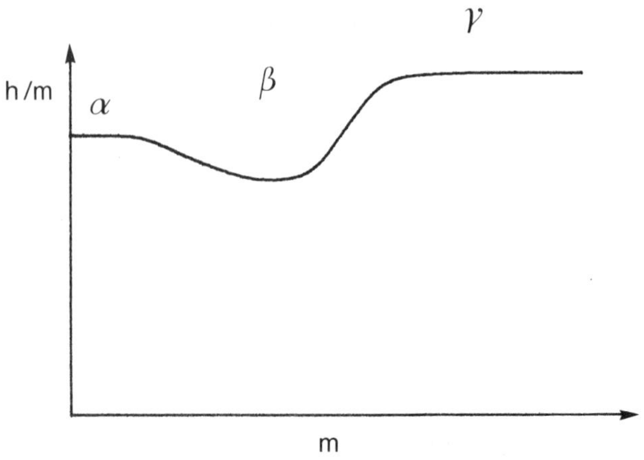

Figure 16. Relative peak height h/m versus the logarithm of the injected amount m.

width and consequently decreases the resolution. Therefore, adsorption effects decrease the resolution and complicate the quantitation in the concentration regions β and γ, where asymmetric peaks occur. However, in the α region, no extra peak broadening is induced and quantitation is not affected. Instead, the adsorption can be exploited to increase resolution in certain cases. This approach will be described in the next section.

D. Gas-Liquid-Solid Chromatography

In this chapter, adsorption in GLC has been largely viewed as a nuisance that decreases the performance of a chromatographic system and that should therefore be avoided as much as possible. In certain cases, however, adsorption can be utilized to increase the separation power of a GLC system, since adsorption forces typically depend on molecular properties other than those determining solubility.

Isomers with different geometric structures often have similar solution properties and are thus not well separated by GLC. In contrast, adsorption on a flat surface will be largely dependent on molecular geometry, providing a clear discrimination between isomers.

The chromatographic technique that utilizes a solid adsorbent covered with a relatively small amount of liquid has been called gas-liquid-solid chromatography (GLSC). A comprehensive review of this method was published by DiCorcia and Liberti [43]. Several examples of greatly increased separation for isomer pairs were reported, such as for *meta*- and *para*-toluic acid [54], 2-methylbutyric and 3-methylbuturic acid [54], and *meta*- and *para*-cresol [55].

In order to obtain adequate chromatographic conditions in a GLSC system, it is necessary to operate in the region of infinite dilution (the α region). Then we get, from Eq. (8),

$$V_N = V_L K_L + A_S K_S \tag{50}$$

Adsorption on uncovered support and liquid surface areas is assumed to be absent. The conditions of linear chromatography apply, leading to symmetrical peaks and constant retention. With too high a concentration, the curvature of the adsorption isotherm will begin (β region) and the chromatographic performance will degenerate.

Thus it is necessary for practical purposes that the α concentration region be as wide as possible. From the preceding discussion (see for examples Figs. 9 and 10), it is evident that this is achieved when the following conditions are met. The heat of adsorption should be relatively low, the surface should be energetically homogeneous, and the surface area of the solid should be relatively large. The most disturbing factor is the presence of even a small number of strongly adsorbing sites.

A useful adsorbent, compatible with the demands of GLSC, is graphitized carbon black. Its properties and use have been thoroughly reviewed by DiCorcia and Liberti [43].

E. Determination of Solution Properties

GLC has been used for many years for the determination of various physicochemical quantities, the most important ones being infinite-dilution activity coefficients and complexation constants. This application of GLC has been extensively reviewed [1,2,10]. In essence, the quantity sought is the bulk partition coefficient K_L, and the demand for accuracy in the experimental measurements is relatively high. The choice of SLP-solute combination is determined by the intended measurement and may be far from optimal in a chromatographic sense. It is therefore essential that adsorption effects be identified and eliminated.

Based on the discussions in this chapter, the following advice and comments on the matter of accurate K_L determination can be given.

When the polarity of the solvent (SLP) and solute are significantly different, the possibility of liquid-surface adsorption always exists and the matter should be investigated.

With polar solutes on nonpolar stationary phases, adsorption effects may be noted as a variation of retention volume with solute concentration.
With nonpolar solutes on polar stationary phases, adsorption effects can be important even if the retention volume is invariant with solute concentration.
The choice of the solid support is important and may influence the magnitude of all types of adsorption effects. Results obtained should preferably be demonstrated to be independent of the support material.
If adsorption effects are detected or suspected, the techniques in Sec. V.A should be applied.

Despite mention of the problem with adsorption in most reviews of the matter, it seems not to be generally acknowledged. In many solution studies described in the literature, the absence of influence of adsorption was not demonstrated, although experimental conditions would not exclude such influence, and elaborate methods were used for further treatment of the measured data. The reliability of these studies is consequently decreased. Unfortunately, considerable experimental effort is needed to safely exclude errors from adsorption effects, and this seems to somewhat decrease the value of GLC for the measurements in question.

F. Elimination of Adsorption

In order to eliminate the influence of adsorption processes on retention in GLC, a number of measures can be taken, with due consideration of the characteristics and goals of the problem at hand.

Polarity of the Stationary Liquid Phase

As we saw in Sec. II.A, adsorption at the gas/liquid interface largely depends on the chemical difference between solute and SLP [or more exactly, the difference in the surface tensions; see Eq. (15)]. From this point of view it is therefore favorable to choose a stationary phase that is chemically similar to the compounds to be separated.

Adsorption effects on the liquid/solid or gas/solid interface are not directly influenced by the polarity of the SLP (see the next paragraph). Since the solubility, and thus the retention due to bulk liquid dissolution, is usually maximal when the polarity of the analyte and the SLP are equal, this situation is also favorable with respect to the relative contribution from support adsorption.

Only in special cases, such as the analysis of mixtures containing very different types of compounds, the application of the original Kováts retention index concept to polar compounds (page

231), or when special efforts for separation are needed, is it necessary to deviate from the general principle of similar polarity of the analyte and the SLP.

Properties of the Solid Support

The properties of the solid support influence the role of adsorption in GLC in two ways: by determining the adsorption coefficients (isotherm) for adsorption on the liquid/solid and gas/solid interfaces and by determining the wetting properties controlling the magnitudes of the various areas in the column. With capillary columns, very similar considerations hold as for porous supports for packed columns.

Considering the first property, it is obvious that supports that exhibit a high degree of inertness are favorable with respect to the elimination of support adsorption. This is accomplished by silanization or polymer coating of diatomaceous supports or capillary column walls, among other procedures. Alternative support materials are also used, such as graphitized carbon black or Teflon.

For certain types of compounds, special support modification principles are used, such as the addition of an alkaline modifier to column packings intended for amine separations. With these types of treatment, a general inertness is not attempted but, rather, elimination of the specific interactions responsible for the adsorption of the compounds in question.

The wetting of the SLP is also of concern. Incomplete wetting leaves some patches of the support uncovered, thus increasing the possibility of gas-solid adsorption effects. The risk of incomplete wetting is greatest with deactivated supports. This risk is somewhat compensated for by the weak adsorption forces of the deactivated surface.

With active supports, such as silica, diatomaceous earth, or glass, wetting is essentially complete for all types of SLP normally used and gas-solid adsorption should be insignificant. Instead, the possibility of adsorption of the analyte at the liquid/solid interface occurs but is balanced by the competitive adsorption of the large excess of SLP.

The extent of adsorption on the gas/solid interface is indirectly influenced by the properties of the support, since the surface area of the liquid varies strongly with its wetting properties. With active supports, complete wetting causes the area of the gas/liquid interface to be stable and approximately equal to the support surface area (see page 210). With deactivated supports, the size of the gas/liquid interfacial area may be very variable (page 211); in some cases it is very small, but occasionally it may be relatively large.

In summary, the choice of a deactivated support versus an untreated support is not straightforward. However, a deactivated support is usually preferred, and definitely so for nonpolar stationary phases. For polar stationary phases, nondeactivated supports are often successfully used.

Amount of SLP

It is apparent that combined adsorption effects relative to the contribution of bulk liquid dissolution retention decrease when SLP loading is increased. Thus, a high loading of SLP is generally preferable from the viewpoint of eliminating adsorption influences. This is in conflict with other aspects of chromatographic performance, such as resolution and time of analysis. Individual consideration is necessary for specific separation problems.

Solute Concentration

On page 213 the typical course of the retention volume variation with solute concentration is discussed in detail. It is clear that, with adsorptive systems, regions of concentration exist in which the total sorption isotherm is linear and, consequently, the adverse effects of adsorption are not perceivable. The discussion on page 227 is relevant here for the description of these regions. Obviously, in some cases adsorption effects can be overcome simply by working in another concentration range.

Practically, the case of high-energy adsorption on small surface areas is the most important. This case is characterized by an onset of adsorption effects at low solute concentrations, although moderate and high amounts of sample permit adequate chromatography. The adsorption effects thus set the lower useful concentration limit for the practical application of the technique. The characteristic of this type of adsorption is of utmost importance in trace analysis, especially of polar compounds. It is amazing how often chromatograms published as examples of gas chromatographic methods intended for trace analysis are either recorded with relatively large sample amounts or presented without any indication of this extremely important parameter.

Other Methods

Some other methods for controlling or bypassing adsorption in GLC can be used in certain cases. A somewhat trivial way to circumvent problems with adsorption (and some other problems also) is to modify the analyte molecules by derivatization, an approach that has been thoroughly reviewed in several texts [56,57]. The use of additives to sample solutions is sometimes shown to be beneficial.

Thus, the addition of ammonia improved trace analyses of free amines [58] by blocking the few highly active adsorption sites remaining after deactivation, both in the column and in sample containers, syringes, and other apparatus. Vapor additives to the carrier gas (for example, water vapor or ammonia) or the use of adsorbing carrier gases have also been suggested [59-62] as offering competition with analyte molecules and thus decreasing support adsorption effects. In spite of the significant improvements in chromatographic performance that are possible, this technique has not gained widespread use. Reasons for this probably include the increased technical complexity of the apparatus needed, the increased number of variables that influence the chromatographic process, and, in some cases, the possibility of detection disturbances.

REFERENCES

1. D. C. Locke, Physicochemical Measurements using Chromatography, *Advances in Chromatography, Vol. 14* (J. C. Giddings, E. Grushka, J. Cazes, and P. R. Brown, eds.), Marcel Dekker, New York, pp. 87-198 (1976).
2. J. R. Conder and C. L. Young, *Physico chemical Measurement by Gas Chromatography*, John Wiley & Sons, Chichester (1979).
3. V. G. Berezkin, Adsorption in gas-liquid chromatography, *J. Chromatogr.*, *159*:359 (1978).
4. J. R. Conder, Existence of gas-liquid interfacial adsorption in solutions of alcohols and ketones in saturated hydrocarbons, *Anal. Chem.*, *48*:917 (1976).
5. A. W. Adamson, *Physical Chemistry of Surfaces*, 3rd Ed., John Wiley & Sons, New York (1976).
6. W. J. Moore, *Physical Chemistry*, 5th Ed., Longman, London (1972).
7. J. W. McBain, *Colloid Science*, D.C. Heath & Co, Boston (1950).
8. C. Eon and G. Guiochon, Surface activity coefficients as studied by gas chromatography, *J. Colloid Interface Sci.*, *45*:521 (1973).
9. D. H. Everett: Thermodynamics of adsorption from solution, Part 2, Imperfect solutions, *Trans. Faraday Soc.*, *61*:2478 (1965).
10. R. J. Laub and R. L. Pecsok, *Physicochemical Applications of Gas Chromatography*, John Wiley & Sons, New York (1978).
11. R. C. Castells, E. L. Arancibia, and A. M. Nardillo, Surface and bulk activity coefficients of non-electrolytic mixtures studied by gas chromatography, *J. Colloid Interface Sci.*, *90*:532 (1982).
12. R. Defay and I. Prigogine (A. Bellemans and D. H. Everett, coauthors), *Surface Tension and Adsorption*, Longman, London (1966).

13. F. Riedo and E. S. Kováts, Effects of adsorption on solute retention in gas-liquid chromatography, *J. Chromatogr.*, *186*:47 (1979).
14. D. E. Martire, R. L. Pecsok, and J. H. Purnell, Static measurements of activity coefficients and surface tensions of some systems at very low mole fraction, *Trans. Faraday Soc.*, *61*:2496 (1965).
15. R. L. Pecsok and B. H. Gump, Gas-liquid interface and solid support effects of polar solute-non polar solvent systems in gas chromatography, *J. Phys. Chem.*, *71*:2202 (1967).
16. J. Å. Jönsson and L. Mathiasson, Mixed retention mechanisms in gas-liquid chromatography, III, Determination of the gas-liquid adsorption coefficient for diiso-propyl ether on n-octadecane, *J. Chromatogr.*, *206*:1 (1981).
17. E. A. Guggenheim and N. K. Adam, The thermodynamics of adsorption at the surface of solutions, *Proc. Roy. Soc. Ser.*, *A*, *139*:218 (1933).
18. B. L. Karger, R. C. Castells, P. A. Sewell, and A. Hartkopf, Study of the adsorption of insoluble and sparingly soluble vapors at the gas-liquid interface of water by gas chromatography, *J. Phys. Chem.*, *75*:3870 (1971).
19. J. W. King, A. Chatterjee, and B. L. Karger, Adsorption isotherms and equations of state of insoluble vapors at the water-gas interface as studied by gas chromatography, *J. Phys. Chem.*, *76*:2769 (1972).
20. L. R. Snyder, *Principles of Adsorption Chromatography*, Marcel Dekker, New York (1968).
21. A. V. Kiselev and Y. I. Yashin, *Gas-Adsorption Chromatography*, Plenum Press, New York (1969).
22. J. Tóth, Gas-(Dampf-)Adsorption auf festen Oberflecken inhomogener Aktivität, III, *Acta Chim. Acad. Sci. Hung.*, *32*:39 (1962).
23. J. Tóth, W. Rudziński, A. Waksmundski, M. Jaroniec, and S. Sokołowski, Adsorption of gases on heterogeneous solid surfaces: The energy distribution function corresponding to a new equation for monolayer adsorption, *Acta Chim. Acad. Sci. Hung.*, *82*:11 (1974).
24. C. H. Eon, Study of liquid-powder interfaces by means of solvent strength parameter measurements, *Anal. Chem.*, *47*:1871 (1975).
25. D. H. Everett, Thermodynamics of adsorption from solution, Part 1, perfect systems, *Trans. Faraday Soc.*, *60*:1803 (1964).
26. W. A. Zisman, Surface energetics of wetting, spreading and adhesion, *J. Paint Technol.*, *44*:41 (1972).
27. G. Brandrup and E. H. Immergut (eds.), *Polymer Handbook*, 2nd Ed., Wiley, New York (1975).

28. J. C. Giddings, Liquid distribution on gas chromatographic support, *Anal. Chem.*, *34*:458 (1962).
29. V. G. Berezkin, D. Kouřilová, M. Krejčí, and V. M. Fateeva, Evaluation of changes in the specific surface area of sorbents based on the volume of stationary phase in gas-liquid chromatography, *J. Chromatogr.*, *78*:261 (1973).
30. J. R. Conder, D. C. Locke, and J. H. Purnell, Concurrent solution and adsorption phenomena in chromatography, I, General considerations, *J. Phys. Chem.*, *73*:700 (1969).
31. D. Dollimore, P. Spooner, and A. Turner, The BET method of analysis of gas adsorption data and its relevance to the calculation of surface areas, *Surface Technol.*, *4*:121 (1976).
32. J. Serpinet, Study of liquid-gas interfacial adsorption by gas chromatography with silanized supports, *Anal. Chem.*, *48*:2264 (1976).
33. J. J. Kirkland, Fluorine-containing polymers as solid supports in gas chromatography, *Anal. Chem.*, *35*:2003 (1963).
34. L. Mathiasson and J. Å. Jönsson, Mixed retention mechanisms in gas-liquid chromatography, II, Contributions of adsorption and partition to the retention volume in systems with polar solutes and non-polar stationary phase on a silanized support, *J. Chromatogr.*, *179*:7 (1979).
35. J. Serpinet, Etude de l'etat physique et du site occupé par des phases stationaires a longues chaines grasses a la surface de diverse supports chromatographiques, *Chromatographia*, *8*:18 (1975).
36. J. Å. Jönsson, L. Mathiasson, and Z. Suprynowicz, Mixed retention mechanisms in gas-liquid chromatography, IV, Comparisons of silanized and Carbowax-modified supports, *J. Chromatogr.*, *207*:69 (1981).
37. J. L. Lorenz and L. B. Rogers, Specification of gas chromatographic behaviour using Kováts' indices and Rohrschneider constants, *Anal. Chem.*, *43*:1593 (1971).
38. G. V. Filonenko, T. I. Dovbush, and A. N. Korol, Relationship between the retention volume and some experimental parameters in gas-liquid chromatography, *Chromatographia*, *7*:293 (1974).
39. J. J. Pesek and J. E. Daniels, Investigation of the retention mechanism of chemically bonded stationary phases in gas chromatography, *J. Chromatogr. Sci.*, *14*:288 (1976).
40. J. Å. Jönsson and L. Mathiasson, Mixed retention mechanisms in gas-liquid chromatography, I, The relation between retention volume and sample size. Determination of bulk liquid contribution, *J. Chromatogr.*, *179*:1 (1979).
41. M. I. Selim, J. F. Parcher, and P. J. Lin, Adsorption of polar solutes on liquid-modified supports, *J. Chromatogr.*, *239*:411 (1982).

42. K. Naito and S. Takei, Adsorption effects in gas-liquid partition chromatography, *J. Chromatogr., 190*:21 (1980).
43. A. DiCorcia and A. Liberti, Gas-liquid-solid chromatography, *Advan. Chromatogr. 14*:305-366 (1976).
44. J. R. Conder, N. K. Ibrahim, G. J. Rees, and G. A. Oweimreen, Wetting transition in squalane on a silanized diatomaceous gas chromatographic support, *J. Phys. Chem., 89*:2571 (1985).
45. J. R. Conder, The derivation of thermodynamic information from asymmetric chromatographic peaks when more than one distribution mechanism contributes to retention, *J. Chromatogr., 39*:273 (1969).
46. Z. Suprynowicz, A. Waksmundzki, and W. Rudziński, The contribution of solution and adsorption phenomena in gas-liquid partition chromatography in which infinite dilution of solute may be assumed, *J. Chromatogr., 67*:21 (1972).
47. E. Kováts, Gas-chromatographische Charakterisierung organischer Verbindungen, Teil 1, Retentionsindices aliphatischer Halogenide, Alkohole, Aldehyde und Ketone, *Helv. Chim. Acta, 41*: 1915 (1958).
48. M. V. Budahegyi, E. R. Lombosi, T. S. Lombosi, S. Y. Mészaros, S. Nyiredy, G. Tarján, I. Timár, and J. M. Takácz, Twenty-fifth anniversary of the retention index system in gas-liquid Chromatography, *J. Chromatogr., 271*:213 (1983).
49. L. Mathiasson, J. Å. Jönsson, A. M. Olsson, and L. Haraldson, Sensitivity of retention index to variation in column liquid loading and sample size, *J. Chromatogr., 152*:11 (1978).
50. J. Bonastre, P. Grenier, and P. Cazenave, Contribution a l'étude des phenomènes de surface en chromatographie gaz-liquide, *Bull. Soc. Chim. Fr.*, 1266 (1968).
51. M. B. Evans and J. F. Smith, Gas-liquid chromatography in qualitative analysis, Part VII, The role of the support in gas chromatographic retention measurement, *J. Chromatogr., 30*:325 (1967).
52. E. Grushka and T. A. Goodwin, The influence of the amount stationary phase and nature of solid support on retention indices, *Chromatographia, 10*:549 (1977).
53. J. Novák, *Quantitative Analysis by Gas Chromatography*, Marcel Dekker, New York (1975).
54. A. DiCorcia, Gas-liquid-solid chromatography of free acids, *Anal. Chem., 45*:492 (1973).
55. A. DiCorcia, Analysis of phenols by gas-liquid-solid chromatography, *J. Chromatogr., 80*:69 (1973).
56. J. Drozd, *Chemical Derivatization in Gas Chromatography*, Elsevier, Amsterdam (1981).
57. K. Blau and G. S. King (eds.), *Handbook of Derivatives for Chromatography*, Heyden, London (1977).

58. M. Dalene, L. Mathiasson, and J. Å. Jönsson, Trace analysis of free amines by gas-liquid chromatography, *J. Chromatogr.*, *207*:37 (1981).
59. N. D. Greenwood and H. E. Nursten, Gas chromatographic studies on basic nitrogeneous drugs, I, Addition of basic compounds or steam to the carrier gas, *J. Chromatogr.*, *92*:323 (1974).
60. A. Nonaka, Gas-solid chromatography of organic compounds using steam as the carrier gas, *Anal. Chem.*, *44*:271 (1972).
61. V. G. Berezkin and L. A. Shkoline, Ammonia as a carrier gas and flame-forming agent in gas chromatography, *J. Chromatogr.*, *119*:33 (1976).
62. E. L. Ilkova and E. A. Mistryukov, The use of ammonia as the carrier gas in gas-liquid chromatography, *J. Chromatogr.*, *54*:422 (1971).

6
Retention in Liquid Chromatography

Władysław Rudziński
Institute of Chemistry
Maria Curie-Skłodowska University
Lublin, Poland

The partition coefficients (capacity ratios) measured in liquid chromatography are affected by many physical phenomena that act in the mobile phase and at the solid/solution interface. Because of the very complicated interactions among these phenomena, it is to be expected that the theoretical and numerical analysis of retention data alone does not provide much opportunity to draw deeper conclusions about the mechanism of retention in liquid-solid chromatography (LSC). For this purpose, some additional physicochemical measurements are necessary to permit us to obtain independent information. Data about the competitive adsorption of solvents onto the solid surfaces of column packings can, for example, be obtained in independent static batch studies. The measurement of vapor pressure data, on the other hand, makes possible the estimation of bulk activity coefficients. It seems that this possibility of obtaining important data in an independent manner has not been sufficiently recognized by those investigating the mechanisms of retetion in liquid-solid chromatography. Accordingly, we have adopted the following policy in presenting the theory of partition coefficient in LSC.

First, we shall consider the competitive adsorption of solvents, which is practically unaffected by solutes present in very small concentrations. We shall then discuss the competitive adsorption of solutes, which is strongly influenced by solvent adsorption. This presentation should help the reader understand the complicated nature of retention processes in LSC and show the proper strategy for further basic research in this field.

I. COMPETITIVE ADSORPTION OF SOLVENTS

A. Adsorption from a Binary Solvent Mixture on a Homogeneous Solid Surface

We shall consider the most general, three-dimensional model of surface solution. This semi-infinite surface solution extends from a planar solid surface and becomes identical with the mobile bulk phase at a sufficiently great distance from the solid surface. Further, we accept the lattice model of this surface solution, assuming that the lattice structure is not perturbed by the presence of the solid phase.

We will consider the surface phase as sliced into L lattice planes parallel to the solid surface and numbered 1, 2, . . ., L, beginning with the lattice plane closest to the solid surface. Each lattice plane contains N_0 lattice sites, which are occupied by N_{1l} and N_{2l} molecules of the first and second solvent, respectively, with l the number of the lattice plane. However, we also assume that the solid-solution forces may produce some specific orientation of the solvent molecules in the first adsorbed layer closest to the solid surface. Let r_1, r_2 denote the numbers of the lattice sites occupied by each of the molecules of the first and the second solvent in the first adsorbed layer or lattice plane. According to our model, r_1, r_2 may be different from unity in the first lattice plane but equal to unity in other adlayers. Thus,

$$r_1 N_{1l} + r_2 N_{2l} = N_0 \qquad l = 1, 2, \ldots, L \tag{1}$$

We complete our model by assuming that a_p and a_v are the numbers of the nearest-neighbors adsorption sites at some arbitrarily chosen site, in the same lattice plane (a_p) and in the lattice plane above or below (a_v). Let z denote the total number of nearest neighbors, equal to $a_p + 2a_v$.

The composition of the quasi-crystalline adsorbed phase is governed by the following set of quesi-chemical reactions,

$$r_2(1)^{(b)} + r_1(2)^{(s)} \rightleftarrows r_2(1)^{(s)} + r_1(2)^{(b)} \tag{2}$$

Here the superscripts (b) and (s) refer to the bulk and the surface phase, respectively. The equilibrium condition for these quasi-chemical reactions reads [1]

$$\frac{1}{r_1}\mu_1 + \frac{1}{r_2}\mu_{21} = \frac{1}{r_1}\mu_{11} + \frac{1}{r_2}\mu_2 \tag{3}$$

where μ and μ_l denote the chemical potentials in the bulk and the surface phase. Appropriate statistical-mechanical consideration yields the following expression for μ_l [2]:

$$\mu_{1l} = \mu_{1l}^0 + RT(1 - r_{12})\phi_{1l} + RT \ln(\phi_{1l}\gamma_{1l}) \tag{4a}$$

$$\mu_{2l} = \mu_{2l}^0 + RT(r_{21} - 1)\phi_{2l} + RT \ln(\phi_{2l}\gamma_{2l}) \tag{4b}$$

where μ_{1l}^0 and μ_{2l}^0 are appropriate standard chemical potentials, $r_{12} = r_1/r_2$, $r_{21} = r_2/r_1$, ϕ_{1l} and ϕ_{2l} are the volume (site) fractions of solvents 1 and 2 in the lth lattice plane, and γ_{1l} and γ_{2l} are the appropriate activity coefficients. The explicit form of these activity coefficients depends not only on the accepted adsorption model but also on the approximation used to account for the intermolecular interactions in the adsorbed phase.

We will use the random mixing approach, known as the Bragg-Williams or Flory-Huggins approach, depending on whether the adsorbed molecules may occupy only one or more adsorption sites. It can easily be shown that the adsorption model described here, together with the Flory-Huggins approach, leads to the following form of the interaction energy U_{in} in the adsorbed phase:

$$U_{in} = U_{in}^{(b)} + U_{in}^{(id)} + U_{in}^{(mix)} \tag{5}$$

where the first of the three terms on the right side of Eq. (5) takes explicit form

$$U_{in}^{(b)} = \frac{N_0}{2} (\phi_1\varepsilon_{11}^{(p)} + \phi_2\varepsilon_{22}^{(p)} + 2\,\alpha_{12}\phi_1\phi_2) \tag{6}$$

which is identical to the appropriate contribution from a single lattice plane to a homogeneous bulk liquid whose composition is identical with the mobile-phase composition. Here, $\varepsilon_{11}^{(p)}$ and $\varepsilon_{22}^{(p)}$ are the interaction energies in the molecular pairs (1-1) and (2-2), whereas α_{12} is the interchange energy,

$$\alpha_{12} = \varepsilon_{12}^{(p)} - \frac{1}{2}(\varepsilon_{11}^{(p)} + \varepsilon_{22}^{(p)}) \tag{7}$$

Further,

$$U_{in}^{(id)} = \frac{N_0}{2}(a_p + a_v)(\phi_{11}\varepsilon_{11}^{(p)} + \phi_{21}\varepsilon_{22}^{(p)}) + \sum_{l=2}^{L} \frac{zN_0}{2}(\phi_{11}\varepsilon_{11}^{(p)} + \phi_{21}\varepsilon_{22}^{(p)}) \tag{8}$$

The last term in Eq. (5) represents the energy of mixing, associated with the nonideality of the adsorbed phase:

$$U_{in}^{(mix)} = \sum_{l=1}^{L-1} N_0\,\alpha_{12}[a_p\,\phi_{11}\phi_{21} + a_v(\phi_{11}\phi_{2(l+1)} + \phi_{1(l+1)}\phi_{21})] + N_0\,\alpha_{12}[a_p\,\phi_{1L}\,\phi_{2L} + a_v(\phi_{1L}\,\phi_2 + \phi_1\phi_{2L} - \phi_1\phi_2)] \tag{9}$$

The activity coefficients γ_{11} and γ_{21} can be obtained from the relationships

$$RT \ln \gamma_{11} = \left(\frac{\partial U_{in}^{(mix)}}{\partial N_{11}}\right)_{N_{21}, N_{1m}, N_{2m}} \tag{10a}$$

$$RT \ln \gamma_{21} = \left(\frac{\partial U_{in}^{(mix)}}{\partial N_{21}}\right)_{N_{11}, N_{1m}, N_{2m}} \tag{10b}$$

where $m \neq l$. After performing these differentiations, we arrive at explicit forms of the surface activity coefficients. For $l = 1, 2, \ldots, L - 1$, we have

$$\ln \gamma_{11} = A_{12}[f_v\phi_{21}(\phi_{2(l-1)} - \phi_{1(l-1)}) + f_p(\phi_{21})^2 + f_v\,\phi_{2(l+1)}] \tag{11a}$$

$$\ln \gamma_{21} = r_{21}A_{12}[f_v\phi_{11}(\phi_{1(l-1)} - \phi_{2(l-1)}) + f_p(\phi_{11})^2 + f_v\,\phi_{1(l+1)}] \tag{11b}$$

where $f_p = a_p/z$ and $f_v = a_v/z$. For the last (Lth) adlayer, we have

$$\ln \gamma_{1L} = A_{12}[f_v \phi_{1L}(\phi_{2(L-1)} - \phi_{1(L-1)})$$
$$+ f_p(\phi_{2L})^2 + f_v(\phi_2)^2] \tag{12a}$$

$$\ln \gamma_{2L} = r_{21}A_{12}[f_v \phi_{1L}(\phi_{1(L-1)} - \phi_{2(L-1)}$$
$$+ f_p(\phi_{1L})^2 + f_v(\phi_1)^2] \tag{12b}$$

In Eqs. (11) and (12), A_{12} stands for $z\alpha_{12}/RT$ and is a convenient measure of the deviation of the surface solution from perfect solution behavior. For a strictly regular solution, A_{12} cannot exceed 2 (according to the Bragg-Williams approach). When $A_{12} > 2$, phase separation will occur within a certain concentration range. This phase separation is the essential phenomenon for liquid-liquid chromatography and will be discussed at the end of this chapter.

Meanwhile, we are concerned with the supercritical behavior of the surface solution when no phase separation takes place in the adsorbed or in the mobile bulk phase. In other words, we are concerned here with liquid-solid chromatography.

According to Eq. (4), the adsorption equation (3) can be written in the compact form

$$\frac{\phi_{11}\gamma_{11}}{\phi_1\gamma_1}\left(\frac{\phi_2\gamma_2}{\phi_{21}\gamma_{21}}\right)^{r_{12}} = K_{12}^{(1)} \tag{13}$$

where the adsorption equilibrium constant $K_{12}^{(1)}$ takes the form

$$K_{12}^{(1)} = e\,\frac{1}{RT}\,[(\mu_1^0 - \mu_{11}^0) - r_{12}(\mu_2^0 - \mu_{21}^0)] \tag{14}$$

Taking the explicit form of the activity coefficients displayed in Eqs. (11) and (12), we arrive at the following detailed form of the adsorption equations:

$$F_1 = \ln \frac{\phi_{11}}{(1 - \phi_{11})^{r_{12}}} - \ln \frac{\phi_1}{(1 - \phi_1)^{r_{12}}} - \ln K_{12}^{(1)}$$
$$+ A_{12}[f_v(1 - 2\,\phi_{1(l-1)} + f_p(1 - 2\,\phi_{11}) + f_v(1 - 2\,\phi_{1(l+1)})$$
$$- (\phi_2)^2 + r_{12}(\phi_1)^2] = 0 \tag{15}$$

whose solution yields the concentration profile $\{\phi_{1l}\}$ in the vicinity of the solid surface.

Now let us consider the meaning of the equilibrium constant $K_{12}^{(1)}$ in more detail. To a good approximation, the difference of the standard chemical potentials $\mu^0 - \mu_1^0$ can be expressed as

$$\mu_1^0 - \mu_{1l}^0 = \varepsilon_{1l} - RT \ln \frac{q_{1l}}{q_1} \tag{16a}$$

$$\mu_2^0 - \mu_{2l}^0 = \varepsilon_{2l} - RT \ln \frac{q_{2l}}{q_2} \tag{16b}$$

where ε_{1l} and ε_{2l} are the adsorption energies of the single molecules 1 and 2, and q_{1l} and q_{2l} and q_1 and q_2 are their molecular partition functions in the adsorbed state and in the bulk phase, respectively. By "adsorption energy" we understand the minimum value of the solid-solvent interaction potential, taken with the reverse sign. The solid-adsorbate forces are usually short-range, so we will neglect their presence in the second and higher layers. Thus, we assume that

$$\varepsilon_{1l} = 0 \qquad \text{and} \qquad \varepsilon_{2l} = 0 \qquad l \geq 2 \tag{17}$$

The terms q_{1l}/q_1 and q_{2l}/q_2 may have an essential effect on $K_{12}^{(l)}$ in the first adsorbed layer, where they are governed mainly by the ratio of the appropriate vibrational partition functions; however, this difference should diminish rapidly in the second and higher adsorbed layers, where the adsorption potential field vanishes. Therefore, we will assume further that the ratios q_{1l}/q_1 and q_{2l}/q_2 are equal to unity in the second and higher adsorbed layers.

These assumptions mean that $K_{12}^{(l)}$ is set equal to unity in the second and higher adsorbed layers. We thus drop the additional index l for ε_{1l}, and ε_{2l}. For the purpose of further investigation, we also define, for the first adsorbed layer, the quantity ε_{12}:

$$\varepsilon_{12} = (\varepsilon_1 - r_{12}\varepsilon_2) - RT \ln \left[\frac{q_{11}}{q_1} \left(\frac{q_2}{q_{21}} \right)^{r_{12}} \right] \tag{18}$$

which we call simply the adsorption energy in the adsorption system (solid + solvent 1 + solvent 2). According to our assumptions, ε_{12} is zero in the second and higher layers.

For the purposes of illustration, we have evaluated the concentration profile $\{\phi_{1l}\}$ for two adsorption systems, characterized by

$\varepsilon/RT = 3$ and $r_{12} = 1$ and showing different positive deviations from Raoult's law, $A_{12} = 1.0$ and 1.95. The value $\varepsilon/RT = 3$ corresponds to a relatively strong preferential adsorption of the solvent 1. Solutions showing negative deviations from Raoult's law are rare and are not used in chromatographic practice. Equation system (15) was solved numerically using an iterative procedure. The homogeneous concentration profile $\{\phi_{1l} = \phi_1\}$ was used as the starting approximation. The result of this calculation is shown in Fig. 1. There, the excess of the preferentially adsorbed solvent 1 ($\phi_{1l} - \phi_1$) is shown using a special three-dimensional representation.

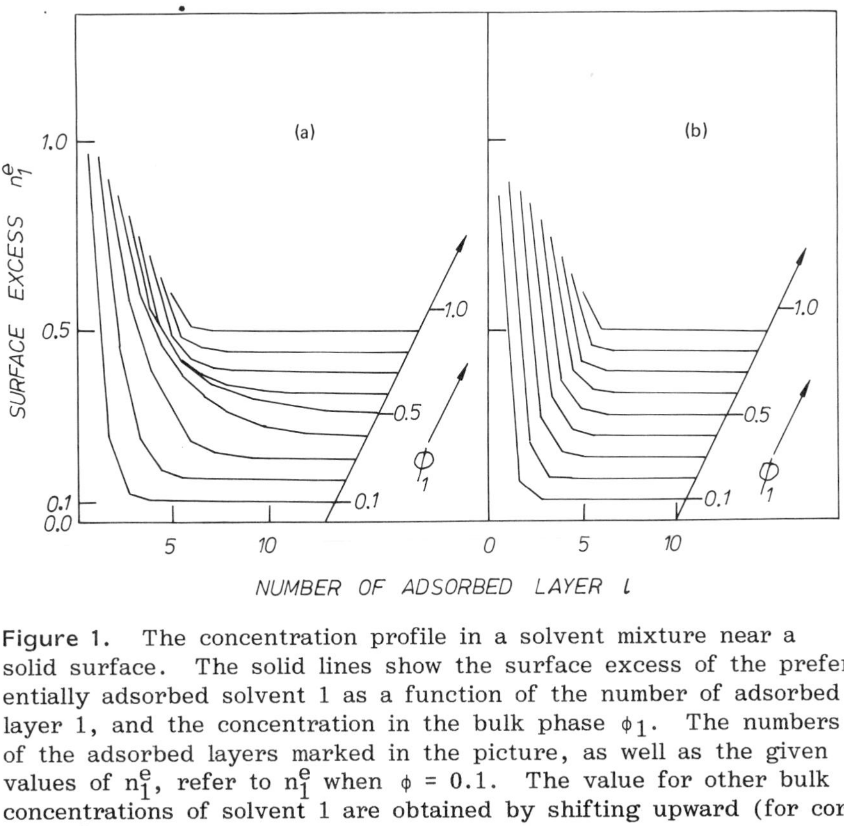

Figure 1. The concentration profile in a solvent mixture near a solid surface. The solid lines show the surface excess of the preferentially adsorbed solvent 1 as a function of the number of adsorbed layer 1, and the concentration in the bulk phase ϕ_1. The numbers of the adsorbed layers marked in the picture, as well as the given values of n_1^e, refer to n_1^e when $\phi = 0.1$. The value for other bulk concentrations of solvent 1 are obtained by shifting upward (for correct n_1^e values) and to the left (for the proper 1 values). (a) Concentration profile for strong positive deviations from Raoult's law ($A_{12} = 1.95$). (b) Concentration profile for moderate positive deviations from Raoult's law ($A_{12} = 1.0$). The monolayer capacity N_0 is assumed to be unity, whereas the adsorption energy ε_{12} is taken as equal to +3RT.

Our calculation shows that, in the case of moderate positive deviations from Raoult's law (i.e., when $A_{12} < 1$), the adsorption has practically a monolayer character. In the case of very strong positive deviations from Raoult's law (i.e., when $A_{12} = 1.95$), formation of the second and higher layers can still be neglected until the concentration of the preferentially adsorbed solvent in the mobile bulk phase does not exceed 0.1. However, even in this extreme case of high positive deviation from Raoult's law, the concentration of the preferentially adsorbed solvent in the second layer is no more than half as great as that in the first layer.

In Fig. 2, the excess adsorption isotherms evaluated according to the correct three-dimensional model are compared with those evaluated under the assumption that adsorption occurs in a purely monolayer fashion. The latter case is easily realized by setting the requirement in the computer program that $\phi_{1l} = \phi_1$ for $l \geq 2$. Our model calculation suggests that, except solutions showing small positive deviations from Raoult's law, the formation of the second and higher layers cannot generally be neglected.

The excess adsorption isotherms can be measured in a simple static experimental method, usually called the batch method, in which a certain amount number of moles n_t of the liquid mixture 1 + 2 is brought into contact with a certain mass of adsorbent. Further, let x_1^0 and x_1 denote the concentrations of solvent 1 before and after contact with the adsorbent, respectively. These concentrations are usually measured refractometrically, interferometrically, or by means of gas chromatography. The measured surface excess n_1^e is then defined as follows [3]:

$$n_1^e = n_t(x_1^0 - x_1) \tag{19}$$

From the material balance condition it follows that

$$n_1^e = n_1 - (n_1 + n_2)x_1 \tag{20}$$

where n_1 and n_2 are the numbers of moles of solvents 1 and 2 in the static adsorption system. Equation (20) can then be rewritten as

$$n_1^e = \frac{N_0}{N} \sum_{l \geq 1} \left[\frac{\phi_{1l}}{r_1} - \left(\frac{\phi_{1l}}{r_1} + \frac{\phi_{2l}}{r_2} \right) x_1 \right] \tag{21}$$

where N is Avogadro's number. Let $n_1^0 = N_0/Nr_1$ be the surface capacity of solvent 1 (i.e., the maximum number of moles of solvent 1 that can be accomodated on the solid surface). Equation (21) can then be written

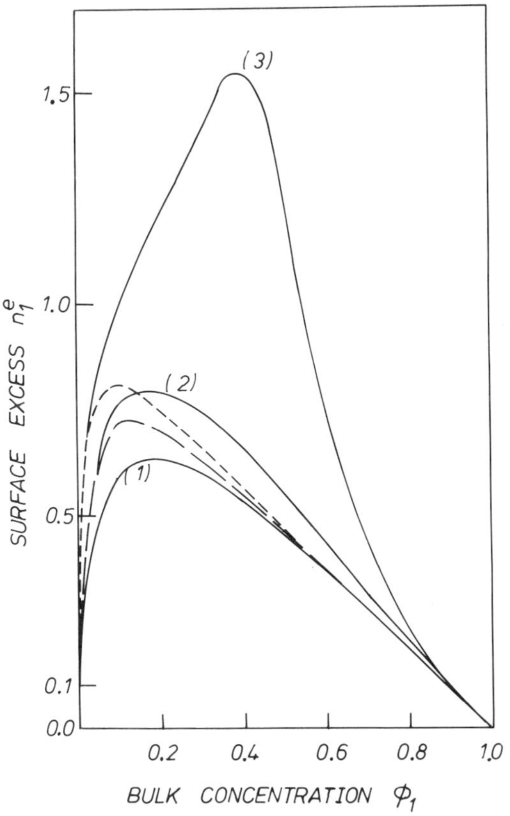

Figure 2. Illustration of the errors made by the incorrect assumption that the adsorption of solvents occurs in a purely monolayer fashion. The solid lines denote n_1^e functions evaluated by assuming a multilayer adsorption model for various values of the nonideality parameter A_{12}: (1) 0.0, (2) 1.0, (3) 1.95. The broken lines denote the n_1^e functions obtained by assuming that adsorption occurs in a purely monolayer fashion. The slightly broken line is for $A_{12} = 1.0$, whereas the strongly broken line is for $A_{12} = 1.95$. The parameters N_0 and ε_{12} are the same as in Fig. 1.

$$n_1^e = n_1^0 \sum_{l \geq 1} \{\phi_{1l} - [r_{12} + (1 - r_{12})\phi_{1l}]x_1\} \tag{22}$$

Clearly,

$$n_1^e + n_2^e = 0 \tag{23}$$

By best fitting Eqs. (15) and (22) to the experimental data for n_1^e, one can numerically extract the adsorption parameters ε, r_{12}, and n_1 and find the concentration profile of the solvents. Such calculations have not yet been performed for the correct three-dimensional adsorption model considered here. This is because they require an advanced computer program. However, they have often been done in basic research on solution adsorption onto solid surfaces, under the simplified assumption that the adsorption occurs in a monolayer fashion [4].

The retention mechanism in LSC is generally a very complicated phenomenon. However, in the hitherto basic research into this problem, the possibility of extracting independent information about the competitive adsorption of solvents has not been sufficiently recognized.

The first attempts along these lines were made by Scott and Kucera [5] and others by Poppe and co-workers [6]. The paper by Slaats et al. [7] contributes an extensive experimental and theoretical study of this problem in LSC. In their paper, extant methods for the experimental determination of excess isotherms are compared and a new one is proposed.

The investigational experience of these authors leads them to the conclusion that the batch method is not very attractive in LSC because the high concentrations lead to a very small difference in the concentration before and after adsorption. These authors also argue that these concentrations are difficult to measure. However, they do not consider the new experimental ideas in this field, presented in the recent work of Everett and co-workers [8]. Their new "high-precision apparatus" yields results that are about 10 times more precise than those obtained by the conventional batch technique. Moreover, the excess isotherm can be measured over a wide temperature range in a single run.

Unfortunately, Slaats et al. [7] did not take into consideration the breakthrough method described in the earlier papers of Poppe and co-workers [6]. In this method, the composition of the mobile phase with respect to one of the components is gradually changed. For the solvent that passes through the column between the solvent change and the step in the recorder output, the amount of solvent 1 adsorbed can be calculated. Thus, having the retention volume V_{1b}^R for the breakthrough from zero to the concentration ϕ_1, the number of moles of component 1 in the column, n_1, can be calculated

$$n_1 = \frac{V_{1b}^R \phi_1}{v_1} \tag{24}$$

where v_1 is the molar volume of component 1 in the bulk solution. The surface excess n_1^e can be calculated as

$$n_1^e = (v_{1b}^R - V_M)\frac{\phi_1}{v_1} \tag{25}$$

where V_M is defined as the total volume of the system, excluding the volume of the adsorbent. There are various ways of estimating V_M, which is of great importance in both practical and theoretical problems in LSC. These are helium pycnometry, calculation of the volume of the packing and subtracting it from the total volume of the column tube, or pycnometry with a liquid. In addition, there are other methods that will be discussed shortly (see also page 35).

Using the breakthrough method, Slaats et al. have found that 2-propanol and ethyl acetate are adsorbed in a multilayer fashion from their liquid mixtures with *n*-heptane onto a silica gel. Their measurements confirm the earlier experimental findings reported by Scott and Kucera.

In general, however, Slaats et al. discard the breakthrough method as less practical and less precise than another method they propose. This is the "minor disturbance" method, in which a small disturbance in the solvent composition is made at the top of the column. With the retention volume resulting from this eluent disturbance, V_{1d}^R, as measured for the entire concentration range of component 1, it is possible to calculate the excess isotherm n_1^e by numerical integration of the differential equation

$$\frac{\partial n_1^e}{\partial \phi_1} = (V_{1d}^R - V_M) \tag{26}$$

by Huber and Gerritse [9]. This equation (see also pages 83 and 108) can be rearranged into

$$n_1^e = \int_0^{x_1} (V_{1d}^R(x_1) - V_M)\frac{d\phi_1}{dx_1}\, dx_1 \tag{27}$$

which is convenient when molar fractions are used. The derivative $d\phi_1/dx_1$ can be calculated by assuming the additivity of the volumes or can be taken from the literature for many solvent mixtures.

In the experiment performed by Slaats et al., the column volume was determined by pycnometry using methanol, acetonitrile, and *n*-heptane as solvents. Then, the excess isotherms for the binary mixtures consisting of acetonitrile + water and methanol + water were measured on two reversed-phase packings. All these isotherms showed an azeotropic point at high concentrations of acetonitrile and methanol.

Similar experiments have been performed at the Institute of Analytical Chemistry in Vienna, according to a slightly different concept in which the volume V_M was estimated according to the relation that results from the definition of n_1^e and Eq. (27) [10]:

$$\lim_{\phi_1 \to 0} \int_0^{\phi_1} V_{1d}^R \, d\phi_1 = V_M \tag{28}$$

Adsorption from two binary mixtures, methanol + water and ethanol + water, onto RPS was measured. The excess isotherms measured in this way also showed azeotropic points at high concentrations of methanol and ethanol. The volume V_M estimated in this way agreed within 15% with its value estimated in a static experiment using various solutes (phenol, benzene, haphthalene, and acenaphthene).

Since these solutes were injected in very small concentrations, their adsorption is assumed to correspond to Henry's law region, where the retention equation (26) takes the simple form

$$V_R = V_M + V_s K \tag{29}$$

where V_R is the total retention volume, K the partition coefficient of solute between the mobile and surface phase, and V_s the volume of the surface phase. It is reasonable to assume that, at very small concentrations of the solute, its adsorption onto the solid surface has a purely monolayer character. The volume of the surface phase V_s is then simply the volume associated with a complete monolayer of the solute on the solid surface. When the dimensions of the solute are not drastically different from those of the solvent molecules, V_s is simply the volume associated with one lattice plane considered in our model of the solid/solution interface. Changing the solutes causes a change in the partition coefficient K in Eq. (29), but the value of V_M remains constant. For various solutes, therefore, the plot of V_R against K should be a straight line with a slope equal to V_s and intercept equal to V_M.

Equation (27) has recently been applied by Busev et al. [11] to obtain the excess isotherm n_1^e from a single elution curve in a single chromatographic experiment. This method is based on the assumption that the broadening effects due to diffusion and mass transfer are absent. From the data reported by Busev et al. [11], it follows that the accuracy of determining n_1^e in this method is comparable to that achieved in the traditional batch experiment.

Finally, let us mention the entire class of methods for determining n_1^e, in which isotopically labeled components are applied. Slaats et al. have recently shown that these methods generally yield slightly different values of n_1^e, than those obtained by the minor disturbance method. This is because, in some LSC systems, the isotopically labeled compounds may interact in a slightly different way with the solid surface.

Interesting results along these lines have recently been obtained by Selim et al. [12]. They have shown that the adsorption of acetone from *n*-hexadecane onto Chromosorb PAW occurs in a multilayer fashion. Wahlund and Beijersten [13] have shown that a simlar multilayer adsorption takes place in the case of 1-pentanol from a phosphate buffer onto Lichrosorb RP-8.

B. Effects of Surface Heterogeneity in Adsorption from Binary Solvent Mixtures on Real Solid Surfaces

The local variations in stoichiometry and crystallography of real solid surfaces cause various surface areas to exhibit different adsorptive properties with respect to molecules of solvents 1 and 2. These different adsorptive properties will first manifest themselves as a variation in both ε_1 and ε_2 across the surface. This variation will be accompanied by a relatively weaker variation in q_1 and q_2. In the literature on adsorption, this phenomenon has long been known as the "energetic heterogeneity" of the solid surface.

Hundreds of published papers have shown the importance of surface heterogeneity in the adsorption of gases on solid surfaces. An exhaustive discussion of the effects of this heterogeneity on various adsorption functions, including adsorption isotherms, isosteric heats of adsorption, and heat capacities, can be found in the recent paper by Rudziński and Jagiełło [14].

Some two or three dozen papers have been devoted to an exact determination of the dispersion of the adsorption energy of the single components ε_1 and ε_2 from experimental gaseous adsorption isotherms. The commonly accepted, quantitative measure of this dispersion is the differential distribution of the number of adsorption sites among various values of the adsorption energy. In most papers devoted to this problem, this differential distribution is denoted $\chi(\varepsilon)$ and simply called the adsorption energy distribution.

Ever more theoretically advanced and numerically sophisticated methods are being published today on the evaluation of $\chi(\varepsilon)$ from various experimental data—largely from experimental adsorption isotherms. Here we will give some unpublished data obtained by using the theoretical-numerical method developed recently by Rudziński et al. [15]. Our calculation has been made for two adsorption systems formed by nitrogen adsorbed on two different silica samples prepared for chromatographic use. The result of our calculation is shown in Fig. 3. It is usually assumed that there are two kinds of surface sites on silica surfaces, "free" and "bonded" hydroxyls. They probably correspond to the two distinct peaks in Fig. 3.

Our analysis suggests that previous views about the adsorption properties of silica should be revised to some extent; describing only two distinct kinds of adsorption sites fails to reflect the broad spectrum of adsorptive properties found on silica surfaces.

In the case of adsorption from a binary liquid mixture 1 + 2, surface heterogeneity will manifest itself through the dispersion of the difference ε_{12} of the adsorption energies of single components.

It is to be expected that, when solvents 1 and 2 are chemically similar, their adsorption energies related to various surface areas will be highly correlated. This means that ε_1 and ε_2 will change in a similar way when moving from one to another surface area. In such a case, the adsorption from a binary liquid mixture 1 + 2 will be like that on a homogeneous solid surface.

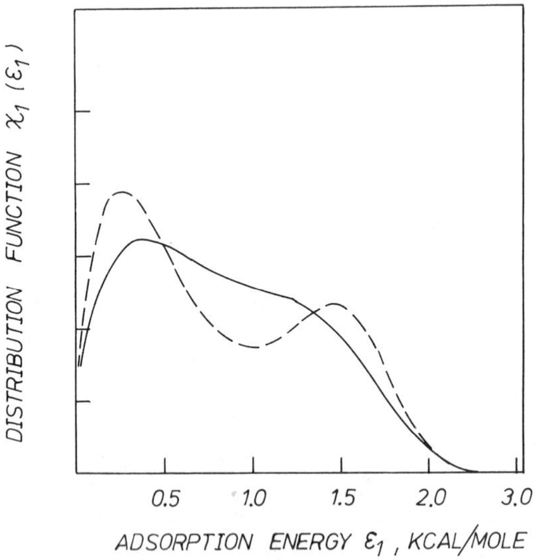

Figure 3. The distribution of the adsorption energy $\chi_1(\varepsilon_1)$ for nitrogen adsorbed on two different silica samples. (Adsorption data obtained courtesy of Prof. K. Unger of the Gutenberg University, Mainz.) The scale for the adsorption energy ε_1 was chosen by assuming that the lowest adsorption energy ε_1^l is equal to zero. This is because only the shape of the distribution function $\chi_1(\varepsilon_1)$ can be determined with the method developed by Rudziński et al. [15] from the adsorption isotherm measured at a single temperature.

It often happens that, although the two solvents 1 and 2 are chemically different, they have some regions that are chemically similar, such as an aliphatic chain. Thus, the dispersion of ε_{12} can generally be expected to be smaller than the dispersion of ε_1 and ε_2 considered separately.

Ościk et al. [16] have analyzed the adsorption data for benzene and cyclohexane on silica, reported by Sircar and Myers [17] and Minka and Myers [18]. These data contain two gaseous isotherms for the adsorption of benzene and cyclohexane from the vapor phase of each, as well as the excess adsorption isotherm for their liquid mixture. The benzene + cyclohexane mixture is very useful for investigation of a theoretical model. While being adsorbed, both the benzene and the cyclohexane molecules collapse onto the solid surface and then occupy an equal surface area of about 30 Å^2. Furthermore, the benzene + cyclohexane mixture shows only slight positive deviations from Raoult's law ($A_{12} \cong 0.5$ at room temperature). Thus, both nonideality and multilayer adsorption effects can practically be neglected in this system. As a result, the calculation of $\chi_1(\varepsilon_1)$, $\chi_2(\varepsilon_2)$, and $\chi_{12}(\varepsilon_{12})$ becomes especially simple. As a result, a curve is obtained for χ_{12} that has two peaks, which are only slightly marked. Thus, in a first approximation, χ_{12} can be determined by a gaussian-like function. In the case of adsorption onto a heterogeneous solid surface from a binary liquid mixture, the concentration profile $\{\phi_{11}^{(h)}\}$ can be evaluated by applying an additional averaging:

$$\phi_{11}^{(h)} = \int_{\Omega_{12}} \phi_{11}(\varepsilon_{12})\chi_{12}(\varepsilon_{12})\, d\varepsilon_{12} \tag{30}$$

where $\phi_{11}(\varepsilon_{12})$ is the solution of equation system (15) for a certain value ε_{12}, whereas Ω_{12} is the physical domain of ε_{12}. For the sake of mathematical convenience, it is often assumed in theories of gas adsorption on heterogeneous solid surfaces that the single adsorption energies ε_1 and ε_2 may vary from zero to plus infinity. A direct consequence of this is the assumption that Ω_{12} is an interval $(-\infty, +\infty)$. Such an assumption is made, in fact, in most papers treating solution adsorption on heterogengeous solid surfaces.

The result of the integration in Eq. (30) depends not only on the form of the energy distribution $\chi_{12}(\varepsilon_{12})$ but also on the topography of a heterogeneous solid surface. Two extreme kinds of surface topography have been considered in our hitherto theoretical investigation. The first is the patchwise model first formulated by Ross and Olivier [19]. It has been assumed in this model that adsorption sites having an equal adsorption energy are grouped into large patches. These patches are sufficiently large that the interactions between admolecules on different or neighbouring patches can be neglected in calculating the state of an adsorption system. Therefore, the adsorption system can be considered a collection of independent subsystems that are in only thermal and material contact. The surface activity coefficients must therefore be defined locally for every patch and will be functions of the concentration profile on each patch.

The other extreme model of surface topography is the random model first formulated by Hill [20]. It is assumed in this model that no spatial correlations exist between adsorption sites having an equal adsorption energy. In other words, adsorption sites having various adsorption energies are distributed completely at random over a heterogeneous surface. In effect, any local concentration in the same layer will be the same throughout the entire solid surface and will be equal to the average concentration in this layer. The surface activity coefficients will therefore have the same form as for a homogeneous solid surface, except that the concentrations ϕ_{11} must be replaced by their averaged values $\phi_{11}^{(h)}$.

It seems that most real solid surfaces should exhibit the random topography. This is because the lack of an energetic order across the solid surface is a consequence of a loss of order in the chemical and crystallographic sense. Graphitized carbons are the solids on which the "patchwise" topography should, on the other hand, give the correct picture of surface topography. Here the patches are identified with graphite domains. The finite dimensions of these graphite domains are sources of long-range longitudinal potential fields whose strength depends on the dimension of a graphite domain. Because of the dispersion of these dimensions in real samples, there will also be a patchwise dispersion of the adsorption energies. Although attempts have been made to apply graphitized carbons in LSC, their weak mechanical properties currently limit their wider application. For this reason, we will focus mainly on surfaces exhibiting random surface topography.

Let us first consider the simple case of adsorption from a binary liquid mixture of molecules of equal size onto a heterogeneous surface exhibiting random surface topography. Then only ϕ_{11} is an explicit function of ε_{12}, and the averaging in Eq. (30) is limited to the first adsorbed layer. Denoting, for simplicity, $\phi_{11}^{(h)}$ by ϕ_{1t}, we write Eq. (30) in the form

$$\phi_{1t} = \int_{-\infty}^{+\infty} \left[1 + e^{(\varepsilon_{12}^{c}-\varepsilon_{12})/RT}\right]^{-1} \chi_{12}\, d\varepsilon_{12} \tag{31}$$

where

$$\varepsilon_{12}^{c} = -RT \ln \frac{\phi_1\gamma_1\gamma_{21}}{\phi_2\gamma_2\gamma_{11}} \tag{32}$$

By simple transformation of the variables ε_{12} and ε_{12}^{c}, one can rewrite Eq. (31) into a Stieltjes transform, as was first shown by Rudziński et al. [21]. Then, using the methods of the Stieltjes transform and assuming that χ_{12} is the gaussian-like function

$$\chi_{12} = \frac{\sin (RT/c_{12})/RTe^{(\varepsilon_{12}'-\varepsilon_{12})/c_{12}}}{1 + 2\cos (RT/c_{12})e^{(\varepsilon_{12}'-\varepsilon_{12})/c_{12}} + e^{2(\varepsilon_{12}'-\varepsilon_{12})/c_{12}}} \tag{33}$$

centered about $\varepsilon_{12} = \varepsilon_{12}'$, one arrives at

$$\phi_{1t} = \frac{K_{12}^{0}(\phi_1\gamma_1\gamma_{21}/\phi_2\gamma_2\gamma_{11})^{m_{12}}}{1 + K_{12}^{0}(\phi_1\gamma_1\gamma_{21}/\phi_2\gamma_2\gamma_{11})^{m_{12}}} \tag{34}$$

where

$$K_{12}^{0} = e^{m_{12}\varepsilon_{12}'/RT} \quad \text{and} \quad m_{12} = \frac{RT}{c_{12}} < 1 \tag{35}$$

In the limit $m_{12} \to 1$, Eq. (34) reduces to the adsorption isotherm $\phi_{11}(\varepsilon_{12}')$, which describes adsorption on a hypothetical homogeneous surface characterized by the adsorption energy ε_{12}'. The energy distribution function (33) then approximates the Dirac delta distribution $\delta(\varepsilon_{12} - \varepsilon_{12}')$ but is not exactly this function. This is because of certain approximations used in establishing the relation between the energy distribution [Eq. (33)] and the averaged adsorption isotherm (34), when using the method of Stieltjes transform. An even more drastic limitation is that the method of the Stieltjes transform cannot be used to describe the adsorption from binary mixtures composed of molecules of different sizes (i.e., when $r_{12} \neq 1$). We will investigate a particular solution for this problem on page 263.

Let us note that, for small values of m_{12} related to a strong dispersion of ε_{12}, the distribution function (33) reduces to

$$\chi_{12} = \frac{(1/c_{12})e^{(\varepsilon'_{12}-\varepsilon_{12})/c_{12}}}{[1 + e^{(\varepsilon'_{12}-\varepsilon_{12})/c_{12}}]^2} \tag{36}$$

In Fig. 4, the shape of the function (36) has been compared with the function χ_{12} evaluated by Ościk et al. [16] for the benzene + cyclohexane mixture adsorbed on silica gel. An integration by parts of Eq. (30) yields

$$\phi_{1t} = \phi_{11}\mathfrak{X}_{12}\Big|_{-\infty}^{+\infty} - \int_{-\infty}^{+\infty} \left(\frac{\partial\phi_{11}}{\partial\varepsilon_{12}}\right)\mathfrak{X}_{12}\, d\varepsilon_{12} \tag{37}$$

where

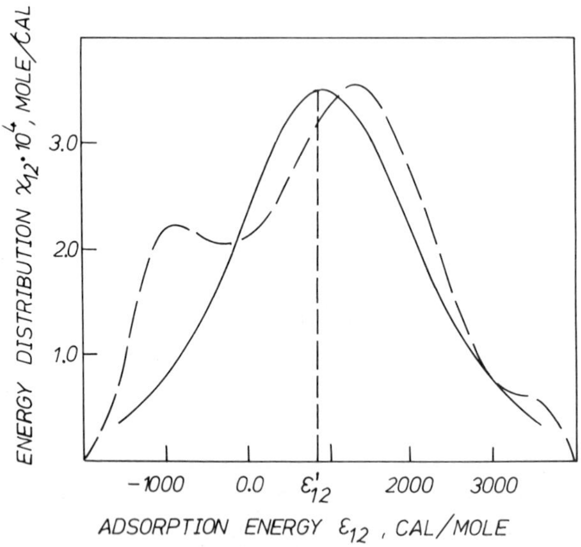

Figure 4. A comparison of the energy distribution functions χ_{12} evaluated by Ościk et al. [16] (broken line) and Rudziński and Partyka [23] (solid line) for the adsorption of a (benzene + cyclohexane) liquid mixture on two different silica samples. The Rudziński-Partyka distribution function was approximated by Eq. (36), and the Ościk distribution function was shifted on the energy scale so that it covers a similar range of adsorption energies. The ratio q_{11}/q_{21} was chosen arbitrarily by Ościk et al.

$$\mathfrak{X}_{12}(\varepsilon_{12}) = \int \chi_{12}(\varepsilon_{12})\, d\,\varepsilon_{12} \tag{38}$$

Because of the infinite integration limits, the first term on the right side of Eq. (37) disappears. The second term can be easily evaluated when m_{12} is small. The function $\mathfrak{X}_{12}$ then takes the form

$$\mathfrak{X}_{12} = \left[1 + e^{m_{12}(\varepsilon'_{12}-\varepsilon_{12})/RT}\right]^{-1} \tag{39}$$

whereas the derivative $\partial\phi_{11}/\partial\varepsilon_{12}$ has the form

$$\frac{\partial\phi_{11}}{\partial\varepsilon_{12}} = \frac{e^{(\varepsilon^c_{12}-\varepsilon_{12})/RT}}{[1 + e^{(\varepsilon^c_{12}-\varepsilon_{12})/RT}]^2} \tag{40}$$

As m_{12} becomes increasingly smaller than unity, the spread of χ_{12} becomes more and more larger than the spread of the derivative $(\partial\phi_{11}/\partial\varepsilon_{12})$. This derivative then behaves as a "sampling" function with respect to χ_{12} during the integration in Eq. (37). The second integral on the right side of Eq. (37) can therefore be evaluated effectively by expanding $\mathfrak{X}_{12}$ into its Taylor series around the point $\varepsilon_{12} = \varepsilon^c_{12}$, at which point the sampling function (derivative) reaches its maximum. Doing so, we obtain

$$\phi_{1t} = -\mathfrak{X}_{12}(\varepsilon^c_{12}) - \sum_{n\geq 1} (\pi RT)^{2n} B_n \left(\frac{\partial^{2n}\mathfrak{X}_{12}}{\partial\varepsilon^{2n}}\right)_{\varepsilon^c_{12}} \tag{41}$$

where B_n is Bernoulli's number. Looking to Eq. (39), we can see that the first term on the right side of Eq. (41) is the averaged isotherm ϕ_{1t} from Eq. (34). The other terms under the sum in Eq. (41) are correction terms that can be neglected when m_{12} is small.

In the more general case in which $r_{12} \neq 1$, the derivative $\partial\phi_{11}/\partial\varepsilon_{12}$ reaches its maximum at the point ε^r_{12}, which is found from the condition

$$\left(\frac{\partial^2\phi_{11}}{\partial\varepsilon^2_{12}}\right)_{\varepsilon^r_{12}} = 0 \tag{42}$$

After performing the differentiation, one arrives at

$$\frac{\partial^2 \phi_{11}}{\partial \varepsilon_{12}^2} = \left(\frac{1}{RT}\right)^2 \frac{\phi_{11}\phi_{21}[(\phi_{21})^2 - r_{12}(\phi_{11})^2]}{(\phi_{21} + r_{12}\phi_{11})^3} \tag{43}$$

From Eqs. (42) and (43) we obtain the following generalization of ε_{12}^c for the case $r_{12} \neq 1$:

$$\varepsilon_{12}^r = RT \ln \frac{(1 + \sqrt{r_{12}})^{r_{12}-1}}{(\sqrt{r_{12}})^{r_{12}}} - RT \ln \frac{\phi_1\gamma_1/\gamma_{11}}{(\phi_2\gamma_2/\gamma_{21})^{r_{12}}} \tag{44}$$

Neglecting the correction terms in Eq. (41), we arrive at the following generalization of the Eq. (34) for the case in which $r_{12} \neq 1$:

$$\phi_{1t} = \frac{K_{12}^r[\phi_1\gamma_1/\gamma_{11}/(\phi_2\gamma_2/\gamma_{21})^{r_{12}}]^{m_{12}}}{1 + K_{12}^r[\phi_1\gamma_1/\gamma_{11}/(\phi_2\gamma_2/\gamma_{21})^{r_{12}}]^{m_{12}}} \tag{45a}$$

where

$$K_{12}^r = \left[\frac{(1 + \sqrt{r_{12}})^{r_{12}-1}}{(\sqrt{r_{12}})^{r_{12}}}\right]^{m_{12}} e^{m_{12}\,\varepsilon'_{12}/RT} \quad \text{and} \quad m_{12} < 1 \tag{45b}$$

Neglecting the nonideality of the adsorbed phase, one can rewrite Eq. (45a) into the compact form

$$\frac{\phi_{1t}}{\phi_{2t}} = \left[\frac{\phi_1\gamma_1}{(\phi_2\gamma_2)^{r_{12}}}\right]^{m_{12}} \left[\frac{(1 + \sqrt{r_{12}})^{r_{12}-1}}{(\sqrt{r_{12}})^{r_{12}}}\right]^{m_{12}} e^{m_{12}\,\varepsilon'_{12}/RT} \tag{46}$$

Let us note here that Eq. (45) is different from the intuitive generalization of Eq. (34) proposed by Jaroniec [22]:

$$\frac{\phi_{1t}}{(\phi_{2t})^{r_{12}}} = \left[\frac{\phi_1\gamma_1}{(\phi_2\gamma_2)^{r_{12}}}\right]^{m_{12}} e^{\varepsilon'_{12}/RT} \tag{47}$$

When $m_{12} \to 1$, our generalization (45) of Eq. (34) becomes less accurate and the intuitive generalization (47) then seems to be more useful.

However, a still more rigorous generalization of Eq. (34) is possible by retaining the correction terms in Eq. (41). By retaining only the first correction term in Eq. (41), we arrive at the following generalization of Eq. (34):

$$\phi_{1t}^{(1)} = -\mathfrak{X}_{12}(\varepsilon_{12}^{r}) - \frac{(\pi RT)^2}{6}\left(\frac{\partial \chi_{12}}{\partial \varepsilon_{12}}\right)_{\varepsilon_{12}^{r}} \tag{48}$$

where the superscript (1) denotes a "first-order" approximation for ϕ_{1t} obtained by retaining the first correction under the sum in Eq. (41). Let $\phi_{1t}^{(0)}$ further denote the zero-order approximation given by Eq. (45). Then Eq. (48) can be rewritten in the form

$$\phi_{1t}^{(1)} = \phi_{1t}^{(0)} + \frac{(\pi m_{12})^2}{6}\ \phi_{1t}^{(0)}(1 + \phi_{1t}^{(0)})(1 + 2\phi_{1t}^{0}) \tag{49}$$

Let us note that Eq. (49) will not work at high concentrations of the first solvent, since ϕ_{1t} evaluated from Eq. (49) may then exceed unity. This is a result of the mathematical approximation used to evaluate ϕ_{1t} and will be discussed in more detail in the forthcoming sections. However, this limitation should not cause trouble in the practical use of Eq. (49), because the preferentially adsorbed solvent 1 will usually play the role of the "moderator," having a greater elution strength. As such, in chromatographic practice it is used in small or moderate concentrations.

The most serious limitation to the use of Eqs. (34), (45), and (49) is that they cannot be used to describe adsorption in systems in which the dispersion of ε_{12} is small (i.e., $m_{12} > 1$). Such systems may be found in reversed-phase chromatography. There the chemical modification of the solid surface may lead to relatively homogeneous surfaces. For such systems, Rudziński and Partyka [23] have recently proposed another approximation for ϕ_{1t}.

When $m_{12} > 1$, the situation becomes the reverse of that considered previously. The distribution function χ_{12} from Eq. (36) behaves like a sampling function with respect to ϕ_{11} during the integration in Eq. (31). Thus, a simple and effective evaluation of ϕ_{1t} can be made by expanding $\phi_{11}(\varepsilon_{12})$ in Eq. (30) into its Taylor series around the point $\varepsilon_{12} = \varepsilon_{12}'$, at which the distribution function χ_{12} reaches its maximum. Doing so and retaining only the first correction, we arrive at the following expression for ϕ_{1t}. For random surface topography we have

$$\phi_{1t} = \phi_{10} + \frac{1}{6}\left(\frac{\pi}{m_{12}}\right)^2 \frac{\phi_{10}\phi_{20}[(\phi_{20})^2 - r_{12}(\phi_{10})^2]}{(\phi_{20} + r_{12}\phi_{10})^3} \tag{50}$$

where ϕ_{10} and ϕ_{20} are the solutions for ϕ_{11} and ϕ_{21} obtained from Eq. (15) by setting $\varepsilon_{12} = \varepsilon_{12}^0$. For patchwise surface topography, this calculation becomes much more complicated, but it has been published elsewhere by Rudziński et al. [24]. Assuming that adsorption occurs in a purely monolayer fashion, one arrives at a relatively simple equation for ϕ_{1t} in the patchwise topography:

$$\phi_{1t} = \phi_{10} + \frac{1}{6}\left(\frac{\pi}{m_{12}}\right)^2 \frac{1 - 2\phi_{10} + (1 - r_{12})(\phi_{10})^2}{(\phi_{10})^2(\phi_{20})^2} \times \left[\frac{1 + (r_{12} - 1)\phi_{10}}{\phi_{10}\phi_{20}} - 2A_{12}f_p\right]^{-3} \tag{51}$$

Because of their complexity, we will postpone writing the explicit forms of the appropriate equations for multilayer adsorption and patchwise surface topography.

Equation (34) has been successfully applied by various authors to describe adsorption from binary liquid mixtures of molecules of similar size, assuming that this adsorption occurs in a purely monolayer fashion [25]. Since the binary mixtures investigated by these authors did not usually exhibit strong deviations from Raoult's law, the assumption of monolayer adsorption should be a source of negligible error only. Equation (50), on the other hand, has been used successfully by Rudziński and Partyka [23] to describe the behavior of excess adsorption isotherms and related heats of immersion in the adsorption from benzene + cyclohexane mixtures on silica gel that has been reported by other authors. It therefore seems reasonable to assume that the gaussian-like energy distributions (33) and (36) well represent the dispersion of ε_{12} in the majority of actual adsorption systems.

Dabrowski and Jaroniec [25] argue that in some cases the expanded gaussian distribution

$$\chi_{12}(\varepsilon_{12}) = \frac{2}{b_{12}}(\varepsilon_{12} - \varepsilon_{12}^1)e^{[-(\varepsilon_{12}-\varepsilon_{12}^1)^2/b_{12}]} \tag{52}$$

may better approximate the actual dispersion of ε_{12}. Here, ε_{12}^1 is the lowest value of ε_{12} in an adsorption system investigated, whereas b_{12} is the appropriate heterogeneity parameter describing the spread of the energy distribution function. When $b_{12} \to 0$, the energy distribution (52) degenerates into the Dirac delta distribution $\delta(\varepsilon_{12} - \varepsilon_{12}^{1})$. When the spread of function (52) is sufficiently large (i.e., the surface is relatively strongly heterogeneous) then from Eq. (41) we obtain

$$\phi_{1t} = e^{[-(\varepsilon_{12}^{r} - \varepsilon_{12}^{1})^2 / b_{12}]} \tag{53}$$

This adsorption equation can be rewritten in the following linear form with respect to $\ln [\phi_1/(\phi_2)^{r_{12}}]$:

$$\left[\ln \frac{1}{\phi_{1t}}\right]^{1/2} = C_{12}^{(b)} - \frac{RT}{(b_{12})^{1/2}} \ln \frac{\phi_1}{(\phi_2)^{r_{12}}} \tag{54}$$

where

$$C_{12}^{(b)} = - \frac{1}{(b_{12})^{1/2}} \left[\varepsilon_{12}^{1} + RT \ln \frac{(r_{12})^{1/2} \gamma_1 \gamma_{21}}{(\gamma_2 \gamma_{11})^{r_{12}}}\right] \tag{55}$$

Let us note that Eq. (45) can also be rewritten in the appropriate linear form with respect to $\ln [\phi_1/(\phi_2)^{r_{12}}]$:

$$\ln \frac{\phi_{1t}}{\phi_{2t}} = C_{12}^{(m)} - m_{12} \ln \frac{\phi_1}{(\phi_2)^{r_{12}}} \tag{56}$$

where

$$C_{12}^{(m)} = -m_{12} \left[\frac{\varepsilon_{12}'}{RT} + \ln \frac{(r_{12})^{1/2} \gamma_1 \gamma_{21}}{(\gamma_2 \gamma_{11})^{r_{12}}}\right] \tag{57}$$

Assuming that $r_{12} = 1$, Eq. (56) is a linear form of Eq. (34), which has been used successfully in many works to correlate experimentally determined data for excess adsorption isotherms. Figure 5 shows an example of these correlations. A recent work by Rudziński et al. [26] shows that, while performing these linear regressions, the differences between the surface areas occupied by molecules of components 1 and 2 should be carefully taken into account.

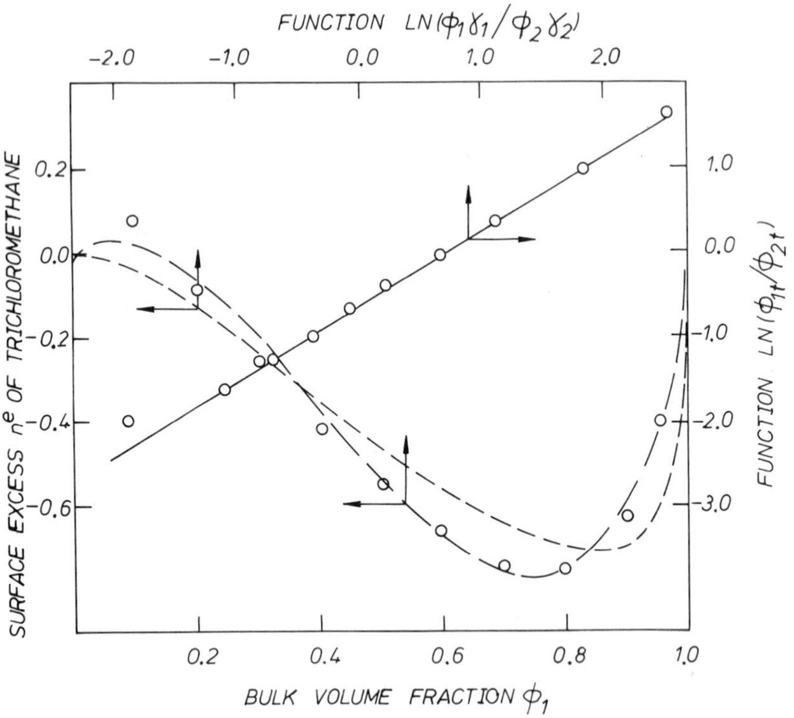

Figure 5. Adsorption of chloroform from a (chloroform + acetone) liquid mixture onto charcoal. (Data obtained by courtesy of Dr. A. Dabrowski of the Institute of Chemistry, UMCS University, Lublin.) The open circles denote the experimental excess isotherm of chloroform, and the strongly broken line is the best possible fit obtained by assuming a unimolecular adsorption on a homogeneous solid surface. The slightly broken line is the best fit obtained by including the effect of surface heterogeneity of charcoal, as in Eq. (34). The circles joined by the solid line show the same best fit but using the linear form of Eq. (34) represented by Eq. (56).

Figure 6 shows how surface heterogeneity affects multilayer adsorption on solid surfaces. Note that surface heterogeneity causes an azeotropic point to appear on the excess isotherm at high concentrations of the preferentially adsorbed solvent.

The approximations represented by Eqs. (50) and (51) and Eq. (41) are essentially different. In Eqs. (50) and (51), the effects of surface heterogeneity are treated as a kind of "perturbation" or a "deviation" from homogeneous surface behavior. In Eq. (41), on the other hand, it is assumed that the form of the energy distribu-

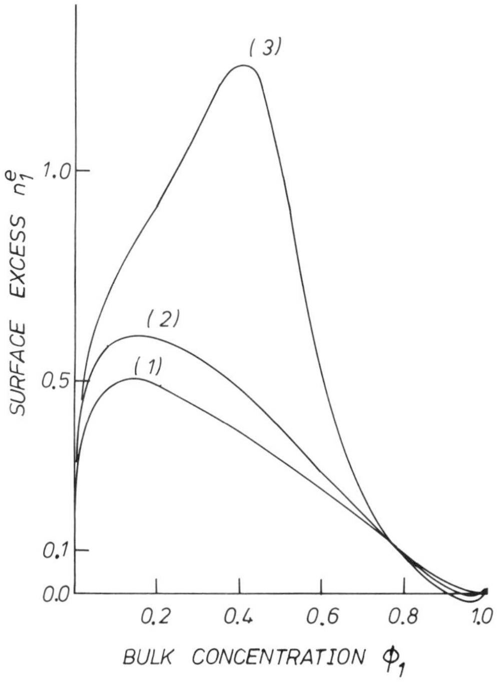

Figure 6. Effects of surface heterogeneity on competitive adsorption from a binary solvent mixture. The most probable adsorption energy ε'_{12} was taken equal to +3RT (i.e., equal to the value of ε'_{12} accepted in Fig. 2). Other parameters and the notation are also the same as in Fig. 2. The only difference lies in assuming that there exists a certain dispersion of the adsorption energy values, described by (36) and characterized by the heterogeneity parameter $m_{12} = 0.5$.

tion function χ is the prime factor governing the behavior of solid/solution interface systems. Other properties of such systems are transmitted only through the form of the function ε^r_{12} into the expression $\mathfrak{X}(\varepsilon^r_{12})$.

There is still another important difference between these two approaches to the problem of surface heterogeneity in solid/solution adsorption. This difference is connected with the behavior of ϕ_{1t} in the limit $\phi_1 \rightarrow 0$. In this case, it can be easily ascertained that the ϕ_{1t} from Eqs. (50) and (51) reduce correctly to Henry's law, in agreement with appropriate basic thermodynamic predictions. However, an inspection of Eq. (54) or (56) brings us to the conclusion that such a reduction does not take place in the case of the approach expressed by Eq. (41), which is to be applied to strongly heterogeneous surfaces. This has not been sufficiently recognized in the hitherto theoretical investigation of adsorption from binary liquid mixtures onto heterogeneous solid surfaces.

However, it can easily be shown [27] that this absence of Henry's law region in the case of strongly heterogeneous surfaces is only apparent and is connected with accepting the nonphysical integration limits in Eq. (37). When using physically meaningful integration limits (ε_{12}^{l} and ε_{12}^{m}) in Eq. (37), we obtain

$$\phi_{1t} = \phi_1 \mathfrak{X} \Big|_{\varepsilon_{12}^{l}}^{\varepsilon_{12}^{m}} - \sum_{r \geq 0} \frac{A_n}{n!} \left(\frac{\partial^n \mathfrak{X}_{12}}{\partial \varepsilon_{12}^n} \right)_{\varepsilon_{12}^{r}} \tag{58}$$

where

$$A_n = \int_{\varepsilon_{12}^{l}}^{\varepsilon_{12}^{m}} \frac{\partial \phi_1}{\partial \varepsilon_{12}} (\varepsilon_{12} - \varepsilon_{12}^{r})\, d\varepsilon_{12} \tag{59}$$

Using the same approximation as in Eq. (41) (i.e., retaining only the first term under the sum in Eq. (58), we arrive at the following expression for ϕ_{1t}:

$$\begin{aligned} \phi_{1t} &= \phi_1(\varepsilon_{12}^{m})\mathfrak{X}_{12}(\varepsilon_{12}^{m}) - \phi_1(\varepsilon_{12}^{l})\mathfrak{X}_{12}(\varepsilon_{12}^{l}) \\ &\quad - \mathfrak{X}_{12}(\varepsilon_{12}^{r})[\phi_1(\varepsilon_{12}^{r}) - \phi_1(\varepsilon_{12}^{l})] \end{aligned} \tag{60}$$

Further, from our definition of $\mathfrak{X}_{12}$,

$$\chi_{12}(\varepsilon_{12}) = \begin{cases} 0 & \text{for} \quad \varepsilon_{12} < \varepsilon_{12}^{l} \\ \chi_{12} & \text{for} \quad \varepsilon_{12}^{l} \leq \varepsilon_{12} \leq \varepsilon_{12}^{m} \\ 0 & \text{for} \quad \varepsilon_{12} > \varepsilon_{12}^{m} \end{cases} \tag{61}$$

there follows the correct definition of $\mathfrak{X}_{12}$:

$$\mathfrak{X}_{12}(\varepsilon_{12}) = \begin{cases} \mathfrak{X}_{12}(\varepsilon_{12}^{l}) & \text{for} \quad \varepsilon_{12} < \varepsilon_{12}^{l} \\ \mathfrak{X}_{12}(\varepsilon_{12}) & \text{for} \quad \varepsilon_{12}^{l} \leq \varepsilon_{12} \leq \varepsilon_{12}^{m} \\ \mathfrak{X}_{12}(\varepsilon_{12}^{m}) & \text{for} \quad \varepsilon_{12} > \varepsilon_{12}^{m} \end{cases} \tag{62}$$

Then, from the normalization condition

$$\mathfrak{X}_{12}(\varepsilon_{12}^{m}) - \mathfrak{X}_{12}(\varepsilon_{12}^{l}) = 1 \tag{63}$$

it follows that the equation for ϕ_{1t} must be written as follows:

1. For $\varepsilon_{12}^{r} < \varepsilon_{12}^{l}$ (i.e., for high concentrations of solvent 1):

$$\phi_{1t} = \phi_1(\varepsilon_{12}^{m}) \tag{64a}$$

2. For $\varepsilon_{12}^{l} \leq \varepsilon_{12}^{r} \leq \varepsilon_{12}^{m}$ (i.e., for moderate concentrations of solvent 1):

$$\phi_{1t} = \phi_1(\varepsilon_{12}^{m})[\mathfrak{X}_{12}(\varepsilon_{12}^{m}) - \mathfrak{X}_{12}(\varepsilon_{12}^{r})] + \phi_1(\varepsilon_{12}^{l})[\mathfrak{X}_{12}(\varepsilon_{12}^{r}) - \mathfrak{X}_{12}(\varepsilon_{12}^{l})] \tag{64b}$$

3. For $\varepsilon_{12}^{r} > \varepsilon_{12}^{m}$ (i.e., for small concentrations of solvent 1):

$$\phi_{1t} = \phi_1(\varepsilon_{12}^{l}) \tag{64c}$$

Equation (64c) is a very important result for LSC systems with strongly heterogeneous solid surfaces. It shows that a species appearing in very small concentrations is adsorbed on a strongly heterogeneous surface just as on a homogeneous surface characterized by the smallest possible difference found on the heterogeneous surface between the adsorption energies of the pure solvents. Let us note here that, because of the small sample sizes of solute used in chromatographic practice, the concentration of solute molecules may always be considered very small.

Thus, among others, Eq. (64c) has an important consequence in the theoretical interpretation of the solvent strength ε_i^0 of a solvent i with respect to another solvent j. This solvent strength is defined by the following equation according to Snyder [28]:

$$\log \frac{k_j}{k_i} = a'\alpha_s(\varepsilon_i^0 - \varepsilon_j^0) \tag{65}$$

where k_i and k_j are the capacity ratios of solute s in solvents i and j. According to Snyder's interpretation, a' is an adsorbent activity function related to the ability of the adsorbent to interact with adjacent molecules of solute or solvent, and α_s is the area occupied by one solute admolecule on the adsorbent surface. According to our adsorption model,

$$k_i = K_{si}^{(1)} \frac{V_s}{V_M} r_s \tag{66}$$

and therefore,

$$\log \frac{k_j}{k_i} = -(\varepsilon_{si} - \varepsilon_{sj}) + \text{constant} \tag{67}$$

where the constant is related to solvent-solute interactions in the adsorbed phase, as well as to the different sizes of the solute and solvent molecules.

In a first crude approximation, the constant can be neglected and the value of $\log k_j/k_i$ gives an idea of how much more strongly solute s will be eluted from the column by solvent j compared with solvent i. Equations (64) and (67) provide us with the correct physical meaning of the elution strength in the case of strongly heterogeneous surfaces.

In the case of a homogeneous adsorbent surface, ε_{ij} may have still another interpretation: it can be related to the interfacial tensions between the solid and solvents j and i, σ_j^0, σ_i^0:

$$\varepsilon_{ij} = -\alpha(\sigma_i^0 - \sigma_j^0) \tag{68}$$

This interpretation makes a further useful approximation possible, as shown recently by Colin and Guiochon [29]. For the particular case of carbon adsorbents, Karger et al. [30] have proposed another useful interpretation based on the concept of solubility parameters.

C. Adsorption from Multicomponent Liquid Mixtures

The experimental and theoretical investigation of adsorption from multicomponent liquid mixtures onto solid surfaces represents a much more complicated problem than the case of adsorption from binary mixtures. It is therefore no surprise that there are few relatively advanced works concerning this problem. These are the papers by Ościk [31], Minka and Myers [18], Brown and Everett [32], and Borówko et al. [33,34].

The multilayer adsorption from an s-component liquid mixture onto a solid can be represented by the following set of quasi-chemical reactions:

$$r_s(i)^{(b)} + r_i(s)_l^{(s)} \rightleftarrows r_s(i)_l^{(s)} + r_i(s)^{(b)} \tag{69a}$$

where

$$l = 1, 2, \ldots, L \qquad \text{and} \qquad i = 1, 2, \ldots, s-1 \tag{69b}$$

In this notation, component s will be the reference substance. The related set of equations yielding the equilibrium conditions for these quasi-chemical reactions is

$$\frac{\phi_{il}\,\gamma_{il}}{\phi_i\,\gamma_i}\left(\frac{\phi_s\gamma_s}{\phi_{sl}\,\gamma_{sl}}\right)^{r_{is}} = K_{is}^{(l)} \tag{70}$$

When

$$r_{is} = 1 \qquad \text{for} \qquad i = 1, 2, \ldots, s-1 \tag{71}$$

is valid, then the equation system (70) reduces to the Langmuir-like form

$$\phi_{il} = \frac{K_{is}^{(l)}(\phi_i\gamma_i\gamma_{sl}/\phi_s\gamma_s\gamma_{il})}{1 + \sum_{j=1}^{s-1} K_{js}^{(l)}(\phi_j\gamma_j\gamma_{sl}/\phi_s\gamma_s\gamma_{jl})} \tag{72}$$

In the general case of $r_{is} \neq 1$, the surface activity coefficients may be evaluated by extending the Flory-Huggins approach outlined at the beginning of this chapter. For the particular case $r_{is} = 1$ and the model of monolayer adsorption, the explicit form of these activity coefficients may be found in the paper by Borówko [34].

Now let us consider the other physical factor affecting the adsorption, which is at least as important as the interactions between the adsorbed molecules. This is the energetic heterogeneity of the actual solid/solution interfaces. Very few papers have been published on the problem of surface heterogeneity in adsorption from multicomponent solutions onto heterogeneous solid surfaces [35,36].

Let $\chi_s(\{\varepsilon_{is}\})$ denote an (s − 1)-dimensional distribution of the adsorption energies ε_{is} normalized to unity:

$$\int_{\Omega_s} \cdots \int \chi_s(\{\varepsilon_{is}\})\, d\varepsilon_{1s} \cdots d\varepsilon_{(s-1)s} = 1 \tag{73}$$

where $\{\varepsilon_{is}\}$ is the set of the variables ε_{is} and Ω_s is the physical domain of these variables. The surface concentrations ϕ_{it} are then found by performing the additional averaging:

$$\phi_{it} = \int_{\Omega_s} \cdots \int \phi_i(\{\varepsilon_{is}\})\chi_s(\{\varepsilon_{is}\})\, d\varepsilon_{1s} \cdots d\varepsilon_{(s-1)s} \tag{74}$$

where $\phi_i(\{\varepsilon_{is}\})$ is the solution of equation system (72) for a particular set of values $\{\varepsilon_{is}\}$.

Experimental and theoretical investigation of adsorption from multicomponent liquid mixtures onto real heterogeneous solid surfaces is still at a very early stage. Consequently, no substantial information has been available until now about the general features of the multidimensional energy distribution $\chi_s(\{\varepsilon_{is}\})$. Only its partially integrated form,

$$\chi_{is}(\varepsilon_{is}) = \int_{\Omega_s} \cdots \int \chi_s(\{\varepsilon_{is}\})\, d\varepsilon_{1s} \cdots d\varepsilon_{(i-1)s}\, d\varepsilon_{(i+s)s} \times \cdots d\varepsilon_{(s-1)s} \tag{75}$$

has been investigated in theories of adsorption from binary mixtures onto heterogeneous solid surfaces. This lack of knowledge has been replaced by some simplifying assumptions that make the problem mathematically tractable.

Let us make a very crude assumption that the competitive adsorption between species i and s is affected mainly by the dispersion of energy ε_{is}. The dispersion of ε_{js} with $j \neq i$ plays only a secondary role and in a first crude approximation can be neglected. With this intuitive assumption, the result of the integration in Eq. (74) should be close to

$$\phi_{it} \simeq \int_{\Omega_s} \phi_i \chi_{is}\, d\varepsilon_{is} \tag{76}$$

Then, assuming that χ_{is} is the function from Eq. (33) or (36), one arrives at the result that equation system (70) must be replaced by

$$\frac{\phi_{it}\gamma_{il}}{(\phi_i\gamma_i)^{m_{is}}} \frac{(\phi_s\gamma_s)^{r_{is}m_{is}}}{\phi_{st}\gamma_{sl}} = K_{is}^r \qquad \text{for } l = 1 \tag{77a}$$

$$\frac{\phi_{il}\gamma_{il}}{\phi_i\gamma_i} \frac{\phi_s\gamma_s}{\phi_{sl}\gamma_{sl}} = 1 \qquad \text{for } l \geq 2 \tag{77b}$$

since according to our model $r_{is} = 1$, $K_{is} = 1$ in the second and higher layers. When the interactions in the adsorbed phase can be neglected, equation system (77a) can be solved with respect to ϕ_{it}, yielding a Langmuir-like expression,

$$\phi_{it} = \frac{K_{is}^r[\phi_i\gamma_i/(\phi_s\gamma_s)^{r_{is}}]^{m_{is}}}{1 + \sum_{j=1}^{s-1} K_{js}^r[\phi_j\gamma_j/(\phi_s\gamma_s)^{r_{js}}]^{m_{js}}} \tag{78}$$

A simplified form of this expression, valid for $r_{is} = 1$, has been proposed by Jaroniec [37].

It should be emphasized that, as in adsorption from binary mixtures, Eqs. (77a) and (78) will be valid until the concentration of one of the solvents becomes very small. However, since the generalization expressed in Eqs. (77a) and (78) is half-intuitive, a rigorous solution of the problem of adsorption at small concentrations obviously cannot be found by following these lines. We will consider this problem again in the next section.

A more rigorous but also very approximate approach to the problem of surface heterogeneity in adsorption from multicomponent liquid mixtures is based on the assumption that there is a high degree of correlation between various energies ε_{is} on various surface sites, holding that they differ only by constant values [35,36]:

$$\varepsilon_{is} = \varepsilon_{1s} + \Delta_{is} \qquad i = 2, 3, \ldots, s - 1 \tag{79}$$

This means that the energy distribution functions χ_{is} have the same shape as χ_{is} and are shifted on the energy scale only by the values Δ_{is}. Further, by assuming the simple case $r_{is} = 1$ with $i = 1, 2, \ldots, s - 1$, one can write Eq. (72) in the form

$$\phi_{i1} = \frac{\Phi_{is}}{\sum_{j=1}^{s-1} \Phi_{js}} \frac{\left(\sum_{j=1}^{s-1} \Phi_{js}\right) e^{\varepsilon_{1s}/RT}}{1 + \left(\sum_{j=1}^{s-1} \Phi_{js}\right) e^{\varepsilon_{1s}/RT}} \tag{80}$$

where

$$\Phi_{js} = \frac{\phi_j \gamma_j \gamma_{s1}}{\phi_s \gamma_s \gamma_{j1}} e^{\Delta_{js}/RT} \tag{81}$$

Assuming also the existence of a random surface topography (i.e., that γ_{j1} and γ_{s1} are functions only of the averaged values ϕ_{it}), we arrive at

$$\phi_{it} = \frac{\Phi_{is}}{\sum_{j=1}^{s-1} \Phi_{js}} \int_{\Omega_{1s}} [1 + e^{(\varepsilon_c^{\Phi} - \varepsilon_{1s})/RT}]^{-1} \chi_{1s}(\varepsilon_{1s})\, d\varepsilon_{1s} \tag{82}$$

where

$$\varepsilon_c^{\Phi} = -RT \ln\left(\sum_{j=1}^{s-1} \Phi_{js}\right) \tag{83}$$

Now we can see that the problem is formally the same as in the case of adsorption from binary mixtures. Therefore, we abandon detailed discussion of Eq. (82). Although new papers continue to appear on the problem of surface heterogeneity in adsorption from multicomponent liquid mixtures, less simplifying assumptions than that in Eq. (79) have not been made yet.

II. COMPETITIVE ADSORPTION OF SOLUTE

A. Simple Theories of Retention in LSC

The term normal-phase liquid chromatography (NPLC) is used to denote solid-liquid chromatography systems in which the competitive adsorption of solvents and solute on a solid surface is the main factor governing the retention mechanism. Historically, these solid-liquid systems were the first to be investigated. Thus it is no surprise that the first theories about the partition coefficient K were developed for these systems.

From a thermodynamic point of view, the theory of the partition coefficient in LSC is nothing but a special version of the theories of adsorption from multicomponent liquid mixtures. This is the case when the concentration of a solute component becomes very small. Consider, for instance, component s from the previous section. Its isotherm equation, ϕ_{sl}, is found from the condition

$$\phi_{sl} = 1 - \sum_{i=1}^{s-1} \phi_{il} \tag{84}$$

Now let us consider the simplest case, $r_{si} = 1$ with $i = 1, 2, \ldots,$ $s - 1$. Assuming in addition that $\phi_s \to 0$, we have from Eqs. (84) and (72),

$$\frac{\phi_{sl}}{\phi_s} = \left(\sum_{i=1}^{s-1} K_{is}^{(1)} \frac{\phi_i \gamma_i \gamma_{sl}}{\gamma_s \gamma_{il}} \right)^{-1} \tag{85}$$

Neglecting interactions in both the surface and bulk phase and limiting our consideration to monolayer adsorption and a binary solvent mixture, we obtain from Eq. (85)

$$K = \frac{x_{sl}}{x_s} = \frac{r_s}{K_{1s}^{(1)}} \left(\phi_1 + K_{21}^{(1)} \phi_2 \right)^{-1} \tag{86a}$$

where the partition coefficient K has the following relation to the capacity factor k:

$$k = K \frac{V_s}{V_m} \tag{86b}$$

In the limit $\phi_1 \to 1$, $K_1 = r_s/K_{1s}$ denotes the partition coefficient in pure solvent 1. Equation (86a) can be rewritten as

$$\log \frac{K_1}{K} = \log \left(\phi_1 + K_{21}^{(1)} \phi_2 \right) \tag{87}$$

Further, let ε'_{12} denote the solvent strength of a binary liquid mixture 1 + 2. Following Snyder [28], [cf. Eq. (65)]

$$\log \frac{K_1}{K} = a' \alpha_2 (\varepsilon_{12}^0 - \varepsilon_1^0) \tag{88}$$

Combining Eqs. (87) and (88), we arrive at the well-known Synder's retention equation,

$$\varepsilon_{12}^{0} = \varepsilon_{1}^{0} + \frac{1}{a'\alpha_2} \log \left(\phi_1 + \phi_2 10^{a'\alpha_2(\varepsilon_2^0 - \varepsilon_1^0)} \right) \tag{89}$$

since the equilibrium constant $K_{21}^{(1)}$ is interpreted by Snyder as

$$K_{21}^{(1)} = 10^{a'\alpha_2(\varepsilon_2^0 - \varepsilon_1^0)} \tag{90}$$

Snyder's equation (89) was for years basic to the theory of elution in LSC. Various authors reported that it gave a relatively good correlation for dozens of systems but it would take far too much time to describe even the most important of them. Many of the experimental data reported were obtained with thin-layer chromatography (TLC). With this technique, the capacity factor k is related to the measurable quantity R_F by the [38]:

$$k = \frac{1 - R_F}{R_F} \tag{91}$$

The increasing accuracy of measurement achieved with ever more modern column chromatography equipment has led to the discovery of discrepancies between the behavior of experimental LSC systems and that predicted by Snyder's equation (89). This should not surprise us, in view of the simplifying assumptions leading to Snyder's equation (89). Recently, some effort has been made by Snyder and co-workers [39,40] to eliminate some of these discrepancies. This will be discussed in the next section.

Now let us note that with the assumptions leading to Snyder's equation (89), Eq. (85) can be rewritten in the equivalent form

$$K = \sum_{i=1}^{s-1} K_i \phi_{i1} \tag{92}$$

which is just another form of Snyder's equation (89). However, it should be strongly emphasized that Eq. (92) is valid for LSC systems with ideal bulk and surface phases, in which the solvent and solute molecules have equal molar volumes in the bulk phase and occupy equal surface areas in the adsorbed phase. Some authors, [41] for instance, tried to account for the nonideality effects in Eq. (92) by evaluating the ϕ_{i1} from the set of equations

$$\frac{\phi_{i1}\gamma_{i1}}{\phi_i\gamma_i} \quad \frac{\phi_{s-1}\gamma_{s-1}}{\phi_{(s-1)1}\gamma_{(s-1)1}} = K^{(1)}_{i(s-1)} \tag{93}$$

Combining two adsorption equations, one of which assumes the total ideality of the adsorption system and the other accounts for the nonideality effects, is obviously inconsistent.

When Eq. (93) is used, Snyder's Eq. (92) must be written in its generalized form,

$$K = \frac{\gamma_s}{\gamma_{s1}} \sum_{i=1}^{s-1} K_i \phi_{i1} \frac{\gamma_{i1}}{\gamma_i} \frac{\gamma^{(i)}_{s1}}{\gamma^{(i)}_s} \tag{94}$$

where $\gamma^{(i)}_s$, $\gamma^{(i)}_{s1}$ are the bulk and surface activity coefficients of solute, respectively, when the mobile phase is pure component i. However, it should again be emphasized that the generalized Snyder's equation (94) is valid only for molecules of equal bulk volume occupying equal surface areas. In the more general case $r_{si} \neq 1$, the equation system (70) cannot be solved analytically and a Langmuir-like expression for ϕ_{i1} is not obtained. Consequently, in this more general case one will not arrive at Eq. (85), leading to Snyder's equation (92).

Now, let us consider the intermediate situation in which solvent molecules occupy an equal surface area, which, however, is different from the area occupied by a solute molecule. In other words, r_{ij} = 1, but $r_{si} = r_{sj} \neq 1$. Then the partition coefficient K must be written as

$$K = r_{si} K^{(1)}_{si} \frac{\gamma_s}{\gamma_{s1}} \left[\frac{\phi_{i1}\gamma_{i1}}{\phi_i\gamma_i}\right]^{r_{si}} \tag{95}$$

where the ϕ_{i1} are found from the Langmuir-like equation

$$\phi_{i1} = \frac{K^{(1)}_{ij}(\phi_i\gamma_i\gamma_{j1}/\phi_j\gamma_j\gamma_{i1})}{1 + \sum_{i=1}^{s-2} K^{(1)}_{ij}(\phi_i\gamma_i\gamma_{j1}/\phi_j\gamma_j\gamma_{i1})} \tag{96}$$

Thus, the partition coefficient K for solute s chromatographed in the (s − 1)-component solvent mixture takes the form

$$\ln K = \ln K_i - r_{si} \ln\left(\sum_{j=1}^{s-1} K_{ji}^{(1)} \phi_j \frac{\gamma_j}{\gamma_{j1}} \right) + \ln \frac{\gamma_s \gamma_{s1}^{(i)}}{\gamma_{s1} \gamma_s^{(i)}} \tag{97}$$

where

$$K_i = r_{si} K_{si}^{(1)} \frac{\gamma_s^{(i)}}{\gamma_{s1}^{(i)}} \tag{98}$$

Thus, from Eq. (98) it follows that

$$\ln \frac{K_j}{K_i} = r_{si}\left[(\varepsilon_j - \varepsilon_i) - RT \ln \frac{q_j}{q_i} \right] \tag{99}$$

In the case of a binary liquid mixture, from Eq. (97) we have

$$\ln \frac{K_1}{K} = r_{s1} \ln \left(\phi_1 + K_{21}^{(1)} \phi_2 \frac{\gamma_2}{\gamma_{21}} \right) + \ln \frac{\gamma_{s1} \gamma_s^{(1)}}{\gamma_s \gamma_{s1}^{(1)}} \tag{100}$$

Thus, according to Snyder's notation, from Eq. (100) we obtain

$$\varepsilon_{12}^0 = \varepsilon_1^0 + \frac{r_s}{a' \alpha_2} \ln \left[\phi_1 \frac{\gamma_1}{\gamma_{11}} + \phi_2 \frac{\gamma_2}{\gamma_{21}} 10^{a' \alpha_2 (\varepsilon_2^0 - \varepsilon_1^0)} \right] + \ln \frac{\gamma_{s1} \gamma_s^{(1)}}{\gamma_s \gamma_{s1}^{(1)}} \tag{101}$$

Equation (101) is a generalization of Snyder's classic equation (89) for LSC systems with binary mobile phases. This generalization accounts for the nonideality effects in both the surface and the mobile phase, as well as for the possibility that solute molecules may occupy a different surface area than that occupied by solvent molecules. Equation (97), on the other hand, is a kind of generalization of Snyder's equation that serves for multicomponent mobile phases. However, it should again be emphasized that neither Eq. (101) nor Eq. (97) can be applied when solvent molecules occupy different surface areas.

Now let us note that Eq. (97) can be rewritten in the equivalent form

$$\frac{1}{K} = \frac{1}{K_i} \left(\sum_{j=1}^{s-1} K_{ji}^{(1)} \phi_j \frac{\gamma_j}{\gamma_{j1}} \right)^{r_{si}} \frac{\gamma_{s1}\gamma_s^{(i)}}{\gamma_s\gamma_{s1}^{(i)}} \tag{102}$$

For a binary solvent mixture ideal in both the surface and the mobile phase, we arrive at the following expression for K:

$$\frac{1}{K} = \frac{1}{K_1} (\phi_1 + K_{21}^{(1)} \phi_2)^{r_{s1}} \tag{103}$$

which is simply a generalization of the Scott-Kucera equation [42] for the case when $r_{s1} = r_{s2} \neq 1$ (i.e., when solute molecules occupy a surface area different than that occupied by solvent molecules). For the still simpler case $r_{s1} = r_{s2} = 1$, we arrive at the classic Scott-Kucera equation [42],

$$\frac{1}{K} = \frac{1}{K_1} + \frac{K_{21}^{(1)} - 1}{K_1} \phi_2 \tag{104}$$

which means that the classic Snyder and Scott-Kucera equations are equivalent.

Now let us again consider Eq. (100). When one of the solvents, say solvent 2, is much more strongly adsorbed than solvent 1 (i.e., $K_{21}^{(1)} \gg 1$), then the term ϕ_1 can be neglected in comparison to the term $(K_{21}^{(1)} \phi_2\gamma_2/\gamma_{21})$. Equation (100) then takes the simpler form

$$\ln K = \ln \frac{K_1}{(K_{21}^{(1)})^{r_{s1}}} - r_{s1} \ln \phi_2 + \ln \left[\left(\frac{\gamma_{21}}{\gamma_2} \right)^{r_{s1}} \frac{\gamma_s\gamma_{s1}^{(1)}}{\gamma_{s1}\gamma_s^{(1)}} \right] \tag{105}$$

When the nonideality effects in both the surface and the mobile bulk phase can be neglected, the last term on the right side of Eq. (105) disappears and we arrive at the well-known equation

$$\ln K = \ln \frac{K_1}{(K_{21}^{(1)})^{r_{s1}}} - r_{s1} \ln \phi_2 \tag{106}$$

first proposed by Soczewiński [43] and investigated independently by Jandera and Churáček [44]. Soczewiński's equation is probably the most popular expression for many investigations who correlate retention data in LSC and TLC systems with mixed mobile phases. Jandera and Churáček successfully used this equation in their impressive work on gradient elution chromatography [45]. Despite the long time that has passed since Soczewiński's equation was published, new experimental data showing its wide applicability continue to be reported.

Now let us consider a ternary solvent mixture,

$$\ln K = \ln \frac{1}{K_{1s}^{(1)}} - \ln (\phi_1 + K_{21}^{(1)} \phi_2 + K_{31}^{(1)} \phi_3) \tag{107a}$$

When solvents 2 and 3 are much more strongly adsorbed than solvent 1 (i.e., $K_{21}^{(1)} \gg 1$ and $K_{31}^{(1)} \gg 1$) the first term in the parentheses can practically be neglected in comparison with the other two. Then, Eq. (107a) recalls the generalized form of Soczewiński's equation, proposed by Paanakker et al. [46]:

$$\ln K = \ln \frac{K_1}{(K_{21}^{(1)})^{r_{s1}}} - r_{s1} \ln (\phi_2 + K_{32}^{(1)} \phi_3) \tag{107b}$$

Another generalization of Soczewiński's equation has been proposed by Haara et al. [47]. They have found experimentally that if for two binary solvent mixtures the following relations hold:

$$\ln K_{12} = C_{12} - D_{12} \ln \phi_2 \tag{108a}$$

$$\ln K_{13} = C_{13} - D_{13} \ln \phi_3 \tag{108b}$$

where the subscripts 12 and 13 in K_{12} and K_{13} refer to binary liquid mixtures 1 + 2 and 1 + 3, then for the ternary solvent mixture 1 + 2 + 3, the following equation can be used to correlate the retention data:

$$\ln K_{123} = C_{123} - D_{123} \ln (\phi_2 + \phi_3) \tag{109}$$

where:

$$C_{123} = C_{12}\phi_2 + C_{13} \phi_3 \tag{110a}$$

$$D_{123} = D_{12}\phi_2 + D_{13} \phi_3 \tag{110b}$$

B. Effects of Surface Heterogeneity in NPLC

Since the competitive adsorption between solvent and solute molecules is the main factor governing the behavior of NPLC systems, the study of the nature of the solid-solution interactions in such systems becomes especially important. First is the dispersion of the adsorptive properties of various solid surface areas, or, the energetic heterogeneity of the solid/solution interfaces.

This problem was first considered by Snyder [28], who proposed writing the partition coefficient for a whole heterogeneous surface K_t in the form

$$K_t = \sum_p g_p K_p \tag{111}$$

where K_p is the partition coefficient for the pth type of adsorption site and g_p is the appropriate weighting factor. In view of what we have said in the preceding Section, the pth kind of adsorption site is simply a collection of sites characterized by the same set of energies $\{\varepsilon_{is}\}$.

Some 10 years after Snyder, Jaroniec et al. [48,49] again raised this important problem in LSC. Below, we refer to the results presented in their three most recent and advanced works concerning this problem [50–52]. For the purpose of clarity we first consider their results for the simpler case of LSC systems composed of molecules of equal sizes. For the same reason, we neglect for the moment the effects of nonideality in the bulk and surface phases.

Jaroniec et al. then write the equation for the adsorption isotherm ϕ_{1t} in the form

$$\phi_{1t} = \frac{(\phi_1/K^0_{s1})^m}{1 + \sum_{i=1}^{s-1} (\phi_i/K^0_{si})^m} \tag{112}$$

According to the discussion on page 275, this equation has an intuitive background but implies certain conclusions about the nature of the solid/solution interface. These are drawn for a system that has two components. Namely, the dispersion of the energies ε_{is} is gaussian-like, and expressed by Eq. (33) or (36). Further, it is assumed that this spread is the same for all the energies ε_{is} (i.e., is characterized by the same heterogeneity parameter m).

Since for a homogeneous solid surface the partition coefficient K can be written as

$$K = \frac{1}{\phi_1} K_{s1}^{(1)} \phi_{11} \tag{113}$$

Jaroniec et al. [50] propose that K_t be expressed as

$$K_t = \frac{1}{\phi_1} \sum_p g_p K_{s1,p} \phi_{11,p} \tag{114}$$

where $\phi_{11,p}$ is the areal concentration of solvent 1 on the pth type of adsorption site. Under these conditions, Jaroniec et al. believe that the summation in Eq. (114) can be well approximated by the expression [51]

$$\sum_p g_p K_{s1,p} \phi_{11,p} \simeq (K_1^0 \phi_{1t})^{1/m} \tag{115}$$

where K_1^0 is an averaged distribution coefficient for the solute s in the pure solvent. The nonideality effects can easily be taken into account, especially in the case of random surface topography. For this, as we know from the preceding section, the surface activity coefficients depend only on the averaged concentrations ϕ_{it}. Then, from Eqs. (114) and (115), it follows that

$$K_t = \frac{1}{\phi_1} \frac{\gamma_s \gamma_{11}}{\gamma_{s1} \gamma_1} (K_1^0 \phi_{1t})^{1/m} \tag{116}$$

In another paper, Jaroniec and Ościk-Mendyk [52] extended their intuitive treatment to the case in which solvent molecules occupy the same surface area, which, however, is different from the surface area occupied by solute molecules. As in their previous work, their starting point was the expression for a homogeneous solid surface,

$$K = \frac{\gamma_s}{\gamma_{s1}} \left(\frac{\gamma_{11}}{\phi_1 \gamma_1} \right)^{r_{s1}} K_{s1}^{(1)} (\phi_{11})^{r_{s1}} \tag{117}$$

In arriving at the model of a heterogeneous solid surface, Eq. (117) is written by Jaroniec et al. in the form

$$K_t = \frac{\gamma_s}{\gamma_{s1}} \left(\frac{\gamma_{11}}{\phi_1 \gamma_1} \right)^{r_{s1}} \sum_p g_p K_{s1,p}^{(1)} (\phi_{11,p})^{r_{s1}} \tag{118}$$

Jaroniec et al. thus believe that the summation in Eq. (118) can be represented by the experession [c.f. Eq. (115)]

$$K_t = \frac{\gamma_s}{\gamma_{s1}} \left(\frac{\gamma_{11}}{\phi_1 \gamma_1} \right)^{r_{s1}} (K_1^0 \phi_{1t})^{r_{s1}/m} \tag{119}$$

where the adsorption isotherm ϕ_{1t} must be evaluated from the expression

$$\phi_{1t} = \frac{(K_{s-1,1} \phi_1 \gamma_1 \gamma_{s-1,1} / \phi_{s-1} \gamma_{s-1} \gamma_{11})^m}{1 + \sum_{i=1}^{s-1} (K_{s-1,i} \phi_i \gamma_i \gamma_{s-1,1} / \phi_{s-1} \gamma_{s-1} \gamma_{i1})^m} \tag{120}$$

Expressions (119) and (120) form the set of equations that were used by Jaroniec and Ośćik-Mendyk [52] to theoretically analyze the capacity coefficients for various dichlorophenol isomers, chromatographed in a benzene + cyclohexane mixture on a silica gel and an aluminum oxide.

Originally, the essential Jaroniec assumption (115) was based on some formal similarities between the problem considered here and the results of a computer simulation of adsorption from gaseous mixtures onto solid surfaces [37]. The recent theoretical results for solution adsorption onto heterogeneous solid surfaces obtained by Rudziński et al. [27] make possible a more rigorous treatment of this problem, at least in the case of strongly heterogeneous solid surfaces.

Thus, let us again consider Eqs. (79) through (83), which assume a high correlation between the energies ε_{is} on various surface sites. The partition coefficient K_t then takes the form

$$K_t = \frac{1 - \sum_{i=1}^{s-1} \phi_{it}}{\phi_s} \tag{121}$$

where according to Eq. (82), we have

$$\sum_{i=1}^{s-1} \phi_{it} = \int_{\Omega_{1s}} \left[1 + e^{\varepsilon_c^{\Phi} - \varepsilon_{1s}/RT} \right]^{-1} \chi_{12} \, d\varepsilon_{1s} \tag{122}$$

Since when solute $\phi_s \to 0$, $\varepsilon_c^{\Phi} \to -\infty$, according to Eq. (64a)

$$\sum_{i=1}^{s-1} \phi_{it} = \frac{\left(\sum_{j=1}^{s-1} \phi_{js}\right) e^{\varepsilon_{1s}^{m}/RT}}{1 + \left(\sum_{j=1}^{s-1} \phi_{js}\right) e^{\varepsilon_{1s}^{m}/RT}} \tag{123}$$

Thus the partition coefficient K_t takes the form of the generalized Snyder equation (94), in which the ϕ_{i1} are replaced by the averaged ϕ_{it}. Further, heterogeneity effects are transmitted through the surface activity coefficients γ_{i1} and γ_{s1}, which are appropriate functions of the averaged ϕ_{it}. Note that Eq. (123) is the solution of the following set of equations:

$$\frac{\phi_{s1}\gamma_{s1}}{\phi_s\gamma_s} \frac{\phi_i\gamma_i}{\phi_{it}\gamma_{i1}} = e^{\varepsilon_{si}^{1}/RT} \tag{124}$$

This means that the competitive adsorption between solute molecules and the molecules of solvent i occurs as on a homogeneous surface characterized by the smallest actual energy ε_{si}^{1} found on the heterogeneous surface investigated.

Consideration of the preceding section suggests that this conclusion should also be valid in the general case $r_{si} \neq 1$. In this general case, the partition coefficient K should be written in the form

$$K = K_{si}^{1} \frac{\gamma_s}{\gamma_{s1}} \left(\frac{\phi_{it}\gamma_{i1}}{\phi_i\gamma_i}\right)^{r_{si}} \tag{125}$$

For better clarity, let us neglect for a while the nonideality effects and limit ourselves to the simple case of a binary mobile phase. Then, from Eq. (45) we obtain

$$\frac{\phi_{2t}}{\phi_2} = \frac{K_{21}^{r}(\phi_2)^{m_{21}-1}}{(\phi_1)^{r_{21}m_{21}} + K_{21}^{r}(\phi_2)^{m_{21}}} \tag{126}$$

because the concentration of the solute on the solid surface is assumed to be negligible when calculating the surface concentrations of solvents 1 and 2.

Now let us consider the region of small concentrations of the more active solvent 2, where Soczewiński's linear relationship is usually found. Let us suppose that as $\phi_2 \to 0$, $\phi_1 \to 1$. Then from Eq. (126) we obtain

$$\lim_{\phi_2 \to 0} \frac{\phi_{2t}}{\phi_2} = K_{21}^{r}(\phi_2)^{m_{21}-1} \tag{127}$$

With the assumption

$$\gamma_s = \gamma_{s1} = \gamma_i = \gamma_{i1} = 1 \tag{128}$$

from Eqs. (125) and (127), we obtain

$$\ln K = \ln\left[K_{s2}^{l}(K_{21}^{r})^{r_{s2}}\right] - (1 - m_{21})r_{s2} \ln \phi_2 \tag{129}$$

which is essentially Soczewiński's kind of relationship between K and ϕ_2 [c.f. Eq. (106)].

The factor $(1 - m_{21})$ is positive but still smaller than unity. For typical solid-solution systems, $0.7 < m_{21} < 0.9$, which means that the term $1 - m_{21}$ lies in the range from 0.1 to 0.3. Thus, it can well happen that the product $(1 - m_{21})r_{s2}$ will be smaller than unity even when r_{s2} itself is much larger than unity.

This is a new derivation of Soczewiński's relationship, based on the concept of the energetic heterogeneity of real solid/solution interfaces. This new derivation explains the following intriguing question. Why, in many cases, is the absolute value of the tangent in the Soczewiśki's plot ln K versus ln ϕ_2 smaller than unity although the usually large solute molecules must obviously occupy a much larger surface area than the small solvent molecules? This is because this tangent (its absolute value) is not equal to r_{s2}, as is suggested by the simple derivation outlined earlier, but is instead equal to the product $r_{s2}(1 - m_{21})$, which in some cases may be smaller than unity.

The generalization of Eq. (129) for the case in which the nonideality of the bulk phase cannot be neglected is straightforward and takes the form

$$\ln K = \ln\left[K_{s2}^{l}(K_{21}^{r})^{r_{s2}}\right] + \ln \frac{\gamma_s}{(\gamma_2)^{r_{s2}}} - (1 - m_{21})r_{s2} \ln \phi_2 \tag{130}$$

Provided that the energetic heterogeneity of actual solid/solution interfaces is the essential source of Soczewiński's linear relationship in the experimental plots ln K versus ln ϕ_2, the exact form of the energy distribution function χ_{12} should play a secondary role in this correlation of experimental data. Therefore, let us investigate the consequences of assuming that ϕ_{it} is evaluated from the Dubinin-Radushkevich equation (54), which is related to the energy distribution (52). We write this equation in the form

$$\ln \phi_{2t} = -\left[C_{21}^{(b)} - \frac{RT}{(b_{12})^{1/2}} \ln \frac{\phi_2}{(\phi_1)^{r_{21}}}\right]^2 \qquad (131)$$

where

$$C_{21}^{(b)} = -\frac{1}{(b_{12})^{1/2}}\left[\varepsilon_{21}^{1} + RT \ln \frac{(r_{21})^{1/2}\gamma_2\gamma_{11}}{(\gamma_1\gamma_{21})^{r_{21}}}\right] \qquad (132)$$

Then,

$$\ln \frac{\phi_{2t}}{\phi_2} = -(C_{21}^{(b)})^2 - \frac{2RT}{(b_{12})^{1/2}} r_{21} \ln \phi_1 - \frac{(RT)^2}{b_{12}} r_{21}^2 \ln \phi_1 + \frac{2(RT)^2}{b_{12}} r_{21} (\ln \phi_1)(\ln \phi_2) - \left[1 - \frac{2RT}{(b_{12})^{1/2}}\right] \ln \phi_2 - \frac{(RT)^2}{b_{12}} \ln^2 \phi_2 \qquad (133)$$

Thus,

$$\lim_{\phi_2 \to 0} \frac{\phi_{2t}}{\phi_2} = -(C_{21}^{(b)})^2 - \left[1 - \frac{2RT}{(b_{12})^{1/2}}\right] \ln \phi_2 - \frac{(RT)^2}{b_{12}} \ln^2 \phi_2 \qquad (134)$$

Consequently, from Eqs. (125) and (134) we obtain

$$\lim_{\phi_2 \to 0} \ln K = A - \left[1 - \frac{2RT}{(b_{12})^{1/2}}\right] r_{s2} \ln \phi_2 - \frac{(RT)^2}{b_{12}} \ln^2 \phi_2 \qquad (135)$$

where

$$A = -\left(C_{21}^{(b)}\right)^2 + \ln K_{s2}^{l} + \ln \frac{\gamma_s}{\gamma_{s1}} + r_{s2} \ln \frac{\gamma_{21}}{\gamma_2} \tag{136}$$

The numerical analysis of various adsorption systems obeying the Dubinin-Radushkevich equation (54) and performed by Dabrowski and Jaroniec [25] has shown that the value of the term $RT/(b_{12})^{1/2}$ varies between 0.08 and 0.3. Taking the highest estimation (i.e., $RT/(b_{12})^{1/2} = 0.3$), we arrive at the conclusion that the term $[1 - 2RT/(b_{12})^{1/2}]$ is still positive but not greater than 0.3. At the same time, the coefficient $(RT)^2/b_{12}$ of the term $\ln^2 \phi_2$ is close to 0.01, which means that the quadratic term $\ln^2 \phi_2$ in Eq. (135) is practically negligible compared with the term $\ln \phi_2$. Thus, we have again arrived at the conclusion that there should be a linear relationship between ln K and $\ln \phi_2$ in a first crude approximation. In other words, we have again arrived at Soczewiński's linear relationship. The quadratic term $\ln^2 \phi_2$ in Eq. (135) may to some extent modify Soczewiński's relationship at very small concentrations of the more active solvent.

Further, the absolute value of the coefficient $[1 - 2RT/(b_{12})^{1/2}]r_{s2}$ may in general be smaller than unity, even when the value of r_{s2} itself is much larger than unity.

Our consideration in this section suggests that the energetic heterogeneity of real solid/solution interfaces is one of the most important factors, if not the main factor, governing the behavior of real LSC systems. It seems that this has not been sufficiently recognized in the past investigation of these systems.

C. Molecular Theories of Retention in LSC

The recent rapid development of column chromatographic equipment has brought an impressive increase in the accuracy of retention measurements in LSC. The simple theories are still very useful in many practical problems as a first quick approximation, but it is also widely realized that essential progress is now needed in the theoretical treatment of partition coefficients. As with the general theories of adsorption from multicomponent liquid mixtures onto solid surfaces, this is achieved by applying the methods of statistical thermodynamics. Their use in adsorption theory is not new; it was initiated in 1950s. The only surprising point is that until now these methods have not been widely applied to the theories relating to partition coefficients in LSC. It seems there are two reasons for this: First, the relatively low accuracy of experiments in LSC, which persisted for some time in both column and thin-layer techniques, did not create the need for more sophisticated and accurate theories for correlating the retention data measured. Second, those involved in chromatography were always mainly impressed by the enormously important practical applications of liquid chromatography; basic research into the retention mechanism was not of great interest for a long time. A gap between experimental and theoretical research is a factor that has retarded the further practical application of LSC. It seems, however, that this is changing.

As we have already said, the theory of partition coefficients in LSC is nothing but a special case of the general theory of adsorption from multicomponent liquid mixtures onto solid surfaces. Impressive progress in the statistical-mechanical treatment of this phenomenon has been made in the past 15 years, in which regard we can mention the excellent series of works by Smirnova [53] and Ash et al. [54] on the multilayer adsorption of different sizes of molecules with orientation effects. Further, the molecules of the mobile solvent mixture are usually small and relatively simple, in contrast to the usually larger and more structurally complicated molecules of solute. Thus, in many typical cases, their adsorption on a support packing in the presence of the solvent molecules represents a theoretical problem similar to that of the adsorption of polymers from solutions. We have no room to discuss even the most advanced theoretical papers.

Meanwhile, the first attempts to apply the methods of statistical mechanics to the theory of partition coefficients in LSC do not seem to have extensively used existing theoretical solutions. By this we mean the works by Martire and Boehm [55,56]. This does not, however, decrease the importance of these works as showing the future trend in theories of the partition coefficient in LSC, and for this reason we will report the data obtained by Martire and Boehm in more detail.

A first step in any statistical-mechanical consideration is to establish a certain molecular picture of the physical phenomena being considered, which is called the model. Next, an appropriate system partition function must be constructed according to the well-known rules of statistical thermodynamics.

Martire and Boehm have to this end applied the canonical partition function Q, although Smirnova, for instance, finds that the grand partition function is more convenient for this purpose. We will stress below the essential results obtained by Martire and Boehm in this way.

Let us consider a monolayer adsorption from a ternary liquid mixture 1 + 2 + 3 of equally sized molecules. It is further assumed that these molecules are cubic and that they make contact with the planar solid surface on one of their six equivalent cubic faces. Each of the remaining five faces interacts with one of the cubic faces of other molecules in the adsorbed phase. Further, the Bragg-Williams approximation is used to calculate the energy of these interactions. This approximation neglects the local correlations in the systems (i.e., it assumes that the molecular environment of a selected molecules is in its neighborhood, the same as the average composition of the system).

Thus, until now, the Martire and Boehm adsorption model was the same as that considered by Borówko [34] when assuming that a fraction (two-thirds) of the nearest-neighbor adsorption sites is located in the same lattice plane (except the first layer, where this fraction is equal to four-fifths).

The essential difference between the model accepted by Borówko and that of Martire and Boehm is that in evaluating the interaction energy U_{in} for the molecules adsorbed directly on the solid surface, the latter authors assume that the composition of the second adsorbed layer is identical with that of the first layer. Borówko, on the other hand, assumes that the composition of the second adsorbed layer is the same as that of the equilibrium bulk phase.

Our model investigation on the multilayer adsorption of solvents (pages 250-252) shows that the extent to which the assumptions of Borówko's or Martire and Boehm are valid depends upon the nature of the solvent solution. Solvent mixtures showing large positive deviations from Raoult's law (e.g., $A_{12} > 1$) form a multilayer in which the concentration in the second layer is closer to that in the first layer rather than to the bulk concentration. For solvent mixtures characterized by $A_{12} < 1$, Borówko's assumption should be more realistic.

In accepting the Martire and Boehm assumptions, one arrives at the following form of the canonical partition function Q_s for the molecules adsorbed on the solid surface:

$$Q_s(\{N_{i1}\}, N_0, T) = N_0! \prod_{i=1}^{3} \left[\frac{(q_{i1} e^{-\varepsilon_i'/kT})^{N_{i1}}}{N_{i1}!} \right] e^{\{-U_{in}/kT\}} \tag{137}$$

where, according to the Martire and Boehm model,

$$U_{in} = \frac{5}{2} \frac{1}{N_0} \sum_{i=1}^{3} \sum_{j=1}^{3} N_{i1} N_{j1} \varepsilon_{ij} \tag{138}$$

and where $\varepsilon_i' = - \varepsilon_i$, according to the convention accepted in statistical mechanics. Similarly, for a lattice plane inside the equilibrium bulk phase, the canonical partition function Q_m takes the form

$$Q_m(\{N_i\}, N_0, T) = N_0! \prod_{i=1}^{3} \left[\frac{(q_i^{N_i})}{N_i!}\right] \exp\left\{\frac{-3}{N_0 kT} \sum_{i=1}^{3} \sum_{j=1}^{3} N_i N_j \varepsilon_{ij}\right\} \tag{139}$$

where q_i is the molecular partition function in the bulk phase. Let solvent 2 be the reference substance. Then the chemical potential differences are readily derived. For the surface phase,

$$\frac{\mu_{i1} - \mu_{21}}{kT} = -\left(\frac{\partial \ln Q_s}{\partial N_{i1}}\right)_{T,N_0} = \ln\left\{\frac{x_{i1}}{x_{21}} \exp\left[\frac{\varepsilon_i' - \varepsilon_2'}{kT} + \ln\frac{q_{21}}{q_{i1}} + \frac{5}{kT}\sum_{j=1}^{3} x_{j1}(\varepsilon_{ij} - \varepsilon_{j2})\right]\right\} \tag{140}$$

where the molar fractions x_i instead of the volume fractions ϕ_i have been introduced to emphasize that this consideration is done for a regular solution model. For the bulk phase, we have

$$\frac{\mu_i - \mu_2}{kT} = -\left(\frac{\partial \ln Q_m}{\partial N_i}\right)_{T,N_0} = \ln\left\{\frac{x_i}{x_2}\frac{q_2}{q_i} \exp\left[\frac{6}{kT}\sum_{j=1}^{3} x_j(\varepsilon_{ij} - \varepsilon_{j2})\right]\right\} \tag{141}$$

At equilibrium, $\mu_{i1} = \mu_i$, and from Eqs. (140) and (141) we obtain

$$\frac{x_{31}}{x_3} = \frac{x_{21}}{x_2} \exp\left\{\frac{1}{RT}\left[-(\varepsilon_3' - \varepsilon_2') + \sum_{i=1}^{3} (6x_i - 5x_{i1})(\varepsilon_{i2} - \varepsilon_{i3})\right]\right\} \tag{142}$$

When $x_3 \to 0$, $x_{31} \to 0$, and $x_{31}/x_3 \to K$, it follows that from Eq. (142) that

$$K_{12} = x_{11}K_1 e^{[-(6x_1 - 5x_{11} - 1)(\alpha_{23} - \alpha_{12} - \alpha_{13})/RT]} + x_{21}K_2 e^{[-(6x_2 - 5x_{21} - 1)(\alpha_{13} - \alpha_{12} - \alpha_{23})/RT]} \tag{143}$$

where K_i is, as usual, the partition coefficient in the pure solvent i:

$$K_i = e^{\{(\varepsilon_i' - \varepsilon_3')/RT\}} \tag{144}$$

At the expense of rapidly increasing the complexity of the statistical-mechanical treatment, the retention of more complicated solute molecules can be calculated. The complexity of the calculation is increases so rapidly that several approximations must be accepted to make the problem analytically tractable. The work of Martire and Boehm has several examples of this. It would take far too long to present here even some selected examples. Therefore, we have confined ourselves to sketching such a statistical-mechanical treatment for the simplest case of an LSC system.

Most investigators working with liquid chromatography are interested in some practical applications of appropriate theoretical expressions for the partition coefficient in their experimental work. For this reason, some compromise must be made between the complexity of appropriate theoretical expressions and the accuracy with which they reproduce the behavior of real experimental partition coefficients. We will present an example of such a compromise.

Let us consider a mixture 1 + 2 + 3 of components of which component (solvent) 1 is assumed to be a monomer occupying one adsorption site, whereas components 2 (solvent) and 3 (solute) collapse onto the solid surface while being adsorbed and occupy adsorption sites r_2 and r_3, respectively. The energy of mixing for this three-component system, $U_{123}^{(mix)}$, may be represented by the approximate expression [18]:

$$U_{123}^{(mix)} = U_{12}^{(mix)} + U_{13}^{(mix)} + U_{23}^{(mix)} \tag{145}$$

where $U_{12}^{(mix)}$ has the same form as $U_{in}^{(mix)}$ in Eq. (9), except that the sum $\phi_{11} + \phi_{21}$ is not unity. Let us also assume that component 3 (solute) is adsorbed in a monolayer fashion. Then the differentiation

$$RT \ln \gamma_{31} = \left. \frac{\partial U_{123}^{(mix)}}{\partial N_{31}} \right|_{N_{11}, N_{21}, N_{1m}, N_{2m}} \tag{146}$$

yields the following explict expression for γ_{31}:

$$\ln \gamma_{31} = r_{31}\{A_{13}[f_v \phi_{12} + f_p \phi_{11}(1 - \phi_{31})] + A_{23}[f_v \phi_{22} + f_p \phi_{21}(1 - \phi_{31})] - A_{12}[f_v \phi_{12}\phi_{22} + f_p \phi_{11}\phi_{21}]\} \tag{147}$$

In the chromatographic limit $\phi_3 \to 0$ and $\phi_{31} \to 0$, from Eq. (147) we obtain

$$\ln \gamma_{31} = r_{31}[A_{13}(f_v \phi_{12} + f_p \phi_{11}) + A_{23}(f_v \phi_{22} + f_p \phi_{21}) - A_{12}(f_v \phi_1 \phi_2 + f_p \phi_{11} \phi_{21})] \quad (148)$$

Similarly, for the equilibrium bulk phase, we have

$$\ln \gamma_3 = A_{13} \phi_1 + A_{23} \phi_2 - A_{12} \phi_1 \phi_2 \quad (149)$$

At the same time, the activity coefficients of solvents 1 and 2 are still given by Eqs. (11) and (12). When the competitive adsorption of solvents can be represented by a monolayer model, then ϕ_{12} and ϕ_{22} in Eq. (148) are replaced by the bulk concentrations ϕ_1 and ϕ_2. Expression (148) may be inserted into various equations for the partition coefficient K, for example, into Eq. (94), (95), (101), (102), or (105). After inserting Eq. (148) into Eq. (94), one arrives at an expression similar in form to the Martire and Boehm equation (143) (when a binary mobile phase is considered). In LSC systems with heterogeneous surfaces, Eq. (148) can again be applied, but the concentrations ϕ_{11} and ϕ_{21} must be replaced by their averaged values ϕ_{1t} and ϕ_{2t} (for the most probable random topographical distribution of energetically different adsorption sites).

D. Theories of the Partition Coefficient in RPLC

The term reversed-phase liquid chromatography (RPLC) is used to denote LSC systems in which the stationary phase is less polar than the mobile liquid phase. This is achieved by applying graphitized carbons as column packings or silica gels with a suitable organic moiety covalently bonded to the silica surface and by using water-methanol-acetonitrile-tetrahydrofurane mixtures as the mobile phase.

The way to obtain graphitized carbons has been known for a long time. The experimental work being carried out in various laboratories is directed at improving the mechanical properties of graphitized carbons. Also well known is the way in which various chemically bonded stationary phases are prepared. There is a large literature on this. We can note, for instance, the excellent monograph by Unger [57] or the paper by Engelhardt and Ahr [58]. More than two-thirds of the analyses being done in world laboratories apply RPLC systems. Thus, knowledge of these systems among the users of chromatography is spreading.

It is commonly accepted that, on the surface of a typical silica not exposed to a temperature above 200°C, the concentration of silanol groups is about 7-8 μmol/m^2 surface. Under optimum conditions the amount of the bonded silane can approach 4 μmol/m^2. Thus, almost half the original silica surface is exposed to the mobile phase. On this basis it is believed that the free space between alkyl chains is large enough to permit penetration by both solvent and the solute molecules, which are usually much larger.

The exact manner in which solvent and solute molecules are adsorbed inside the surface phase is, however, still far from well understood. Morel and Serpinet [59], for instance, have discovered that the system of alkyl chains bonded to a silica surface may undergo one or even two phase transitions whose nature is still unknown. It has only been postulated that these are transitions from the ordered rigid structure characteristic of crystals to a liquid-like structure. Further, these phase transitions take place at room temperature, at which most RPLC experiments is done. There is no doubt that these phase changes may affect the retention mechanism in RPLC systems. It therefore seems that the theory of the partition coefficient in RPLC is still at an initial stage and should be matter of further theoretical study.

A characteristic and commonly accepted feature of RPLC systems is the relatively small contribution of surface phenomena to the retention mechanism. This is especially so for chemically bonded phases. Solute-solvent interactions in the mobile bulk phase are assumed to be the main factor governing the behavior of RPLC systems. Let us consider this problem in more detail.

All the equations for the partition coefficient K developed in the three previous sections can be written in the general form

$$\ln K = \ln K^{(in)} + \ln K^{(comp)} \tag{150}$$

where

$$\ln K^{(in)} = \ln \frac{\gamma_s}{\gamma_{s1}} \tag{151}$$

the interaction contribution to ln K due to solute-solvent interactions in the mobile and the surface phases. Further, ln $K^{(comp)}$ is a term connected with the competitive adsorption of solvents on a solid surface. This term may take various forms, depending on the model of the solid/solution interface. We will therefore call this term the competition contribution to ln K. There is a certain interference between these two terms, and therefore a sharp separation of interaction and competition effects is not possible.

In the absence of a strong preferential adsorption of one of the components of a solvent mixture, the term ln $K^{(comp)}$ is practically constant over the entire range of solvent concentrations. Also constant is the term γ_{s1}, with the result that the change in ln K is due to changes in the bulk activity coefficient γ_s. This simple model was the basis on which two simple theories of retention in RPLC were developed.

The first of these was the theory of Schoenmakers et al., formulated in the language of solubility-parameter theory. Below, we refer briefly to the results of their second paper [60], in which this theory was extended to the case of ternary mobile phases. The term ln $K^{(in)}$ is then identified with ln K, and the activity coefficient of solute in phase f is expressed by the equation

$$RT \ln \gamma_{s(f)} = v_s(\delta_s - \delta_{(f)})^2 \tag{152}$$

where entropy effects are neglected. Further, f = s, m, where s and m refer to the surface and mobile phases, respectively, δ_s and $\delta_{(f)}$ are the solubility parameters for the solute and for the phase f, respectively. The combination of Eqs. (151) and (152) yields

$$\ln K = \frac{v_s}{RT}[(\delta_s - \delta_{(m)})^2 - (\delta_s - \delta_{(s)})^2] \tag{153}$$

Assuming that

$$\delta_{(m)} = \sum_{i=1}^{3} \phi_i \delta_i \quad \text{and} \quad \sum_{i=1}^{3} \phi_i = 1 \tag{154}$$

Eq. (153) can be written as

$$\ln K = \frac{v_s}{RT}\{[\delta_s - \phi_1\delta_1 - \phi_2\delta_2 - (1 - \phi_1 - \phi_2)\delta_3]^2 - (\delta_s - \delta_{(s)})^2\} \tag{155}$$

Assuming further that $\delta_{(s)}$ is constant, after a rearrangement we obtain

$$\ln K = A_1\phi_1 + B_1\phi_2 + A_2\phi_1^2 + B_2\phi_2^2 + 2\sqrt{A_2B_2}\,\phi_1\phi_2 + C \tag{156}$$

where

$$A_1 = \frac{2v_s}{RT}(\delta_s - \delta_3)(\delta_3 - \delta_1) \tag{157a}$$

$$B_1 = \frac{2v_s}{RT}(\delta_s - \delta_3)(\delta_3 - \delta_2) \tag{157b}$$

$$A_2 = \frac{v_s}{RT}(\delta_1 - \delta_3)^2 \tag{157c}$$

$$B_2 = \frac{v_s}{RT}(\delta_2 - \delta_3)^2 \tag{157d}$$

$$C = \frac{v_s}{RT}[(\delta_3 - \delta_s)^2 - (\delta_{(s)} - \delta_s)^2] \tag{157e}$$

The theoretical research of Schoenmakers et al. [60] was accompanied by considerable experimental work to prove the validity of Eqs. (156) and (157). The behavior of over 30 solutes in the ternary phases water + methanol + acetonitrile and water + methanol + tetrahydrofuran was investigated. Using the Gibbs triangle representation, Schoenmakers et al. demonstrated the behavior of the so-called isoeulotropic lines connecting the concentration points for which an equal retention time was found experimentally. These isoeulotropic lines were then compared with the theoretically calculated lines. Generally, the agreement between the experimentally measured and theoretically calculated isoeluotropic lines was good at high concentrations of water and methanol. This agreement became poorer at higher concentrations of acetonitrile and tetrahydrofuran.

In the case of a binary liquid mixture 2 + 3 (i.e., when $\phi_1 \to 0$) Eq. (156) reduces to

$$\ln K = C + B_1\phi_2 + B_2\phi_2^2 \tag{158}$$

predicting a quadratic dependence of ln K upon the mobile-phase composition.

A similar result was obtained by Jandera et al. [61], whose basic assumptions are exactly the same as those accepted by Schoenmakers et al.; however, their further theoretical treatment follows that first outlined by Snyder for NPLC. Thus, Jandera et al. assume that the transfer of a solute molecule from the mobile to the stationary phase is mainly the result of polar interactions in the mobile phase. The standard free energy change for this process ΔF_S is given by the expression

$$-\Delta F_S = \Delta F_{c,S} + \Delta F_{m,S} \tag{159}$$

where $\Delta F_{c,S}$ is the free energy change required for the creation of a cavity in the mobile phase to accomodate the solute molecule, whereas $\Delta F_{m,S}$ is the energy change from the interaction between the solute molecule and the surrounding mobile phase. The further essential assumption in the theoretical treatment of Jandera et al. is that the solvent mixture is considered to occupy the cavity volume as a chemical individual interacting with the neighboring solvent mixture.

Jandera et al. believe that, in the case of polar interactions between molecules A and B, the energy of the interaction ε_{AB} can be represented by the product

$$\varepsilon_{AB} = C_A I_A C_B I_B \tag{160}$$

where I_A and I_B are interaction indexes of A and B and C_A and C_D are coefficients that correct for the scale chosen for I. According to Jandera et al., "This originates from the fact that a given molecule A can interact with another molecule B in different ways, depending on the relative chemical functionalities of A and B. This cannot be fully accounted for by the indexes I_A and I_B and thus requires the introduction of the coefficients C_A and C_B. Because a given solute is characterized by a unique value of I, the distinction between the different selectivity classes will appear at the level of the coefficient C."

Thus, with the definition in Eq. (160), Eq. (159) can be written in the following form:

$$-\Delta F_s = C_m^2 I_m^2 + C_m C_s I_m I_s \tag{161}$$

where subscripts s and m refer to the molecules of the solute and mobile phase, respectively. For a binary mobile phase, I_m is represented by

$$I_m = \phi_1 I_1 + \phi_2 I_2 \tag{162}$$

After inserting Eq. (162) into Eq. (161) and rearranging the obtained expression, we have

$$\ln K = -\frac{\Delta F_s}{RT} = c + b_1 \phi_2 + b_2 \phi_2^2 \tag{163}$$

where

$$c = \frac{1}{RT} C_m I_1 (C_m I_1 + C_s I_s) \tag{164a}$$

$$b_1 = \frac{1}{RT} C_m (I_2 - I_1)(C_s I_s + 2C_m I_1) \tag{164b}$$

$$b_2 = \frac{1}{RT} C_m^2 (I_2 - I_2)^2 \qquad (164c)$$

As a matter of fact, the quasi-linear dependence of ln K upon the mobile-phase composition in RPLC has been widely confirmed experimentally by Jandera et al. [62]. Sometimes, however, for example in the separation of Aza-arenes on Hypersil C8 in the water + methanol mixture described by Colin et al. [63], deviations from linear behavior are readily visible.

The theory of Jandera et al. referred to above is an interesting combination of Snyder's concept of polarity indexes and the solvophobic theory of retention in RPLC developed some years earlier by Horváth et al. [64]. The theory of Horváth et al. is still the most advanced theoretical treatment of partition coefficients in RPLC and is based on the idea first put forth by Sinanoǵlu [65].

In the model accepted by Horváth et al., the interaction between the solute and the stationary phase is considered a reversible association of the solute molecules S with the hydrocarbonaceous ligand L at the surface:

$$S + L \rightleftharpoons SL \qquad (165)$$

where the complex SL is assumed to be formed by solvophobic interactions. The process is characterized by the equilibrium constant K:

$$K = \frac{[SL]}{[S][L]} \qquad (166)$$

According to Horváth et al. (whose solvophobic theory is restricted to unionized solutes), "... the molecular association in solution can be conceptually broken down into two processes. One is the interaction of the molecules S and L, to yield SL in a hypothetical gas phase without any intervention of the solvent. The other more involved process entails the interactions of the associating species and the complex individually with the solvent proper. ... the association in the gas phase is assumed to occur by van der Waals forces only and the free energy change of the process is denoted by $\Delta F_{vdw,assoc.}$"

The standard unitary free energy change for the overall association in solution $\Delta F^0_{assoc.}$, is given by

$$\Delta F^0_{assoc.} = \Delta F_{vdw,assoc.} + (\Delta F_{c,SL} + \Delta F_{m,SL}) - (\Delta F_{c,s} + \Delta F_{m,s})$$
$$- (\Delta F_{c,L} + \Delta F_{m,L}) - RT \ln \frac{RT}{P_0 v_m} \qquad (167)$$

where the last term, which contains the mole volume v_m and the atmospheric pressure P_0, accounts for the entropy change arising from the change in "free volume."

Following Sinanoğlu [65], Horváth et al. assume that the free energy of the cavity formation for some species j can be expressed as

$$\Delta F_{c,j} + \kappa_j^e \alpha_j^{(v)} \sigma_m N \left[1 - \left(1 - \frac{\kappa_j^s}{\kappa_j^e} \right) \left(\frac{\partial \ln \sigma_m}{\partial \ln T} + \frac{2}{3} \alpha_j^{(v)} T \right) \right] \quad (168)$$

where $\alpha_j^{(v)}$ is the molecular surface area of species j, σ_m is the surface tension of the solvent mixture, and the product $\alpha_j^{(v)} \sigma_m$ approximately represents the energy required for the formation of the cavity with the area $\alpha_j^{(v)}$. Further, κ_j^e is a correction factor that expresses the ratio between this energy and the energy required to expand the planar surface of the solvent mixture by the same area $\alpha_j^{(v)}$ and κ_j^s is the corresponding function for entropy. The value κ_j^e can be estimated as follows:

$$\kappa_j^e = 1 + (\kappa^e - 1) \left(\frac{v_m}{v_j} \right)^{2/3} \quad (169)$$

where κ^e is related to the internal energy change associated with vaporization of the solvent and can be calculated from the heat of vaporization. A similar relationship is used to estimate κ_j^s.

The terms $\Delta F_{m,j}$ in Eq. (167) are assumed to be the sum of a van der Waals component, $\Delta F_{m,j,vdw}$, and an electrostatic term, $\Delta F_{m,j,es}$:

$$\Delta F_{m,j} = \Delta F_{m,j,vdw} + \Delta F_{m,j,es} \qquad j = S, L, SL \quad (170)$$

The van der Waals term can be calculated from the ionization potentials, refractive indices, and molecular volumes of the solvent and species j. The electrostatic term, on the other hand, can be calculated from the static dielectric constant of the solvent and from the static dipole moment and polarizability of the species j.

With a number of additional simplifying assumptions, the term $\Delta F^0_{assoc.}$ can be fully estimated theoretically and the partition coefficient K evaluated as follows:

$$\ln K = -\frac{\Delta F^0_{assoc.}}{RT} \quad (171)$$

The simplifying assumptions mentioned above are

$$\kappa_j^s = \kappa_j^e = 1 \qquad \text{for } j = L, S, SL \tag{172a}$$

$$\Delta F_{m,SL,vdw} = \Delta F_{m,L,vdw} \tag{172b}$$

$$\mu_{SL} = \mu_S \tag{172c}$$

$$\mu_L = 0 \tag{173d}$$

A great advantage of the theory developed by Horváth et al. is the possibility of ab initio calculation of K using thermodynamic quantities and physical constants that are often tabulated in the literature. However, it is also known that purely ab initio calculations in physics and chemistry rarely yield a result in quantitative agreement with experimental data. The theory of Horváth et al. is not an exception. Figure 7 shows a comparison between the theoretically calculated and experimentally measured partition coefficients for one solute chromatographed in the water + methanol mixture investigated by Horváth et al. The theory of Horváth et al. is very rigorous in that it takes into account the interactions in both the mobile and the surface phase. It is especially important in RPLC systems in which graphitized carbons are used as column packings. A real advance in applying the solvophobic theory to RPLC systems with graphitized carbons was recently made by Ciccioli et al. [66].

However, since the theory of Horváth et al. requires difficult and time-consuming calculations, it is not very convenient for a rapid prediction of retention data. This is probably why this advanced theory is not widely applied. The simple theories of Schoenmakers et al. and of Jandera et al. are much more useful, but unfortunately they neglect the contribution to retention made by changes in the activity coefficient of solute in the surface phase γ_{s1}.

Looking for a compromise between these two situations, we focus our attention on the theory developed by Ościk as long ago as 1965 [67]. The starting point in Ościk's theoretical treatment was the equation

$$\ln K = \ln \gamma_s - \ln \gamma_{s1} \tag{173}$$

Following the consideration of Buchowski [68], based on the Scatchard-Hildebrand theory of solutions, the bulk activity coefficient γ_s can be expressed as

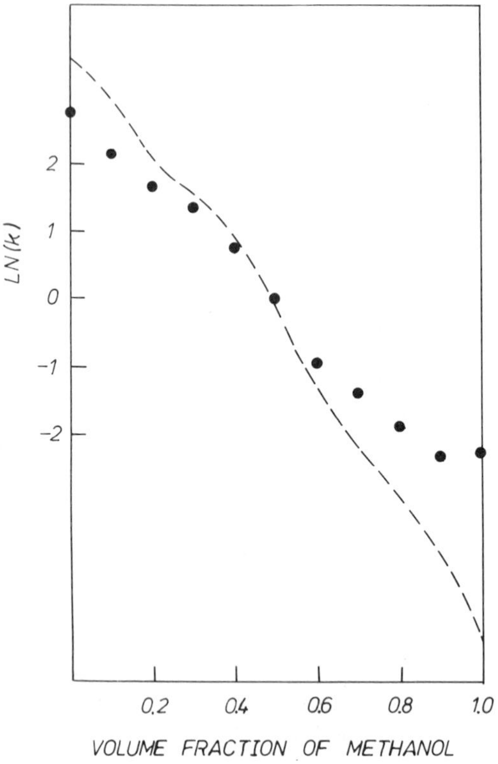

Figure 7. A comparison of the experimentally measured and theoretically calculated capacity factors for *o*-toluic acid chromatographed in a (water + methanol) mixture on an octadecyl silica column. The filled circles show the experimental values of ln k found by Horváth et al. [64]; the line represents the result of their theoretical calculation. The theoretical values of ln k were increased by a constant value, such that they could be equal to the experimental values at the bulk concentration of methanol equal to $\phi_1 = 0.5$. (The theoretical curve was calculated by computer after digitizing the original data presented in Fig. 15 of the paper by Horváth et al. [64] and their subsequent summation.)

$$\ln \gamma_s = \sum_{i=1}^{s-1} \phi_i \mu_s^{(i)} - \frac{v_s}{2} \sum_{i=1}^{s-1} \sum_{j=1}^{s-1} \phi_i \phi_j B_{ij} \tag{174}$$

where B_{ij} is the concentration-independent coefficient and $\mu_s^{(i)}$ is the chemical potential of solute in a pure bulk solvent i. Ościk then proposes to write γ_{s1} in a similar form,

$$\ln \gamma_{s1} = \sum_{i=1}^{s-1} \phi_{i1}\mu_{s1}^{(i)} - \frac{v_s}{2} \sum_{i=1}^{s-1} \sum_{j=1}^{s-1} \phi_{i1}\phi_{j1}B_{ij} \tag{175}$$

where $\mu_{s1}^{(i)}$ is the chemical potential of solute in the pure solvent i on the solid surface. After inserting Eqs. (174) and (175) into Eq. (173) and rearranging the resulting expression, one obtains

$$\ln K = \sum_{i=1}^{s-1} \phi_i \ln K_i + \sum_{i=1}^{s-1} \mu_{s1}^{(i)} n_i^e + \frac{v_s}{2} \sum_{i=1}^{s-1} \sum_{j=1}^{s-1} B_{ij}(\phi_{i1}\phi_{j1} - \phi_i\phi_j) \tag{176}$$

where

$$\ln K_i = \mu_s^{(i)} + \mu_{s1}^{(i)} \tag{177}$$

Thus, when the preferential adsorption of any of the solvents is small (i.e., the n_i^e are small), then ln K should be a linear function of the mobile-phase composition. With a binary mobile phase, the possible deviations from linearity should be proportional to the surface excess of one of the solvents.

There is a basic misunderstanding in the literature that is connected to Ościk's theory. This is the attempt made by various authors [69,70] to apply Ościk's theory to typical NPLC systems, in which the term ln $K^{(comp)}$, not ln $K^{(in)}$, plays the major role.

III. LIST OF SYMBOLS

Latin

a'	Snyder's adsorbent activity function, defined in Eq. (65)
a_p, a_v	numbers of the nearest neighbors-adsorption sites in the same lattice plane (p) and the plane above (v)
A	function defined in Eq. (136)
A_1, A_2	constants defined in Eqs. (156) and (157)
A_{ij}	nonideality parameter, equal to $z\alpha_{ij}/RT$
b_{ij}	heterogeneity parameter defined in Eq. (52)
B_1, B_2	constants defined in Eqs. (156) and (157)
B_{ij}	nonideality parameter defined in Eq. (174)
c_{12}	heterogeneity parameter defined in Eqs. (33) and (36)
C	constant defined in Eqs. (156) and (157)
C_i	scale correction coefficient defined in Eq. (160)

C_{ij}	constant defined in Eq. (108)
C_{123}	constant defined in Eqs. (109) and (110)
$C_{12}^{(b)}$	function defined in Eq. (55)
$C_{12}^{(m)}$	function defined in Eq. (57)
D_{ij}	constant defined in Eq. (108)
D_{123}	constant defined in Eqs. (109) and (110)
f_p, f_v	lattice parameters equal to a_p/z and a_v/z
F_l	function defined in Eq. (15)
g_p	fraction of surface sites of pth type
k	capacity ratio
k_i	capacity ratio of solute in the pure solvent i
K_i	partition coefficient in the pure solvent i
K_p	partition coefficient for the pth type of surface sites
K_t	partition coefficient of solute for a total heterogeneous surface
K_{ij}	partition coefficient for a binary solvent mixture i + j
$K^{(in)}$	interaction factor in K, defined in Eq. (150)
$K^{(comp)}$	competition factor in K defined in Eq. (150)
K_{ij}^0	equilibrium constant defined in Eq. (35)
$K_{ij}^{(l)}$	equilibrium constant for the adsorption in the l-th lattice plane
l, m	indexes of lattice planes
L	number of lattice planes in the adsorbed phase
m_{12}	heterogeneity parameter defined in Eq. (35)
n_i^e	molar surface excess of solvent i
n_i^0	surface capacity for component i
N	Avogadro's number
N_0	number of adsorption sites in one lattice plane
N_{il}	number of molecules of component i in the lth lattice plane
q_i	molecular partition function of component i in the bulk phase
q_{il}	molecular partition function of component i in the lth lattice plane
Q_m, Q_s	canonical system partition functions for one lattice plane in the mobile m and surface s phases
r_i	number of lattice sites in the first lattice plane, occupied by one molecule i
r_{ij}	ratio equal to r_i/r_j
R	gas constant
T	absolute temperature

U_{in}, $U_{in}^{(b)}$, $U_{in}^{(id)}$, $U_{in}^{(mix)}$	interaction energies defined in Eqs. (5), (6), (8), and (9)
$U_{123}^{(mix)}$	interaction energy for a ternary solution, defined in Eq. (145)
v_i	molecular volume of component i in the bulk phase
v_m	molar volume of solvent mixture
V_M	total volume of a chromatographic column, excluding the volume of the adsorbent
V_R	total retention volume
V_s	volume of the surface phase
V_{ib}^R	retention volume for the breakthrough from zero to concentration ϕ_i
V_{id}^R	retention volume of the concentration disturbance of component i
x_i	molar concentration of i in the mobile bulk phase
x_{il}	molar concentration of i in the lth lattice plane
x_i^0	concentration of a solvent mixture before contact with adsorbent
z	total number of nearest-neighbors adsorption sites

Greek

α_i	surface area occupied by one admolecule i
$\alpha_j^{(v)}$	molecular surface area of species j
α_{12}	"interchange energy" defined in Eq. (7)
γ_i	activity coefficient of component i in the bulk phase
γ_{il}	activity coefficient of component i in the lth lattice plane
δ_i	solubility parameter of substance i
ε_i	adsorption energy of a molecule i
ε_i^0	elution strength of pure solvent i
ε_{ij}^0	elution strength of a binary solvent mixture 1 + 2 defined in Eq. (88)
ε_{ij}	difference between ε_i and ε_j, defined in Eq. (18)
$\varepsilon_{ij}^!$	most probable value of ε_{ij}

ε_{12}^{c} concentration function defined in Eq. (32)

ε_{ij}^{l}, ε_{ij}^{m} minimum and maximum value of ε_{ij} on a given heterogeneous surface

ε_{12}^{r} generalized function ε_{12}^{c}, defined in Eq. (44)

$\varepsilon_{ij}^{(p)}$ interaction energy between a molecule i and molecule j, which are nearest neighbors

κ^{e} parameter in Eq. (169) related to the internal energy change and associated with the vaporization of solvent

κ_{j}^{e} ratio of the energy required for the formation of a cavity in the solvent to the energy required to expand a planar solvent surface by the area equal to the cavity area

κ_{j}^{s} ratio of the same kind as κ_{j}^{e} but connected with entropy production

μ_i chemical potential of component i in the bulk phase

μ_{il} chemical potential of component in the lth lattice plane

μ_{i}^{0} standard chemical potential of component i in the bulk phase

μ_{il}^{0} standard chemical potential of component i in the lth lattice plane

σ_m interfacial tension between air and solution

σ_{i}^{0} interfacial tension between solid and solvent i

ϕ_i volume fraction of solvent i in the bulk phase

ϕ_{il} volume (areal) fraction of i in the lth lattice plane

ϕ_{i0} value of the function $\phi(\varepsilon_{ij})$ at the point $\varepsilon_{ij} = \varepsilon_{ij}'$

ϕ_{it} the same as $\phi_{il}^{(h)}$

$\phi_{il}^{(h)}$ averaged value of ϕ_{il} defined in Eq. (30)

$\phi_{1t}^{(n)}$ nth-order approximation for ϕ_{1t}

χ_i differential distribution of surface sites among various values of ε_i

χ_{ij} differential distribution of adsorption sites among various values of ε_{ij}

$\mathfrak{X}_{ij}$ indefinite integral of χ_{ij}

Ω_{ij} physical domain of the variable ε_{ij}

Combined Symbols

$\gamma_{s(f)}$ activity coefficient of solute in phase f

ΔF_s standard free energy change for the transfer of a solute molecule from the mobile to stationary phase defined in Eq. (159)

ACKNOWLEDGMENT

The author would like to express his warm thanks to Dr. Jolanta Narkiewicz-Michalek for her technical assistance in preparing the manuscript.

REFERENCES

1. D. H. Everett, Thermodynamics of adsorption from solutions, Part 1, Perfect systems, *Trans. Faraday Soc.*, *60*:1803 (1964). Thermodynamics of adsorption from solutions, Part 2, Imperfect systems, *Trans. Faraday Soc.*, *61*:2478 (1965).
2. R. Defay and I. Prigogine (with collaboration of A. Bellemans and translated by D. H. Everett), *Surface Tension and Adsorption*, Longmans, London (1966).
3. G. Schay, On the thermodynamics of physical adsorption from solutions of non-electrolytes at the surfaces of solid adsorbents, *J. Colloid Interface Sci.*, *42*:478 (1973).
4. G. Schay and L. G. Nagy, Critical discussion of the use of adsorption measurements from liquid phase for surface area estimation, *J. Colloid Interface Sci.*, *38*:302 (1972).
5. R. P. Scott and P. Kucera, Solute-solvent interactions on the surface of silica gel, *J. Chromatogr.*, *149*:93 (1978).
6. E. H. Slaats, J. C. Kraak, W. J. T. Brugman, and H. Poppe, Study of the influence of competition and solvent interaction on retention in liquid-solid chromatography by measurement of activity coefficients in the mobile phase, *J. Chromatogr.*, *149*: 255 (1978).
7. E. H. Slaats, W. Markowski, J. Fekete, and H. Poppe, Distribution equilibria of solvent components in reversed-phase liquid chromatographic columns and relationship with the mobile phase volume, *J. Chromatogr.*, *207*:299 (1981).
8. S. G. Ash, R. Bown, and D. H. Everett, A high-precision apparatus for the determination of adsorption at the interface between a solid and a solution, *J. Chem. Thermodynamics*, 5:239 (1973).
9. J. F. K. Huber and R. G. Gerritse, Influence of static and dynamic effects on the reproducibility of retention data in gas-liquid chromatography, *J. Chromatogr.*, *80*:25 (1973).
10. J. F. K. Huber, W. Rudziński, and A. Dawidowicz, Unpublished work.
11. S. A. Busev, S. I. Zverev, O. G. Larionov, and S. Jakubov, Study of adsorption from solutions by column chromatography, *J. Chromatogr.*, *241*:287 (1982).

12. M. I. Selim, J. F. Parcher, and P. J. Lin, Adsorption of polar solutes on liquid-modified supports, *J. Chromatogr.*, *239*:411 (1982).
13. K. G. Wahlund and I. Beijersten, Adsorption of 1-pentanol on alkyl-modified silica and its effect on the retention mechanism in reversed-phase liquid chromatography, *Anal. Chem.*, *54*:128 (1982).
14. W. Rudziński and J. Jagiełło, Low-temperature adsorption of gases on heterogeneous solid surfaces: Effects of surface topography, *J. Low Temp. Phys.*, *48*:307 (1982).
15. W. Rudziński, J. Jagiełło, and Y. Grillet, Physical adsorption of gases on heterogeneous solid surfaces: Evaluation of the adsorption energy distribution from adsorption isotherms and heats of adsorption, *J. Colloid Interface Sci.*, *87*:478 (1982).
16. J. Ościk, A. Dabrowski, S. Sokołowski, and M. Jaroniec, Adsorption on heterogeneous surfaces: Analogy between adsorption from gases and solutions, *J. Res. Inst. Catalysis, Hokkaido Univ.*, *23*:91 (1976).
17. S. Sircar and A. L. Myers, A thermodynamic consistency test for adsorption from binary liquid mixtures on solids, *AIChE J.*, *17*:186 (1971).
18. C. Minka and A. L. Myers, Adsorption from ternary liquid mixtures on solids, *AIChE J.*, *19*:453 (1973).
19. S. Ross and J. P. Olivier, *On Physical Adsorption*, Interscience, London (1964).
20. T. L. Hill, Statistical mechanics of adsorption, VI, Localized unimolecular adsorption on a heterogeneous surface, *J. Chem. Phys.*, *17*:762 (1949).
21. W. Rudziński, J. Ościk, and A. Dabrowski, Adsorption from solutions on patchwise heterogeneous solids, *Chem. Phys. Lett.*, *20*:444 (1973).
22. M. Jaroniec, A simple model for the adsorption of mixtures on solids involving energetic heterogeneity of the surface and differences in molecular sizes of the components, *Thin Solid Films*, *81*:L97 (1981).
23. W. Rudziński and S. Partyka, Adsorption from solutions on solid surfaces: Effects of surface heterogeneity on adsorption isotherms and heats of immersion, *J. Chem. Soc. Faraday Trans. 1*, *77*:2577 (1981).
24. W. Rudziński, J. Narkiewicz-Michalek, and S. Partyka, Adsorption from solutions onto solid surfaces: Effects of topography of heterogeneous surfaces on adsorption isotherms and heats of immersion, *J. Chem. Soc. Faraday Trans. 1*, *78*:2361 (1982).
25. A. Dabrowski and M. Jaroniec, Application of Dubinin-Raduskevich and Freundlich equations to characterize the adsorption from non-ideal binary liquid mixtures on heterogeneous solid surfaces,

Acta Chim. Acad. Sci. Hung., *99*:255 (1979). Effects of surface heterogeneity in adsorption from binary liquid mixtures: Analysis of experimental data by using Langmuir-Freundlich type equations, *J. Colloid Interface Sci.*, *73*:475 (1979). A. Dabrowski and M. Jaroniec, Application of Dubinin-Radushkevich type equation for describing the adsorption from non-ideal solutions on real solid surfaces, *Acta Chim. Acad. Sci. Hung.*, *104*:183 (1980).

26. W. Rudziński, L. Lajtar, J. Zajac, E. Wolfram, and I. Pászli, Ideal adsorption from binary liquid mixtures on heterogeneous solid surfaces: Equations for excess isotherms and heats of immersion, *J. Colloid Interface Sci.*, *96*:339 (1983).
27. W. Rudziński, J. Narkiewicz-Michalek, R. Schöllner, H. Herden, and W. D. Einicke, Adsorption from binary liquid mixtures on heterogeneous solid surfaces: Analytical approximations for excess isotherms in the case of ideal adsorbed phases, *Acta Chim. Acad. Sci. Hung.*, *113*:207 (1983).
28. L. R. Snyder, *Principles of Adsorption Chromatography*, Marcel Dekker, New York (1968).
29. H. Colin and G. Guiochon, The solvent eluotropic strength on carbon adsorbents, *Chromatographia*, *15*:133 (1982).
30. B. L. Karger, L. R. Snyder, and C. Eon, Expanded solubility parameter treatment for classification and use of chromatographic solvents and adsorbents, *Anal. Chem.*, *50*:2126 (1978).
31. J. Ościk, Selected problems of the thermodynamics of adsorption from multicomponent solutions (in Russian), in *Physical Adsorption from Multicomponent Phases* (M. M. Dubinin, ed.), Nauka, Moscow (1972).
32. C. E. Brown and D. H. Everett, Colloid science, *Specialist Periodical Rep.*, *2*:52 (1975).
33. M. Borówko and M. Jaroniec, General formulation of adsorption from liquid mixtures on heterogeneous solid surfaces, *Rev. Roum. Chim.*, *25*:475 (1980).
34. M. Borówko, Adsorption from multicomponent solutions on homogeneous solid surfaces, *J. Colloid Interface Sci.*, *85*:540 (1982).
35. M. Borówko, M. Jaroniec, and W. Rudziński, Adsorption from multicomponent mixtures on heterogeneous solid surfaces, *Z. Phys. Chem.* (Leipzig), *260*:1027 (1979). Statistical thermodynamics of adsorption from multicomponent liquid mixtures on heterogeneous surfaces, *Monatsh. Chem.*, *112*:59 (1981).
36. M. Borówko, M. Jaroniec, J. Ościk, and R. Kusak, Adsorption of multicomponent liquid mixtures on heterogeneous surfaces, *J. Colloid Interface Sci.*, *69*:311 (1979).
37. M. Jaroniec, Kinetics of adsorption from liquid mixtures on heterogeneous solid surfaces, *J. Res. Inst. Catalysis, Hokkaido Univ.*, *26*:155 (1979).
38. E. Soczewiński and J. Jusiak, A simple molecular model of adsorption chromatography, XIV, R_F or R_M? Secondary retention

effects in thin layer chromatography, *Chromatographia, 14*:23 (1981).

39. L. R. Snyder and J. L. Glajch, Solvent strength of multicomponent mobile phases in liquid-solid chromatography: Binary-solvent mixtures and solvent localization, *J. Chromatogr., 214*:1 (1981). Solvent strength of multicomponent mobile phases in liquid-solid chromatography: Mixtures of three or more solvents, *J. Chromatogr., 214*:21 (1981).
40. L. R. Snyder, Mobile phase effects in liquid-solid chromatography: Importance of adsorption-site geometry, adsorbate delocalization and hydrogen bonding, *J. Chromatogr., 255*:3 (1983).
41. M. Jaroniec, B. Klepacka, and J. Narkiewicz, Liquid adsorption chromatography with a two-component mobile phase, I, Effects of non-ideality of the mobile phase, *J. Chromatogr., 170*:299 (1979).
42. R. P. Scott and P. Kucera, Solute interactions with the mobile and stationary phases in liquid-solid chromatography, *J. Chromatogr., 112*:425 (1975).
43. E. Soczewiński, Solvent composition effects in thin-layer chromatography systems of the type silica gel-electron donor solvent, *Anal. Chem., 41*:179 (1969).
44. P. Jandera and J. Churáček, Gradient elution in liquid chromatography, I, The influence of the composition of the mobile phase on the capacity ratio (retention volume, band width, and resolution) in isocratic elution—theoretical considerations, *J. Chromatogr., 91*:207 (1974).
45. P. Jandera and J. Churáček, Liquid chromatography with programmed composition of the mobile phase, in *Advances in Chromatography*, Vol. 19 (Giddings et al., eds.), Marcel Dekker, New York (1980), p. 125.
46. J. E. Paanakker, J. C. Kraak, and H. Poppe, Some aspects of the influence of the mobile phase composition in normal-phase liquid-solid chromatography, with special attention to the role of water present in binary organic mixtures, *J. Chromatogr., 149*:111 (1978).
47. S. Haara, K. Kunihiro, H. Yamaguchi, and E. Soczewiński, Ternary solvent system design for liquid-solid chromatography, *J. Chromatogr., 239*:687 (1982).
48. M. Jaroniec, J. Narkiewicz, and M. Borówko, Dependence of the capacity ratio on mobile phase composition in liquid adsorption chromatography, *Chromatographia, 11*:581 (1978).
49. M. Borówko and M. Jaroniec, Dependence of the distribution coefficient on the mobile phase composition in liquid adsorption chromatography, II, Analytical equations for the distribution coefficient involving non-ideality of the mobile phase and heterogeneity of the adsorbent surface, *Chromatographia, 12*:672 (1979).

50. M. Jaroniec, J. Różylo, and W. Golkiewicz, Liquid adsorption chromatography with mixed mobile phases, II, Effects of adsorbent heterogeneity, *J. Chromatogr.*, *178*:27 (1979).
51. M. Jaroniec and A. Patrykiejew, Theory of adsorption from multicomponent liquid mixtures on solid surfaces and its application to liquid adsorption chromatography, *J. Chem. Soc. Faraday Trans. 1*, *76*:2486 (1980).
52. M. Jaroniec and B. Ościk-Mendyk, Application of excess adsorption data measured for components of the mobile phase for characterizing chromatographic systems, *J. Chem. Soc. Faraday Trans. 1*, *77*:1277 (1981).
53. N. A. Smirnova, Lattice model for the surface region of solutions consisting of different-sized molecules with orientation effects, *Fluid Phase Equilibria*, *2*:1 (1978).
54. S. G. Ash, D. H. Everett, and G. H. Findenegg, Multilayer theory for adsorption from solution. A general theory for the adsorption of r-mers and its application to flexible tetramers and trimers, *Trans. Faraday Soc.*, *66*:708 (1970).
55. R. E. Boehm and D. E. Martire, A unified theory of retention and selectivity in liquid chromatography. Liquid-solid (adsorption) chromatography, *J. Phys. Chem.*, *84*:3620 (1980).
56. D. E. Martire and R. E. Boehm, Molecular theory of liquid adsorption chromatography, *J. Liquid Chromatogr.*, *3*:753 (1980).
57. K. Unger, *Porous Silica*, Elsevier, Amsterdam (1979).
58. H. Engelhardt and G. Ahr, Properties of chemically bonded phases, *Chromatographia*, *14*:227 (1981).
59. D. Morel and J. Serpinet, Gas chromatographic evidence for phase transitions in very compact octadecyl bonded silicas, *J. Chromatogr.*, *200*:95 (1980). Influence of the liquid chromatographic mobile phase on the transitions of alkyl-bonded silicas studied by gas chromatography, *J. Chromatogr.*, *214*:202 (1981).
60. P. J. Schoenmakers, H. A. H. Billiet, and L. de Galan, Systematic study of ternary solvent behavior in reversed-phase chromatography, *J. Chromatogr.*, *218*:261 (1981).
61. P. Jandera, H. Colin, and G. Guiochon, Interaction indexes for prediction of retention in reversed-phase liquid chromatography, *Anal. Chem.*, *54*:435 (1982).
62. P. Jandera, J. Churáček, and H. Colin, Gradient elution in liquid chromatography, XIV, Theory of ternary gradients in reversed-phase liquid chromatography, *J. Chromatogr.*, *214*:35 (1981).
63. H. Colin, J.-M. Schmitter, and G. Guiochon, Liquid chromatography of azaarenes, *Anal. Chem.*, *53*:625 (1981).
64. C. Horváth, W. Melander, and I. Molnár, Solvophobic interactions in liquid chromatography with nonpolar stationary phases, *J. Chromatogr.*, *125*:129 (1976).

65. I, Sinanoğlu, Solvent effects on molecular associations, in B. Pullman (ed.), *Molecular Associations in Biology*, Academic Press, New York (1968), p. 427.
66. P. Ciccioli, R. Tappa, and A. Liberti, An experimental method for testing the solvophobic theory by using graphitized carbon black in GC and LC, *Chromatographia, 16*:330 (1982).
67. J. Ościk, Adsorption from multicomponent solutions (in Polish), *Przem. Chem.*, *44*:129 (1965).
68. H. Buchowski, Standard chemical potentials in a multicomponent system in relation to excess thermodynamic functions of mixing, *Bull. Acad. Polon. Sci. Ser. Sci. Chim.*, *10*:195 (1962).
69. J. Ościk, J. K. Różylo, B. Ościk-Mendyk, and M. Jaroniec, Adsorption TLC with binary mobile phases, *Chromatographia*, *14*:95 (1981).
70. M. Jaroniec and J. Ościk, Liquid-solid chromatography. Recent progress in theoretical studies concerning the dependence of the capacity ratio upon the mobile phase composition, *J. HRC&CC*, 5:3 (1982).

7

Retention and Detection in Ion Exchange Chromatography

Dennis R. Jenke
Travenol Laboratories, Inc.
Morton Grove, Illinois

Gordon K. Pagenkopf
Montana State University
Bozeman, Montana

As a practical analytic methodology, ion chromatography (IC) is a relatively new technique, having been introduced in the chemical literature in 1975 by Small et al. [1]. Despite its youth, the technique has achieved a rapid acceptance and wide application in the areas of environmental analysis, quality and process control, clinical chemistry, microelemental analysis, and geologic analysis, primarily owing to the versatility, sensitivity, and simplicity of its various commercial embodiments.

In its most general sense, ion chromatography may be defined as a technique whereby mixtures of ions are separated and quantitated. Its original intent, and widest application to date, is in the quantitation of ions (both organic and inorganic) that have dissociation constants less than 7. These species had traditionally been a problem in chromatographic analysis for two reasons: (1) many of these species contain no chromophores or electrophores and are therefore immune to detection via photometric, fluorescent, or electrochemical means; and (2) the highly ionic chracter of these species required the utilization of ion-exchange resins for their separation, which were inconvenient owing to problems of swelling and a slow rate of mass transport [2]. Since separation in IC is accomplished by ion exchange [that is, the sample is passed through a column containing an ion-exchange material, and separation is accomplished via the dynamic competition between the analyte(s) and ionic component(s) of the mobile phase (eluent ion(s))], one observes that IC represents a popularized and more highly evolved form of the more generalized method of ion-exchange chromatography.

The two distinct technical advances that have served to stimulate the evolution of the IC methodology are (1) the development of a technology capable of reproducibly manufacturing resin materials that offer a low exchange capacity, good mass transport properties, freedom from swelling, and effective ion selectivity; and (2) the identification and development of detection methodologies capable of distinguishing between a charged analyte and a mobile phase that also contains charged ions serving as the eluent species. Continued technical advances in both column and detector design and coupling have served to broaden the application of IC from that described previously to include the quantitation of organic ions, alkali, alkaline earth, transition, and lanthanide metals and weakly dissociated inorganic anions.

It is the purpose of this chapter to examine separation and detection as two important facets of IC methodology from a theoretical standpoint throught the description and development of theoretical formulations that model the separation process, and to consider the theory of operation of the most widely used detectors that are indigenous to IC methodology. For a more practical discussion of ion chromatography methodologies, a comparison of their relative merits, and a guide to potential applications, the reader is referred to the many excellent reviews available on this subject [2-8].

I. MODELING OF ANALYTE ELUTION

A. General

The qualitative observation that the retention characteristics of analytes in IC were in some way or another related to the ionic nature of the mobile phase employed was shared by many researchers during the early development of IC [1,9-12]; however, few attempts have been made to quantitatively model separation behavior based on a theoretical evaluation of the ion-exchange process. Two such models have been proposed. Both treat the separation process in terms of classic thermodynamics, that is, that the separation of analytes by an ion-exchange resin can be considered to be controlled mechanistically by the thermodynamic equilibrium that exists at the individual resin sites between the analyte species and any ionic component of similar charge type in the mobile phase (eluent ions). These models differ slightly in derivation, but most prominently with respect to the number of mobile-phase eluent ions with which each is equipped to deal. Gjerde and associates have developed a model that is applicable to a situation in which the mobile phase contains a single active component [9], and Jenke and Pagenkopf have modified and applied a more generalized approach, originally described by Hoover [13], to the various subclasses of IC [14-16]. Both are equivalent in their foundation (if somewhat different in derivation), and their successful application is mobile-phase dependent. Each will be considered in greater detail in terms of derivation and application in the following discussion.

B. Mobile Phase Containing Single Eluent Ion

One can consider the equilibrium that exists between two species of similar charge type on a strong ion-exchange resin to have the form

$$xA^y + y[B - R_x] \rightleftharpoons yB^x + x[A - Ry] \quad (1)$$

where R represents the ion exchanger, A and B are the competing ions, and x and y are the absolute values of their associated charges. In this derivation, species A represents the analyte ion and species B is the mobile-phase eluent. The equilibrium constant for this expression ($K_B{}^A$) becomes

$$K_B^A = \frac{[A - Ry]^x [B]^y}{[A]^x [B - R_x]^y} \quad (2)$$

where the brackets indicate the activities of the various species. It is noted that although both [A] and [B] are solution-phase activities, $[A - R_y]$ and $[B - R_x]$ both refer to the activity of a resin-bound ion. If the sample size is small, the quantity of the eluent ion on the equilibrated resin, $[B - R_x]$, is essentially constant and equal to the column capacity Q. The ratio of the amount of A on the resin to that in the solution phase in contact with the resin ($[A - R_y]/[A]$) is a distribution ratio D, which classically can be related to a reduced retention volume V_e' for analyte A by the equation

$$D = \frac{V_R'}{w} \quad (3)$$

where the reduced retention volume is the difference between the measured retention volume for analyte A and the column void volume, and w is the weight of the resin in the column.

Substituting D and Q into Eq. (2) yields

$$K_B^A = D^x \left(\frac{[B]}{Q}\right)^y \quad (4)$$

which, when combined with Eq. (3), produces

$$K_B^A = \left(\frac{V'_R}{w}\right)^x \left(\frac{[B]}{Q}\right)^y \tag{5}$$

Rearranging and taking the logarithm of both sides:

$$\log K_B^A w^x = x \log V'_R + y \log [B] - y \log Q \tag{6}$$

Now if the weight of the resin in the column and the capacity of the column are constant, and if the selectivity coefficient is truly independent of the mobile-phase composition, then Eq. (6) simplifies to

$$\log V'_R = \frac{-y}{x} \log [B] - \text{constant} \tag{7}$$

which is consistent with the Gjerde derivation [9].

C. Mobile Phase Containing Multiple Eluent Ions

Table 1 lists common mobile phases used in ion chromatography in terms of their salt content, appropriate acid (or base) dissociation constants, and common pH. One can see that, especially in anion analysis, the mobile phase can often contain one (or more) weak acids dissociated among their various charged forms, each of which may contribute to the elution process. As an extreme case, consider the mixed carbonate-bicarbonate eluent used for most common suppressed applications. In this system, the mobile phase contains the bicarbonate, carbonate, and hydroxide species, all of which may contribute to analyte migration. Clearly, in such a situation, the single-species derivation can only roughly approximate analyte behavior, even if the column selectivity is much larger for one of the eluent ions than for the others.

In order to consider the interaction that can occur in a situation in which multiple eluents and the analyte compete for resin sites on chromatographic columns, the conventional theory of liquid chromatography can be coupled to a thermodynamic treatment of ion-exchange equilibrium to produce a viable model. Such an approach has been explored by Hoover [13] and later refined and applied to the various forms of ion chromatography by Jenke et al. [14-16] and has proven to be moderately successful in terms of predicting retention characteristics. The mathematical development of such models will be discussed in detail below.

One of the important assumptions made in this approach (which is borrowed from conventional chromatographic theory) is that the reduced retention volume of the analytes is equal to the volumetric distribution coefficient [see Eq. (3)]. Additional assumptions made in this derivation are as follows.

1. Stoichiometric exchange occurs between ionic species of different formal charge (electroneutrality requirement).
2. At any given time the exchange sites on the column are dominated by eluent species, and the total column capacity Q can be written

$$Q = \Sigma_i [E_i] \tag{8}$$

   where $[E_i]$ is the concentration of the ith eluent species, having a formal charge i.
3. For every ionic pair in the system there is a selectivity coefficient corresponding to a concentration equilibrium for the exchange reaction. That is, for the exchange reaction

$$A_i + \frac{i}{j} R_j B_j \rightleftharpoons R_i A_i + \frac{i}{j} B_j \tag{9}$$

   the selectivity coefficient k_{AB} can be written

$$k_{AB} = \frac{[A_i]}{(A_i)} \left(\frac{(B_j)}{[B_j]} \right)^{(i/j)} \tag{10}$$

   where the brackets indicate a concentration of species in the resin and the parentheses concentration of species in solution.

Let us note in passing that implicit in these statements are the assumptions that column capacity is independent of species and eluent composition, that the nature of the column and solution chemical environment is such that ionic strength effects can be ignored, and that exchange isotherms are essentially linear. Let us also note that selectivity coefficients for both analyte and eluent species can be expressed in terms of a single species; in the following derivation, selectivity coefficients will be symbolized by either K or X, depending upon whether an analyte or eluent species is being considered. Thus, selectivity coefficients referring to the relationship between two eluent ions are denoted by the symbol X, and coefficients relating to an eluent ion and an analyte are given by K.

For derivation of the model designed to simulate the behavior of a nonsuppressed chromatographic system, consider the case of a

Table 1. Eluents Commonly Used in Ion Chromatography

Eluents	Conjugate acid	Dissociation constant(s), pK_a	pH range
Anion analysis			
Suppressed methods, salts			
$Na_2B_3O_7$	$H_2B_4O_7$	9.1	9-11
NaOH	H_2O	14	9-11
$NaHCO_3$-Na_2CO_3	H_2CO_3	6.3, 10.3	9-11
Nonsuppressed methods, salts			
Na benzoate	$C_6H_5CO_2H$	4.18	4-6
Na phthalate	$C_6H_4(CO_2H)_2$	3.10, 5.40	4-6.5
Na phthalate-Na borate	$C_6H_4(CO_2H)_2$-H_3BO_3	3.10, 5.40, 9.24	8-10
Na citrate	$HO_2CC(OH)(Ch_2CO_2H)_2$	3.07, 4.75, 6.40	5-7
Na malonate	$HO_2CCH_2CO_2H$	2.80, 6.10	4-6
NaOH	H_2O	14	9-11
Na tartrate	$H_2C(CHOH)_2CO_2H$	3.0, 4.4	3-4
Cation analysis			
Suppressed methods, reagents			
HNO_3			2-3
Phenylene diamine-HNO_3			2-3
Zn^{2+}-HNO_3			2-3
Nonsuppressed methods, reagents			
HNO_3			5-7
Ethylene diamine		9.93, 6.86	5-7

polyprotic acid H_2P serving as the eluent source. At low pH, the contribution of OH^- to analyte elution is small, and thus the column capacity can be written

$$Q = [E_1] + 2[E_2] \tag{11}$$

where E_1 refers to the HP^- species and E_2 is the P^{-2} species. By definition of the selectivity coefficient,

$$X_{21} = \frac{[E_2]}{[E_2]}\left(\frac{(E_1)}{[E_1]}\right)^2$$

or, by rearrangement,

$$[E_2] = \frac{X_{21}(E_2)[E_1]^2}{(E_1)^2} \tag{12}$$

Combination of Eqs. (11) and (12) yields

$$Q = [E_1] + \frac{2X_{21}(E_2)[E_1]^2}{(E_1)^2} \tag{13}$$

Solution of Eq. (13) by the quadratic formulation for $[E_1]$ results in Eq. (14):

$$[E_1] = \frac{\sqrt{1 + \dfrac{8(E_2)X_{21}Q}{(E_1)^2}} - 1}{4(E_2)X_{21}/(E_1)^2} \tag{14}$$

Now the volumetric distribution coefficient for analyte i can be written

$$D_i = K_{ij}D_j^{\,i/j} \tag{15}$$

where

$$D_j = \frac{[E_j]}{(E_j)}$$

and, for a divalent analyte,

$$D_2 = \frac{K_{21}[E_1]^2}{(E_1)^2} \tag{16}$$

Combination of Eq. (16) with Eq. (14) produces a result that, via Eq. (3), defines the relationship between analyte behavior and eluent composition:

$$D_2 = \frac{K_{21}}{(E_1)^2}\left(\frac{\sqrt{1 + \dfrac{8(E_2)X_{21}Q}{(E_1)^2}} - 1}{\dfrac{4(E_2)X_{21}}{(E_1)^2}}\right)^2 \tag{18}$$

$$D_2 = \frac{K_{21}(E_1)^2}{16X_{21}^2(E_2)^2}\left(\sqrt{1 + \frac{8(E_2)X_{21}Q}{(E_1)^2}} - 1\right)^2 \tag{18}$$

It should be noted that rarely is eluent composition expressed in terms of the absolute distribution of the eluent among its various protonated forms; generally the eluent is described in terms of pH and total acid present (E_T). Considering the acid dissociation equilibrium $HP^- \rightleftharpoons H^+ + P^{-2}$, we note that

$$\frac{(P^{-2})(H^+)}{(HP^-)} = K_{a2}$$

or

$$\frac{(E_2)}{(E_1)} = \frac{K_{a2}}{(H^+)} \tag{19}$$

and since

$$(E_1) + (E_2) = E_T$$

$$(E_1) = \frac{E_T}{1 + K_{a2}/(H^+)} \tag{20}$$

Substitution of Eqs. (19) and (20) into Eq. (18) produces

$$D_2 = \frac{K_{21}(H^+)^2}{16X_{21}^2K_{a2}^2}\left(\sqrt{\frac{1 + 8K_{a2}X_{21}Q[1 + K_{a2}/(H^+)]}{(H^+)E_T}} - 1\right)^2 \qquad (21)$$

Until now, the derivation ignores the possibility that the eluent pH is sufficiently low that (H_2P) is significant. For phthalate eluents with $pk_{a1} = 3.2$ and a normal operating pH range from approximately 4.2 to 6.5, this effect can become important. In this case, E_T can be replaced by $(1 - \alpha_{H_2P})E_T$, where

$$\alpha_{H_2P} = \frac{(H^+)^2}{K_{a1}K_{a2} + K_{a1}(H^+) + (H^+)^2} \qquad (22)$$

Similarly, for a univalent analyte,

$$D_1 = \frac{K_{11}[E_1]}{(E_1)}$$

and

$$D_1 = \frac{K_{11}(H^+)}{4X_{21}K_{a2}}\left(\sqrt{1 + \frac{8K_{a2}X_{21}Q[1 + K_{a2}/(H^+)]}{(H^+)E_T}} - 1\right) \qquad (23)$$

Equations (41) and (43) can be generalized to define the behavior of an analyte of any change n by the expression

$$D_n = K_{n1}\left(\frac{(H^+)}{4X_{21}K_{a2}}\right)^n\left(\sqrt{1 + \frac{8K_{a2}X_{21}Q[1 + K_{a2}/(H^+)]}{(H^+)(E_T)}} - 1\right)^n$$

(24)

Although the theoretical treatment for a three-component system (for example, the OH^-/HCO_3^--CO_3^{2-} eluent used in suppressed systems) is strictly analogous to that for the case considered above, the mathematics become somewhat more complicated and therefore will not be given here. However, the appropriate mathematical derivation produces the relationship between elution volume and eluent composition for divalent and monovalent analytes shown below.

$$D_2 = \left(\frac{K_{21}}{(X_{21})^2}\right)\left(\frac{(E_1 + X_{31}E_3)^2}{8E_2^2}\right)\left[1 + \frac{8X_{21}QE_2}{2(E_1 + X_{31}E_3)^2} - \sqrt{1 + \frac{8X_{21}QE_2}{(E_1 + X_{31}E_3)^2}}\right] \tag{25}$$

$$D_1 = \frac{K_{11}}{X_{21}}\left(\frac{E_1 + X_{31}E_3}{4E_2}\right)\left(\sqrt{1 + \frac{8X_{21}QE_2}{(E_1 + X_{31}E_3)^2}} - 1\right) \tag{26}$$

where

$E_1 = (HCO_3^-)$

$E_2 = (CO_3^{2-})$

$E_3 = (OH^-)$

Utilization of the relationships defined above requires determination of the selectivity coefficients for all analyte and eluent species. These species can be determined by batch-type experimentation [17] or by calculation from appropriate mathematical theories [18-23], but neither approach is sufficiently precise to fulfill the requirements of the formulation in this chapter. Batch-type extractions performed with relatively long equilibration times have no chance at approximating the kinetic behavior of the column exchange process and are generally specific for the experimental conditions in which they are measured; therefore, their validity over varying analyte-eluent concentration ranges is questionable. Mathematical formulations that attempt to estimate selectivity coefficients from independent data include consideration of the thermodynamic nature of the exchange process or a statistical evaluation of resin structure. These formulations generally consider the exchanger phase to be a concentrated aqueous electrolyte and couple the Gibbs-Duhem relationship for osmotic pressure with the Donnen model of ion exchange; all have been applied with some degree of success in modeling cation exchange. However, it is clear that this type of mathematical development cannot account for all selectivity effects exhibited by various resins and pairs of ions [24]. It is also observed that the application of this type of treatment to determining selectivities in ion chromatography is limited, in that the theories can only be applied to more concentrated systems (in which the osmotic coefficients can be accurately measured). A third approach involves the calculation of selectivity coefficients based on a minimum number of experiments performed using the system to be modeled. Retention times and calculated eluent compositions are put into the appropriate expression for each experimental condition evaluated, producing a series of simultaneous equations with only the selectivity coefficients as variables. These equations can be solved for the coefficients by a variety of strategies; the use of simplex optimization [13] and iterative minimization of the calculated error [15] have both been reported in the literature.

As a final note, we observe that, although in the previous derivations the term "concentration" has been employed in the most rigorous sense solution and resin-phase activities should be used throughout. This substitution has no effect on the nature of the derivation, but eluent concentrations (C) in the generalized equations should be replaced by activity (A); thus,

$$C_E = \gamma_E A_E \tag{27}$$

where γ_E is the activity coefficient. The importance of this activity correction is confirmed when one observes that (1) the ionic strengths of common ion-chromatographic eluents may be quite large (in excess of 0.01 N), and thus that γ_E may be significantly less than 1, and (2) the effect of ionic strength on the activity coefficient for a multivalent ion is greater than that for a singly charged species.

D. Variations of the Multiple-Species Approach

Single-Species Eluent

As noted earlier, mobile phases in ion chromatography can, over a limited range of composition for some eluents and over the entire useful composition range for others, be considered to consist of one component that controls the elution process. In this case, the mathematical process used to generate the multiple-species model can also be applied. In the case of a single-eluent species of charge m, the column capacity becomes

$$Q = m[E]$$

For an analyte of charge n, the analyte (A) distribution coefficient is equivalent to the reduced retention volume and related to the analyte-eluent distribution via their selectivity coefficient (K_{nm}); thus,

$$D_A = K_{nm} \left(\frac{[E]}{E}\right)^{n/m}$$

Therefore, in general,

$$D_A = K_{nm} \left(\frac{Q}{mE}\right)^{n/m} \tag{28}$$

It is observed that this equation has the same form as Eq. (7), the single eluent equation derived by Gjerde [9]. For an eluent containing a divalent active species (e.g., phthalate above pH 4.5), the appropriate retention equations for both a monovalent and divalent analyte become

$$D_1 = K_{12}\left(\frac{E}{2Q}\right)^{1/2}$$

$$D_2 = K_{22}\left(\frac{E}{2Q}\right)$$

Single-Site Interaction

The resins used in ion chromatography generally have either a silica or a polymeric support. For silica-based resins, it has been estimated that the average surface area per exchange site is of the order of 90 Å^2 [15], which is similar to the 60 Å^2 area reported for a poly(styrene-divinylbenzene) anion exchanger [25]. The area occupied by typical analyte and eluent species is significantly less than this, at approximately 30 Å^2, and thus if the exchange sites are evenly distributed on the resin, it is spatially impossible for any species, regardless of its charge, to interact simultaneously with more than one adjacent site at any given time. In this situation, species charge will not affect the total number of column-species interactions that can occur but affect only the strength or stability of the interaction. Under these conditions, the column capacity becomes (for an eluent containing two species, one monovalent and the other divalent)

$$Q = [E_1] + [E_2] \tag{29}$$

which, when coupled with the derivation process used in the original model (remembering that any analyte, regardless of change, also can only interact within a single site at a given time), leads to

$$D_n = \frac{K_{n1}Q}{E_1 + E_2/X_{21}} \tag{30}$$

where n is the analyte charge. Such a treatment has been used with limited success for the modeling of ion behavior in ion chromatography [15]; however, Gjerde et al., in their study of polymeric resins containing fewer than 1 quaternary functional group per 1000 benzene rings, conclude that the even distribution of exchange sites is unlikely [9]. They conclude that since monovalent and divalent analytes react differently in response to changes in eluent composition, the sites on such resins are not evenly distributed throughout the column but are instead clustered together near the more accessible surfaces.

Column Capacity Effects

Ion exchange resins as a class have a tendency to swell when taking up solvent during the process of exchange; as a general rule, the size of the individual resin pores is directly related to the total ionic content of the solution with which they are in contact (i.e., the mobile phase). The nature of the relationship between mobile-phase ionic strength and resin size is inverse in that an increasing ionic strength causes a decrease in pore size. As a resin swells within the confined space of a column, some sites may become unavailable for interaction with either the eluent or analyte species, thus causing the column capacity to decrease. Alternatively, Diamond and Witney have observed in their study of ion-exchange resins that, as eluent concentrations increase, the interaction between the solution and the resin is no longer strictly ion exchange, but that nonexchange adsorption or resin invasion by the aqueous ions may occur [26]. Qualitatively, then, one would expect that the effective capacity of a column (i.e., the number of sites actually available for interaction) is inversely related to the ionic strength of the mobile phase. In its most correct sense, then, the net capacity (Q), as it appears in the model equations given to this point, should be replaced by the effective capacity (Q_e), where Q_e is a function of both Q and mobile-phase ionic strength. This relationship has not been quantitated to date, but the presence of such a concentration-related effect has been observed in applications of these models [15]. Although an actual change in capacity as a function of eluent strength has not been demonstrated, Diamond and Whitney have observed as much as a 13% change in resin volume (determined as a decrease in column void volume) for silica-based anion exchanges as the ionic strength of an eluent is increased by a factor of 10 from 1×10^{-3} M [27].

E. Window Diagrams for Optimization

The equations derived to this point are effective in describing the behavior of a single species under a given set of conditions, but in a practical sense it is sometimes more useful to know how a group of analytes (or interferents) will behave. This is especially true when one wants to optimize a particular assay in terms of both interspecies resolution and total analysis time. Graphically, one can visualize analyte behavior with the aid of window diagrams that describe the relationship between the retention characteristics of a pair of analytes and some operational variables of a chromatographic system. Such diagrams have been used to characterize and optimize separations in gas [28,29], high-performance liquid [30-33], and ion [14,16,34] chromatographies. The retention relationship used herein is the reduced retention ratio for two analytes, A and B, which is written

$$\alpha_{A,B} = \frac{t_A - t_M}{t_B - t_M} = \frac{D_A}{D_B} \tag{31}$$

where A and B are chosen so that $t_A > t_B$, $\alpha > 1$, and the t refers to retention times; t_M is the void volume equivalent. One notes that at $\alpha = 1$ peak overlap occurs and that as the value of α gets larger, the resolution between the two analytes also improves. The operational variable of interest is eluent composition, which for a two-species mobile phase can be expressed either by individual species concentration or by total salt content and pH (if the two species are both dissociation products of a weak acid). The interpretation of the resulting three-dimensional diagrams is made simpler by considering two-dimensional contours of these response surfaces.

Figure 1 documents the two most common types of diagram encountered in ion chromatography and considers the separation of three species; species A and B of similar charge and species C, which has a somewhat greater charge. Interpretation of a window diagram proceeds as follows. The lowest line in the diagram identifies the analyte pair with the poorest resolution, and the highest line identifies the pair that is best resolved. Thus, the lower line limits optimization in terms of resolution, whereas the top line limits the technique in terms of total analysis time. In order to interpret the diagrams, the analyst must determine initially the minimum resolution the system can exhibit; for this example let us use a value of $\alpha = 1.5$. In case A in Fig. 1, all three analyte pairs have an α value greater than 1.5 for all the conditions documented, and therefore separation efficiency is limited by the analysis time. The optimum eluent composition is the one that provides the shortest analysis time, shown in this diagram as the lowest point reached by the C/A

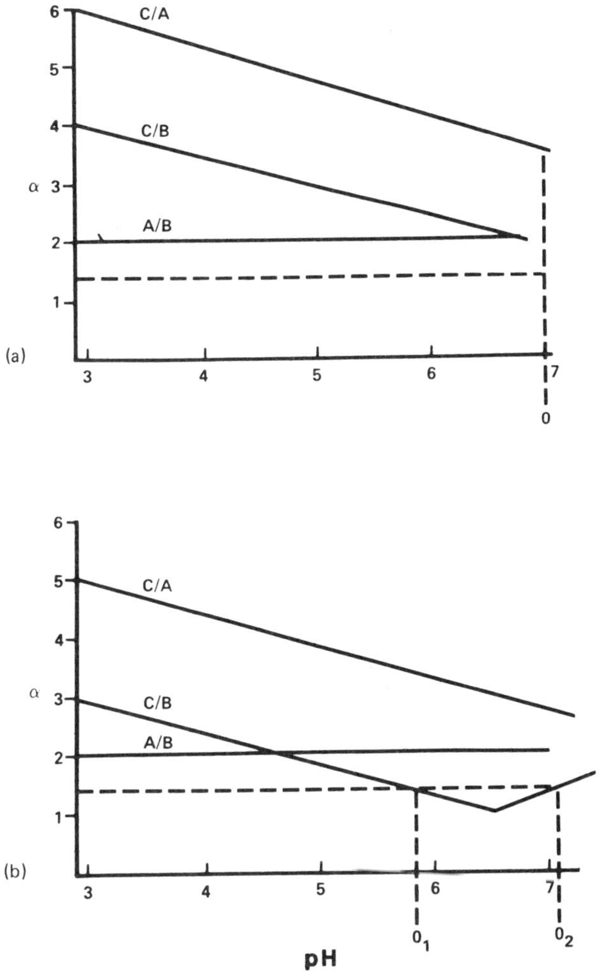

Figure 1. Interpretation of typical window diagrams. Species A and B are monovalent anions; species C is divalent. The area below the dotted line represents the resolution limit set by the analyst. (Reprinted with permission from *Analytical Chemistry*, *56*, p. 2575. Copyright 1984, American Chemical Society.)

(upper) line (point 0). In case B, species C is less strongly retained by the resin than in the previous case; the net result of this is a decrease in $\alpha_{C,A}$ and $\alpha_{C,B}$. If the resin used in this case is silica based and thus limited to an eluent pH 7 by its own stability, eluent composition 0_1 defines the only point on the diagram where the resolution between the most poorly resolved species pair just meets the analyst-defined resolution limit. Thus 0_1 represents the optimum eluent composition in terms of resolution and analysis time. One observes that at pH 6.5, $\alpha_{C,B} = 1$, and these species coelute. As the pH continues to increase, species C is affected to a greater extent than species B (owing to the charge difference) with the result that their α values again increase but their elution sequence shifts. To wit, at a pH lower than 0_1, the elution order is B followed by C, but at 0_2 the order is reversed. At point 0_2 the resolution between the most poorly resolved species again reaches the defined limit, making this eluent composition analytically acceptable. Indeed, it is observed that 0_2 represents a more desirable eluent composition than does 0_1, since total analysis time is shorter.

A third possibility, not considered in Fig. 1, is that in which the α value for one of the species pairs is always less than 1.5. In this case the desired separation could not be accomplished under any analytic conditions.

Perhaps the most significant feature of these diagrams is the relationship between the slope of a line defining the behavior for a given analyte pair and their relative charge ratio. For species that are similarly charged, there is no eluent composition effect on α and the slope of the α line is zero. For species of dissimilar charge, α changes inversely with increasing eluent ionic strength. The slope of the α line (or the rate of the change in α) is directly related to the magnitude of analyte charge ratio. These observations are conceptually consistent with the multispecies elution model described earlier. For similarly charged species, the ratio of retention volumes [e.g., from Eq. (21)] simplifies to the ratio of the selectivity coeffiients of the individual species. Since the magnitude of each selectivity coefficient is independent of eluent composition, the ratio should be constant. For dissimilarly charged species, the ratio of reduced retention volumes obtained from the model reduces to an expression that retains some eluent composition dependence. For example, for two species A and B, where A is divalent and B is monovalent, which are eluted with a mobile phase containing monovalent and divalent active species, their retention volume ratio becomes [see Eqs. (21) and (23)]

$$\alpha_{A,B} = \frac{D_A}{D_B} = \frac{K_{A1}(H^+)}{4K_{B1}X_{21}K_{a2}}\left(\sqrt{1 + \frac{8K_{a2}X_{21}Q[1 + K_{a2}/(H^+)]}{(H^+)E_T}} - 1\right) \tag{32}$$

Thus, α for these two species would remain a function of eluent composition.

F. Example of Model Application

The multiple-eluent species model has been successfully used to predict the behavior of anionic analytes for numerous column-eluent couples [14-16,27,34]. In the example presented here, the response of five analytes (Cl^-, NO_3^-, Br^-, SO_4^{2-} and $S_2O_3^{2-}$) to changing pH in a phthalate-type eluent with a Hamilton PRP-X100 column was studied. Specifically, the mobile phase contained 2×10^{-3} M potassium hydrogen phthalate and was adjusted to pH 4.0, 4.25, 4.50, 5.00, and 5.50. Given that the dissociation constants for phthalic acid are 3.10 and 5.40, one observes that this pH range corresponds to an eluent dominated by the monovalent phthalate ion (HP^-, pH 4.0) to one in which the divalent species (P^{-2}, pH, 5.5) comprises 60% of the total dissolved salt. As shown in Table 2, the resulting composition and ionic strength of these mobile phases are vastly different, and thus this example should be an excellent test of the model's utility. In terms of column characteristics, PRP-X100 is a strong anion exchanger prepared by the chemical functionalization of a PRP-1 macroporous co-poly(styrene-divinylbenzene) reversed-phase material [25]. The exchange capacity of this material is 0.17 mEq/g, and it contains 1.15 mEq sites per gram.

Since the eluents studied contain both dissociated forms of phthalate over the pH range studied, Eqs. (21) and (23) are applicable to this example. It is observed from Table 2 that the ionic strengths of these mobile phases are substantial (10^{-3} M or higher) and therefore that the solution's composition is more correctly represented in terms of species activities rather than species concentrations. This is especially true for the divalent eluent P^{-2}, since the magnitude of the activity coefficient is directly related to the square of the ionic charge. We observe from the model equations that the column's capacity is a constant that need not be determined in order to produce a set of relative selectivity coefficients. Consequently, in this example Q is assigned a value of 1 for convenience. Selectivity coefficients are calculated from all the available retention time-eluent composition data via a process of iterative minimization of the calculation error and are then used to predict analyte behavior for the eluent compositions studied.

Selectivity coefficients calculated for each analyte are summarized in Table 3. The values for the X_{21} selectivity coefficient, which defines the relative selectivity between the two eluent species P^{-2} and HP^-, agree very well for similarly charged analytes; one expects there to be a difference in the X values calculated for each analyte type (i.e., monovalent versus divalent), due to the difference in Eqs. (21) and (23). Each analyte produces a slightly different value because the minimization process considers each analyte separately, not all analytes of a given charge simultaneously. The K values for the selectivity coefficient relating analytes to the HP^-

Table 2. Eluent Compositions Used in the Study

Eluent	pH	Composition ($\times 10^{-3}$ M)[a] Total phthalate	HP^-	P^{-2}	Ionic strength
1	4.00	2	1.650	0.0657	0.995
2	4.25	2	1.675	0.1186	1.163
3	4.50	2	1.629	0.2052	1.286
4	5.00	2	1.302	0.5182	1.838
5	5.50	2	0.7633	0.9610	2.667

[a]Total phthalate and ionic strength are expressed as concentrations; HP^- and P^{-2} are expressed as activities.

eluent mimic the elution order, with a larger K value indicative of a longer elution time. The minimized error in calculation of the K values is moderately good, varying from a high of 6.13% for SO_4^{2-} to a low of 2.57% for Cl^-. These values for the relative error are only slightly above the estimated analytic errors of 5 and 2.5% for divalent and monovalent analytes, respectively.

Figures 2 and 3 document the ability of the model, armed with the calculated relative selectivity coefficients, to predict the behavior of analytes over the range of eluent compositions studied. One observes that the agreement is quite good; as noted in Table 3,

Table 3. Selectivity Coefficients and Model Fit Parameters

Analyte	Selectivity coefficients X_{21}	K_{n1}	% RSD in K_{n1}	Model fit,[a] average deviation |% RSD|	% RSD
SO_4^{2-}	3.14×10^{-3}	3.59×10^{-5}	6.13	2.65	−0.55
$S_2O_3^{2-}$	2.95×10^{-3}	5.89×10^{-5}	5.78	3.22	+0.43
NO_3^-	1.70×10^{-3}	4.61×10^{-3}	2.86	3.22	−0.07
Br^-	1.63×10^{-3}	3.90×10^{-3}	3.12	1.50	−0.07
Cl^-	1.70×10^{-3}	2.65×10^{-3}	2.57	1.61	−0.75

[a]Curve fit parameters: slope, 0.980; intercept, 0.07; R^2, 0.9973.

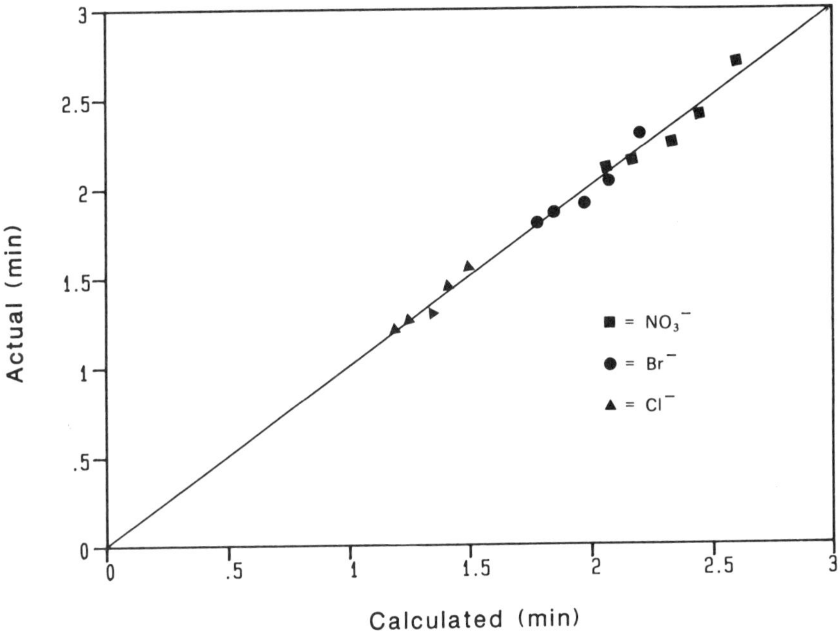

Figure 2. Comparison of observed versus calculated retention behavior: Monovalent analytes.

linear regression analysis of the plot of actual versus calculated retention times produces a line with a slope of 0.980, an intercept of 0.07, and a correlation coefficient of 0.9973. This behavior agrees quite well with the values of 1.0, 0.0, and 1.0, respectively, for the slope, intercept, and R^2 value, which are indicative of an exact fit. Focusing on individual analytes, one observes that the average error in the determination of retention times for the monovalent analytes is approximately 1.5%, but that for the divalent species is roughly double this value, at 3.0%. Again, the magnitude of this error is within the expected analytic precision. If one observes the magnitude of the relative deviation (that is, the signed average), it is noted that the model tends to underestimate the observed behavior to a small degree.

Figure 4 shows the constant total salt contour (2×10^{-3}M) of the window diagram defining the relationship between the reduced retention ratio and eluent pH. The shape of the diagram is completely consistent with the observations made earlier; all lines relating the behavior of similarly charged analytes are horizontal, indicating no pH dependence. Although the relationship between α and pH is not linear, smooth curves of very similar shape ade-

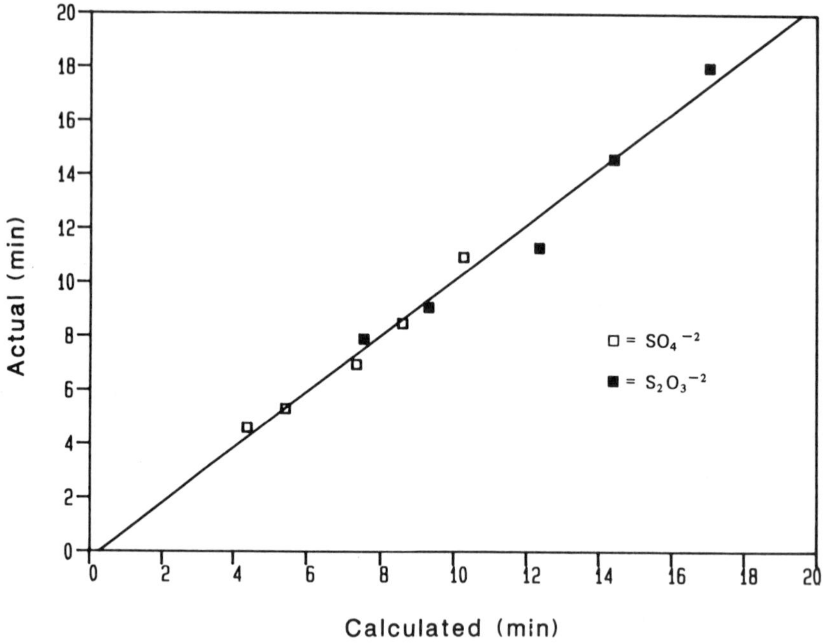

Figure 3. Comparison of observed versus calculated retention behavior: Divalent analytes.

quately define the behavior of dissimilarly charged ion couples. Interpretation of the diagram in terms of optimization is exceedingly simple; in all cases the lower resolution limit is defined by the NO^-_3-Br^- separation and the upper limit of analysis time is defined by the $S_2O_3^{2-}$-Cl^- pair. No crossover of analyte peaks (i.e., no change in elution order) occurs under the conditions studied. Clearly, the condition that optimizes the separation in terms of both resolution and total analysis time is identified as the highest pH studied. One observes that if pH were raised to 6 (or, for that matter, to as high as 6.4), resolution would still have been maintained but total analysis time would have decreased even more. We note, however, that this trend eventually breaks down as pH increases past 6.4. Since after this pH phthalic acid speciation is dominated by P^{-2}, increasing pH past this point cannot increase eluent strength (and affect analyte retention) until the hydroxide concentration becomes so high (e.g., pH 11) that it itself becomes an effective eluent.

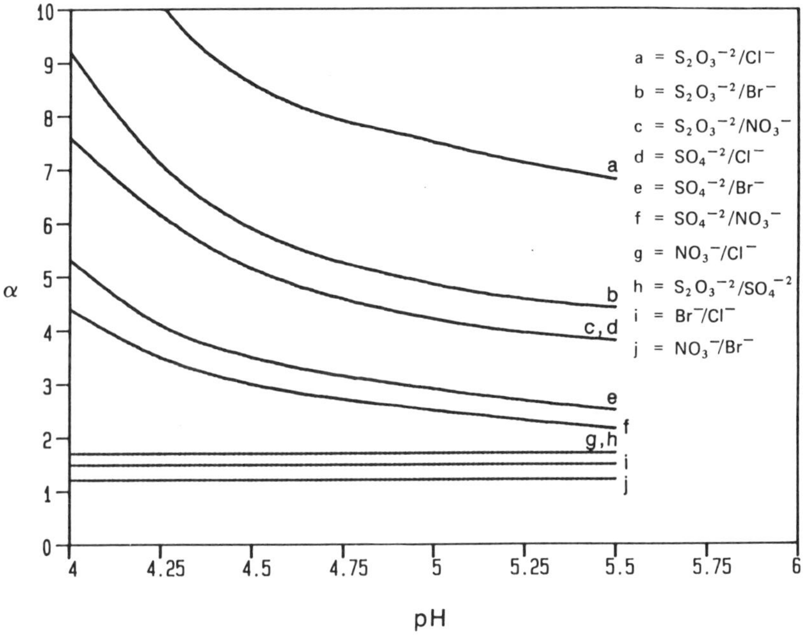

Figure 4. Constant total salt content contour of the window diagram for the Hamilton column-phthalate eluent system.

II. DETECTION THEORY

As noted in the introduction, the commercial development of ion chromatography required not only progress in column manufacturing techniques, but also the identification of appropriate analyte detection methodologies. In a theoretical sense the only requirement of an IC detector is its ability to respond to the presence of ionic species in a flowing solution, but in a practical sense the detector must have other operating characteristics as well. Sensitivity, reproducibility (and linearity) of response, simplicity of design, and ease of operation are important practical concerns, but of primary interest in evaluating detectors for their utility in ion chromatography is their selectivity (i.e., the ability to distinguish between analyte and eluent ions) and universality (i.e., the ability to respond to all analytes). Since the user is concerned with quantitating ionic species in the various applications of IC, all these analytes have in common the ability to conduct electricity across two electrodes between which an electric field has been applied. However, note that, since the eluent ions themselves are also ionic and therefore also have the ability to conduct electricity, a conductivity detector used in IC applications must have the ability to distinguish between these two types of charged components in the mobile phase. This is especially difficult to accomplish practically, since the background conductivity of most common IC eluents is relatively high. Essentially, then, the key to the commercial development of IC (in terms of detection) has been to devise a means of dealing with a high level of background conductance. Two approaches have been devised to date; in their original work, Small et al. [1] reported a method in which the eluent is chemically modified (or suppressed) by passage through a suppressor device. In this device the background conductance of the mobile phase is drastically reduced as the weak acid (or base) eluent is converted to its undissociated form. Alternatively, other researchers have successfully coupled low-capacity columns and weakly conducting eluents (generally organic acids) with detectors capable of electronically coping with the somewhat higher background conductivities inherent in these systems [35]. Methods using this strategy have been called nonsuppressed or single-column systems in the literature.

Another property shared by the ionic analytes in IC is that they displace an equivalent concentration of eluent ions from the mobile phase as they elute from the separator columns. Thus, indirect detection of analytes is possible if the eluent ions themselves have some property that can be monitored. In this situation, the detector responds to the decrease in eluent ion concentration instead of the increase in analyte ion concentration. However, owing to electroneutrality constraints, these two quantities are equivalent in IC.

Many other types of detectors have been used in conjunction with ion chromatographic separation methods. These include electrochemical (amperometric [36,37], potentiometric, e.g., ion selective electrodes [36-41], and coulometric [42]), atomic emission-absorption spectrometric [43-46], refractive index [47,48], and permitivity [49] detectors. Inorganic analytes have been quantitated with conventional spectrophotometric detectors either directly, owing to their moderate absorptivities in the lower wavelength region (190-205 nm) of the ultraviolet (UV) [50-52], or after either pre- or postcolumn derivatization to form absorbing complex ions [53-55]. Despite their appropriateness in specific applications, these detectors are more generic in terms of their application to liquid chromatography in general (as opposed to IC in particular), and their performance has been considered in detail from a theoretical standpoint elsewhere in the literature. In this chapter, we will limit ourselves to a consideration of the conductivity and indirect detection methodologies.

A. Conductivity

Electrolytic conductance is, by definition, the ability of an electrolytic solution to conduct electricity across two electrodes between which an electric field is applied. The specific conductance of a solution (Λ) can be written

$$\Lambda = (\lambda^{+} + \lambda^{-}) \frac{\alpha C}{1000\ K}$$

where α is the degree of dissociation of a given solute, C is the concentration of that solute, K is the cell constant of the specific detector, and λ^{+} and λ^{-} are the limiting equivalent conductances of the cationic and anionic portions of the solute. Following a derivation similar to that described by Okada and Kumamoto [56] and Wilson et al. [57] for an eluent containing a single moiety (and thus one cationic and anionic component), the background conductivity (Λ) measured by the detector becomes

$$\Lambda_b = \frac{\Sigma(\eta\lambda_H + \lambda_{H_{m-n}A})\, \alpha_n C_E}{1000K} \tag{34}$$

where H_mA is the eluent salt (and thus m is the charge of its completely dissociated anion A) and α_n is the degree of dissociation of the eluent salt with respect to the specific $H_{m-n}A$ species. Note that the summation (Σ) is made over all values of n from 1 to n in all the equations derived in this section. When an analyte A elutes from the column, it displaces an equivalent concentration of eluent species from the mobile phase. Depending upon the nature of the mobile phase (the dissociation behavior of its salt and the mobile-phase pH) and the relative selectivities of all species involved in the exchange process, the analyte may displace one or more of the eluent's various dissociated forms. Thus, we define C_a^n as the amount of analyte eluted with or displacing eluent ion $H_{m-n}A$. The conductivity of the mobile phase containing the eluent ion can thus be expressed as

$$\Lambda_a = \frac{[\lambda_a C + \Sigma\, n\, \lambda_H + \alpha_n C_E + \lambda_{H_{m-n}A}(\alpha_n C_E - C_a)]^n}{1000K} \tag{35}$$

Note that the counterion associated with the analyte does not contribute to the conductance of that portion of the mobile phase containing the analyte (after separation), since the retention behavior of the analyte and its counterion should be drastically different. Thus, the total contribution of positively charged species to the measured conductance is expressed by the second term in the above expression. One also observes that

$$\sum C_s^n = C_s$$

the absolute concentration of the analyte. Since the conductivity detector is differential in nature (i.e., it measures changes in the conductance of a flowing stream), the response to the analyte ($\Delta\Lambda$) becomes:

$$\Delta\Lambda = \Lambda_a - \Lambda_b \tag{36}$$

Substituting Eqs. (34) and (35) into this expression, followed by the appropriate mathematical simplifications (especially the realization that the contribution of the positive species in the mobile phase to its conductance is the same in the presence and absence of the analyte), produces the following generalized expression describing the magnitude of the detector response:

$$\Delta\Lambda = \frac{\lambda_s C_s - \Sigma \lambda_{H_{m-n}A} C_a^n}{1000K} \tag{37}$$

Nonsuppressed or single-column ion chromatography refers to a method in which a conductivity detector is directly coupled to the separator column. It should be noted that this approach is, as a whole, instrumentally less complicated than the suppressed approach, which will be discussed later, but it was not the first approach to be exploited commercially. The time lag between the introduction of the two primary methods of analyte detection reflects that accurate and reproducible measurement of relatively small changes in a highly conductive mobile phase required the development of appropriately designed conductivity monitors. We again repeat that it is not the purpose of this chapter to provide a direct and quantitative comparison of the two methodologies, although qualitatively it is observed that, at present, the suppressed method is theoretically and practically capable of roughly an order of magnitude better sensitivity than an otherwise equivalent nonsuppressed procedure [2,3,8]. Note also that the detection limits obtained with the nonsuppressed technique may be somewhat poorer than those achieved with suppressed methods, at least in part because of its greater sensitivity to short-term temperature fluctuations. This increased sensitivity to temperature is due to the somewhat higher background conductivities common for mobile phases used in nonsuppressed systems, coupled with the temperature dependence of conductivity (roughly 2%/°C [2]). Temperature compensation or control features, or both, have been added to many detectors used in nonsuppressed methods to minimize this type of effect.

B. Suppressed Conductivity

Eluent suppression is a chemical process by which the background conductance of an IC mobile phase is changed by changing either the degree of dissociation or the total ionic character of the mobile phase. The key to the suppression reaction is that the counterions associated with the active eluent species are replaced in the mobile phase by either hydrogen or hydroxide ions via the process of ion exchange. The net result of this process is neutralization of the active eluent species. Consider, for example, a typical weak acid eluent used in the separation of anionic analytes. The mobile phase contains not only the negatively charged active components but also positive counterions. In eluent suppression, these counterions are replaced by an equivalent concentration of hydrogen ions, with the net result that the dissociated weak acids become protonated and the total ionic character (and thus the background conductivity) of the mobile phase is decreased. Since many of the analytes that are commonly quantitated by IC are themselves strong acids, their dissociation is not affected by the suppression process. Diagrammatically, this process, accomplished with a fiber-type reactor, is shown in Fig. 5. Conversely, if a strong acid (e.g., HCl) is used to produce a cation separation, eluent suppression will involve the exchange of the counterions (e.g., Cl^-) for OH^-. Essentially, this produces the neutralization of the hydrogen eluent species to form water, which once again results in a decrease in the ionic content and conductance of the mobile phase. As was true of the anion methods, most potential cationic analytes (alkali and alkaline earth metals) are unaffected by the suppression process and can remain readily detectable. In general, then, the suppression reaction takes the forms

$$N - H + (X^+, E^-) \rightleftharpoons N - X + HE$$

for anion analysis and, for cation analysis

$$N - OH + (E^+, X^-) \rightleftharpoons N - X + EOH$$

where N represents the suppression reactor, X is the eluent counterion, and E is the active eluent species.

Since its original application, the suppression process has essentially been accomplished by three different mechanisms. Initially,

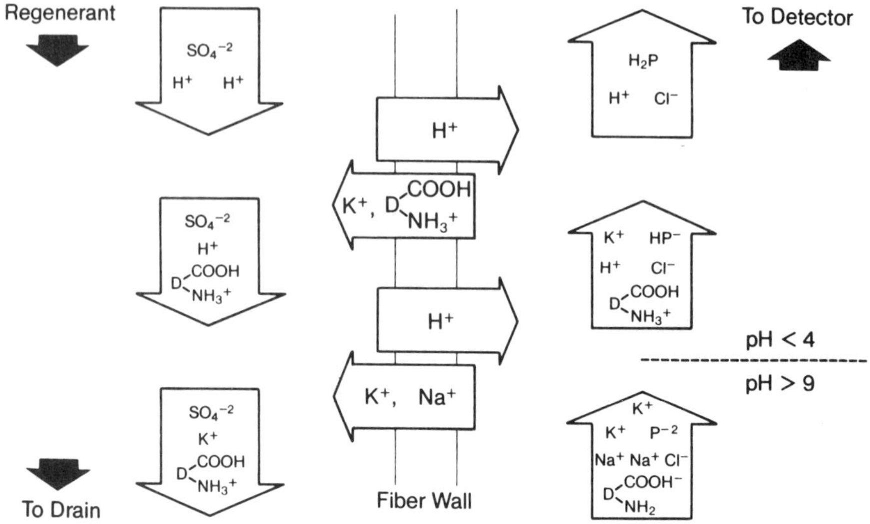

Figure 5. Operation of the fiber suppressor.

the suppressor consisted of a high-capacity ion-exchange column loaded with the appropriate neutralization ion [1]; this accounts for the classic distinction between suppressed methods using two columns (the separator and suppressor) and nonsuppressed or single-column (separator alone) methods. Although analytically useful, the suppressor column was characterized by a number of undesirable features, which included (1) the need for regeneration as the neutralizing ion was depleted, (2) decreased chromatography performance (e.g., band broadening), and (3) variable sensitivity as a function of column depletion. Eventually the column was replaced by a hollow fiber device capable of providing continuous suppression. The operation of such a fiber reactor is shown in Fig. 5, specifically as related to anion applications. The sulfonated polyethylene fiber wall allows the transport of cationic species (i.e., H^+ from the regenerating solution and the counterions from the mobile phase), but the anions (eluent, analyte, and regenerant) will not permeate the wall owing to Donnan exclusion [58]. Thus, suppression is accomplished in the classic sense. Although various researchers have proposed alternate designs for fiber-type suppressors [59-62], the major limitation of this type of device is a somewhat limited exchange capacity because of the mass transport processes required. Recently, Dionex, Inc., has introduced a micromembrane suppressor (MMS), which reportedly has nearly a 100-fold greater suppression capacity than the conventional fiber [7]. Essentially, the MMS consists of alternating layers of ion-exchange "screens" and ultrathin ion-exchange membranes. The mobile phase flows countercurrently in the suppression screens located above and below the eluent screen. Ion exchange between the screens is accomplished through the membranes separating them. Thus, the ability to chemically suppress a mobile phase has improved drastically as IC matures as an analytic technique, but the advances have been technical (in terms of facilitating the ion-exchange process), as oppossed to theoretical changes in the concept of suppression. It is noted that, besides their ability to modify and control background conductance, suppressors may also be useful in removing potential interferents in IC separations as part of the supppression-exchange process [63].

C. Indirect Photometry

As mentioned previously, as an analyte elutes from the separator column in IC, it displaces an equivalent concentration of eluent ion from the mobile phase. Thus, if the eluent ions themselves have some property that can be monitored, the presence of the analyte can be confirmed and its concentration quantitated by indirectly observing the behavior of the eluent ions. A most useful property of the weak organic acid eluents commonly used in the nonsuppressed anion methods is their UV absorptivity; thus, indirect photometric chromatography, first mentioned by Laurent and Bourdon [64] and brought into practice by Small and Miller [65], has received much attention in the recent literature. The theoretical principle of this method, alternatively referred to as "vacancy" chromatography, can be qualitatively explained as follows (FIg. 6). Actual detection involves measuring the absorbance decrease as the analyte displaces an ultraviolet-active component in the mobile phase. Thus, the UV detector responds to the presence of the analyte by producing a negative peak in the baseline absorbance plot. The magnitude of this inverse peak is directly related to the difference in concentration and molar absorptivity between the analyte and eluent species. Thus,

$$R = C_s(A_s - A_e) \tag{38}$$

where R is the detector response, C is the concentration, A is the molar absorptivity, and the subscripts s and e refer to the analyte and eluent species, respectively. For a transparent analyte, $A_s = 0$ and

$$R = -C_sA_e \tag{39}$$

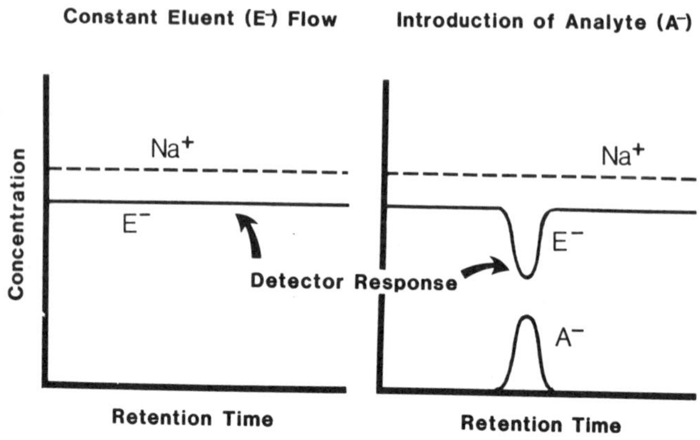

Figure 6. Principle of indirect photometric chromatography.

The detector response is thus independent of analyte identity. Coupling this observation with an experiment designed to determine the magnitude of retention time shifts as the eluent composition (salt concentration and pH) is changed allows one to quantitate the molar concentration of an unidentified analyte without calibration curve or linear regression standardization [66]. Measuring the peak area response for a single analyte of known molar concentration (and charge) makes it possible to calculate a response factor that is constant for a specific eluent composition. The equivalent concentration of any analyte can then be determined as the product of this response factor and the absolute detector response. By measuring retention times for both the known and unknown analytes in two distinct mobile phases, the analyst can calculate a reduced retention ratio via Eq. (31) for both elution conditions. The magnitude and nature of the change in the retention ratio produced as a result of the two elution conditions allows one to determine the charge of the unknown analyte, which can then be used to convert equivalent concentrations to their molar counterparts. A similar process for use in the more general case in which the analyte itself may or may not adsorb radiation has been described by Wilson and Yeung [67].

The indirect approach as a whole has the potential advantages of widespread applicability, good sensitivity and selectivity, instrumental simplicity, and ready adaptability to conventional HPLC equipment. Its potential power is underscored by recent work that describes the simultaneous determination of cations and anions in a single system [68]. The mobile phase used in this application contains cationic and anionic components that are both UV absorbing, and each analyte type is detected at a distinct (and separate) wavelength. The indirect analytic concept can also be applied to other properties of a suitable mobile phase; the use of both radioactivity [69] and fluorescence [70] in indirect methods has been reported in the literature. These last two methods share some of the advantages inherent in the indirect approach. But neither clearly exhibits the applicability and adaptability that are characteristic of the photometric approach.

REFERENCES

1. H. Small, T. S. Stevens, and W. C. Bauman, Novel ion exchange chromatographic method using conductimetric detection, *Anal. Chem.*, *47*:1801-1809 (1975).
2. C. A. Pohl and E. L. Johnson, Ion chromatography—the state-of-the-art, *J. Chromatogr. Sci.*, *18*:442-452 (1980).
3. J. S. Fritz, D. T. Gjerde, and C. Pohlandt, *Ion Chromatography*, Huthig Verlag, New York (1982).
4. C. Schmuckler, Recent developments in ion chromatography, *J. Chromatogr.*, *313*:47-57 (1984).
5. J. S. Fritz, Ion chromatography: A review of methods and recent developments, *L. C. Magazine*, *2*(6):446-452 (1984).
6. J. R. Benson, Modern ion chromatography, *Am. Lab.*, *17*(6): 30-40 (1985).
7. G. O. Franklin, Development and applications of ion chromatography, *Am. Lab.*, *17*(6):65-80 (1985).
8. R. A. Wetzel, C. A. Pohl, J. M. Riviello, and J. C. MacDonald, Ion chromatography, *Chem. Anal. (N.Y.)*, *78*:355-416 (1985).
9. D. T. Gjerde, G. Schmuckler, and J. S. Fritz, Anion chromatography with low conductivity eluents, II, *J. Chromatogr.*, *187*: 35-45 (1980).
10. T. B. Hoover and G. D. Yager, *Ion Chromatography of Anions*, U. S. Environmental Protection Agency, Report #EPA-600/4-80-029 (1980).
11. D. Jenke, Anion peak migration in ion chromatography, *Anal. Chem.*, *52*:1535-1536 (1981).
12. J. A. Glatz and J. E. Girard, Factors affecting the resolution and detectability of inorganic anions by non-suppressed ion chromatography, *J. Chromatogr. Sci.*, *20*:266-273 (1982).
13. T. B. Hoover, Multiple eluent and pH effects on ion chromatography of phosphate and arsenate, *Sep. Sci. Technol.*, *17*:295-305 (1982).
14. D. R. Jenke and G. K. Pagenkopf, Modeling of analyte response to changing eluent composition in suppressed ion chromatography, *J. Chromatogr.*, *269*:202-207 (1983).
15. D. R. Jenke and G. K. Pagenkopf, Models for prediction of retention in non-suppressed ion chromatography, *Anal. Chem.*, *56*: 88-91 (1984).

16. D. R. Jenke, Modeling of analyte behavior in indirect photometric chromatography, *Anal. Chem.*, *56*:2674-2681 (1984).
17. R. Kanin, *Elements of Ion Exchange*, Reinhold Publishing, New York (1960).
18. A. W. Davidson and W. J. Argersinger, Jr., Equilibrium constants of cation exchange processes, *Ann. N.Y. Acad. Sci.*, *57*: 105-115 (1953).
19. E. Glueckauf, A theoretical treatment of cation exchangers, I, The prediction of equilibrium constants from osmotic data, *Proc. R. Soc.* (London), *A214*:207-225 (1952).
20. D. H. Freeman, Thermodynamics of binary ion-exchange systems, *J. Chem. Phys.*, *35*:181-191 (1961).
21. B. A. Soldana, Q. V. Larson, and G. E. Meyers, The osmotic approach to ion-exchange equilibrium, II, Cation exchangers, *J. Amer. Chem. Soc.*, *77*:1338-1344 (1953).
22. J. A. Marinsky, M. M. Reddy, and S. Amdur, Prediction of ion exchange selectivity, *J. Phys. Chem.*, *77*:2128-2132 (1973).
23. H. P. Gregor, Gibbs-Donnan equilibria in ion exchange resin systems, *J. Amer. Chem. Soc.*, *73*:642-649 (1951).
24. Y. Marcus and A. S. Kertes, *Ion Exchange and Solvent Extraction of Metal Complexes*, Wiley-Interscience, New York (1969).
25. D. P. Lee, A new anion exchange phase for ion chromatography, *J. Chromatogr. Sci.*, *22*:327-331 (1984).
26. R. M. Diamond and D. C. Witney, in *Ion Exchange* (J. Marissky, ed.), Marcel Dekker, New York, pp. 227-274 (1966).
27. D. R. Jenke, Observations on and modeling of analyte elution in ion chromatography, Ph.D. Thesis, Montana State University, Bozeman, Montana (1983).
28. R. J. Laub and J. H. Purnell, Criteria for the use of mixed solvents in gas-liquid chromatography, *J. Chromatogr.*, *112*:71-79 (1975).
29. R. J. Laub and J. H. Purnell, Optimization of gas chromatographic analysis of complex mixtures of unknown composition, *Anal. Chem.*, *48*:1720-1724 (1976).
30. S. N. Deming and M. L. Turoff, Optimization of reverse-phase liquid chromatographic separation of weak organic acids, *Anal. Chem.*, *50*:546-548 (1978).
31. W. P. Price, Jr., R. Edens, D. C. Hendrix, and S. N. Deming, Optimized reverse-phase high-performance liquid chromatographic separation of cinnamaric acids and related compounds, *Anal. Biochem.*, *93*:233-237 (1979).
32. W. P. Price, Jr., and S. N. Deming, Optimized separation of scopdetin and feralic and *p*-coumaric acids by reverse-phase high-performance liquid chromatography, *Anal. Chim. Acta*, *108*:117-231 (1979).

33. B. Sachok, J. J. Stranahan, and S. N. Deming, Two factor minimum alpha plots for the liquid chromatographic separation of 2,6 disubsituted anilines, *Anal. Chem.*, *53*:70-74 (1981).
34. D. R. Jenke and G. K. Pagenkopf, Optimization of anion separation by non-suppressed ion chromatography, *Anal. Chem.*, *56*: 85-88 (1984).
35. D. T. Gjerde and J. S. Fritz, Effects of capacity on the behavior of anion-exchange resins, *J. Chromatogr.*, *176*:199-206 (1979).
36. P. Edwards and K. K. Hoak, A pulsed amperometric detector for ion chromatography, *Am. Lab.*, *15*(4):78-87 (1983).
37. R. D. Rocklin and E. L. Johnson, Determination of cyanide, sulfide, iodide and bromide by ion chromatography with electrochemical detection, *Anal. Chem.*, *55*:4-7 (1983).
38. H. Hershcovitz, C. Yarnitzky, and G. Schmuckler, Ion chromatography with potentiometric detection, *J. Chromatogr.*, *252*: 113-119 (1982).
39. K. Suzuki, H. Aruga, and T. Shirai, Determination of monovalent cations by ion chromatography with ion selective electrode detection, *Anal. Chem.*, *55*:2011-2013 (1983).
40. P. R. Haddad, P. W. Alexander, and M. Trojanowicz, Ion chromatography of Mg, Ca, Sr and Ba ions using a metallic copper electrode as a potentiometric detector, *J. Chromatogr.*, *294*: 397-402 (1984).
41. P. R. Haddad, P. W. Alexander, and M. Trojanowicz, Ion chromatography of inorganic anions with potentiometric detection using a metallic copper electrode, *J. Chromatogr.*, *321*:363-374 (1985).
42. J. E. Girard, Ion chromatography with coulometric detection for the determination of inorganic ions, *Anal. Chem.*, *51*:836-839 (1979).
43. G. R. Ricci, L. S. Shepard, G. Colovos, and N. E. Hester, Ion chromatography with atomic absorption spectometric detection for determination of organic and inorganic arsenic species, *Anal. Chem.*, *53*:610-613 (1981).
44. D. Chakraborti, D. Hillman, K. J. Irgolic, and R. A. Zingaro, Hitachi Zeeman graphite furnace atomic absorption spectrometer as a selenium specific detector for ion chromatography, *J. Chromatogr.*, *249*:81-92 (1982).
45. S. W. Downey and G. M. Hieftje, Replacement ion chromatography with flame photometric detection, *Anal. Chim. Acta*, *153*:1-13 (1983).
46. J. M. Pettersen, Determination of free chromate in lignosulfonate dispersants by ion chromatography with atomic absorption spectrometric detection, *Anal. Chim. Acta.*, *160*:263-266 (1984).
47. F. A. Buytenhuys, Ion chromatography of inorganic and organic ionic species using refractive index detection, *J. Chromatogr.*, *218*:57-64 (1981).

48. D. R. Jenke, P. K. Mitchell, and G. K. Pagenkopf, Use of multiple detectors and stepwise elution in ion chromatography without suppressor columns, *Anal. Chim. Acta, 155*:279-285 (1983).
49. J. F. Alder, P. R. Fielden, and A. J. Clark, Simultaneous conductivity and permitivity detector with a single cell for liquid chromatography, *Anal. Chem., 56*:985-988 (1984).
50. R. J. Williams, Determination of inorganic anions by ion chromatography with ultra-violet absorbance detection, *Anal. Chem., 55*:851-854 (1983).
51. J. P. Ivey, Novel eluent for the UV and conductometric detection of anions in unsuppressed ion chromatography, *J. Chromatogr., 267*:218-221 (1983).
52. G. P. Ayers and R. W. Gillett, Sensitive detection of anions in ion chromatography using UV detection at wavelengths less than 200 nm, *J. Chromatogr., 284*:510-514 (1984).
53. M. Zenki, Determination of alkaline earth metals by ion-exchange chromatography with spectrophotometric detection, *Anal. Chem., 53*:968-971 (1981).
54. S. Matsushita, Simultaneous determination of anions and metal cations by single column ion chromatography with ethylenediaminetetraacetate as eluent and conductivity and ultra-violet detection, *J. Chromatogr., 312*:327-336 (1984).
55. D. Yan and G. Schwedt, Trace analysis of aluminum and iron by ion chromatography and post-column derivatisation, *Fresenius' Z. Chem., 320*:252-257 (1985).
56. T. Okada and T. Kuwamoto, Sensitivity of non-suppressed ion chromatography using divalent organic acids as eluents, *J. Chromatogr., 284*:149-156 (1984).
57. S. A. Wilson, E. S. Yeung, and D. R. Bobbitt, Quantitative ion chromatography without standards by conductivity detection, *Anal. Chem., 56*:1457-1460 (1984).
58. T. S. Stevens, J. C. Davis, and H. Small, Hollow fiber ion-exchange suppressor for ion chromatography, *Anal. Chem., 53*:1488-1492 (1981).
59. Y. Hanaoka, T. Murayrama, S. Muramoto, T. Matsuura, and A. Nanba, Ion chromatography with an ion-exchange membrane suppressor, *J. Chromatogr., 239*:537-548 (1982).
60. P. K. Dasgupta, Linear and helical flow in a perfluorosulfonate membrane of annular geometry as a continuous cation exchanger, *Anal. Chem., 56*:96-103 (1984).
61. P. K. Dasgupta, Ion chromatographic separation of anions with ion interaction reagent and an annular helical suppressor, *Anal. Chem., 56*:769-776 (1984).
62. P. K. Dasgupta, R. Q. Bligh, and M. A. Mercurio, Dual membrane annular helical suppressor in ion chromatography, *Anal. Chem., 57*:484-490 (1985).

63. R. Payton, D. Brown, and D. Jenke, Elimination of matrix interferences in indirect photometric chromatography, *Anal. Chem.*, 57:2264-2267 (1985).
64. A. Laurent and R. Bourdon, Assay of anions by ion-exchange chromatography, *Anal. Pharm. Fr.*, *36*:453-460 (1978).
65. H. Small and T. E. Miller, Jr., Indirect photometric chromatography, *Anal. Chem.*, *54*:462-469 (1982).
66. D. R. Jenke, Standardization of transparent analyte response in indirect photometric chromatography, *Anal. Chem.*, *56*:2468-70 (1984).
67. S. A. Wilson and E. S. Yeung, Qualitative ion chromatography with an ultra-violet absorbance detector without standards, *Anal. Chim. Acta*, *157*:53-63 (1984).
68. Z. Iskandarni and T. E. Miller, Jr., Simultaneous independent analysis of anions and cations using indirect photometric chromatography, *Anal. Chem.*, 57:1591-1594 (1985).
69. S. Banergee and J. R. Steimers, Indirect ion chromatography with radioactive eluents, *Anal. Chem.*, 57:1476-7 (1985).
70. S. Mho and E. S. Yeung, Detection method for ion chromatography based on double-beam laser-excited indirect fluorimetry, *Anal. Chem.* 57:2253-2256 (1985).

8
Retention in Steric Exclusion Chromatography

Lave Fischer
AB Sangtec Medical
Bromma, Stockholm, Sweden

Neutral gels exhibit very interesting properties as liquid chromatographic materials: they often give separations related to the size of the substances being separated. This was found independently for polymers, particularly those used for plastics in organic solvents, and biomacromolecules (proteins, carbohydrates, nucleic acids, and so on) in aqueous solution.

Both polymer chemistry and biochemistry have come to rely to a very large extent on steric exclusion chromatography (SEC) for the separation of substances under mild, nondestructive conditions and for characterization of molecular weights and molecular weight distributions. Important for the success of this technique have been its reproducibility and its independence of the composition of the eluent: conditions can be chosen independently of the separation technique to suit the molecules to be separated. This is, however, also related to a limitation of the method: there is no possibility of using gradients of varying eluting power. The development of packing materials with better mechanical properties and selectivity than the materials originally used have brought the technique close to its potential capacity.

Although steric exclusion chromatography may sometimes be difficult to distinguish from some types of liquid chromatography based on adsorption (specific or unspecific), it has the following characteristics:

1. The stationary phase consists of a gel in which the solute molecules can penetrate through the entire macroscopic volume of the particles of the packing material.
2. Conditions are chosen so that adsorption (specific or nonspecific) is avoided or reduced. If adsorption does not take place, the elution volume depends on the size of the sample molecules in a way that would be expected from a molecular sieving mechanism, and elution volumes higher than the total bed volume do not occur.
3. Solutes are ideally eluted in the order of the size of the molecules, the larger molecules being eluted first.

Chromatographic systems with these characteristics were first described by Porath and Flodin [1,2]. The materials and techniques developed by them for use with aqueous eluents were mainly used for separation and characterization of proteins, polysaccharides, nucleic acids, and other substances of biologic interest. The separation obtained was explained qualitatively by a size-exclusion mechanism: a larger molecule was unable to penetrate as completely into the gel matrix as a smaller molecule. Therefore, a larger molecule was partitioned more into the liquid phase than a smaller molecule, thereby eluting earlier from the column. This mechanism was visualized in terms of pores of different sizes and with varying diameters. Such a crude mechanical assumption of the gel structure also underlay the first attempt at a quantitative description of the separation [1,2]. This description of the mechanism also forms the basis for the name these authors gave the technique, gel filtration (the name was probably originally suggested by Arne Tiselius).

Independently of the use of this technique in biochemistry, it was reinvented by Moore and co-workers [3,4] for characterization of polymers in organic solvents. Their description of the mechanisms underlying the separations was practically identical to that proposed by Porath and co-workers, although the concept of "pores" and their importance is not introduced by Moore and co-workers.

Although the fundamental identity of "gel filtration," as described by Porath and co-workers, and the "gel permeation chromatography" of Moore and co-workers, was soon realized, the two "schools" have continued to develop independently, with different terminology and different approaches to data reduction. The name used here, "steric exclusion chromatography," is that recommended by ASTM (ANSI/ASTM E 682-79).

The use of SEC was stimulated by the commercial introduction of gel materials. This made it possible to perform SEC separations reproducibly and without extensive organic synthesis to produce the gel materials. The first materials available consisted of cross-linked dextran (Sephadex, from Pharmacia, Uppsala, Sweden) for the separation of biochemically important substances in aqueous solution and cross-linked polystyrene gels (Styrogel, made by Dow Chemical and sold by Waters Associates) for the separation and characterization of polymers. Other commercial gel materials have followed, with better mechanical properties, a wider choice of fractionation ranges, more uniform particle size, and different chemical properties. This has led to a universal adoption of SEC as a standard technique, both in biochemistry and in polymer chemistry. Much of the success of the SEC technique can also be related to the fact that the separation is carried out without direct interaction or binding. The composition of the eluent and the temperature need not be chosen to give the right amount of binding, and can therefore be adjusted to give maximum stability or solubility to the substances being analyzed. The solutes also do not have contact with environments with unfavorable polarity. There is no saturation of the binding capacity of the gel phase, and the partition isotherms are therefore linear up to very high solute concentrations. These characteristics contribute to give high recoveries (most close to 100%) and a high column capacity.

The widespread use of SEC has led to the publication of a large number of review articles [5-19].

I. BASIC MECHANISM OF THE SEPARATION

SEC can be regarded as a special case of partition chromatography in the sense that the substance distributes through the whole volume of the gel phase (in contrast to adsorption chromatography, in which the active surface is the parameter that influences the separation). In SEC, the fundamental equation for partition chromatography without a gradient is written

$$V_e = V_0 + K_{av}V_g \tag{1}$$

where V_e is the elution (retention) volume of the substance, V_0 is the void volume in the column (see Fig. 1) [20-22], and V_g is the volume of the gel. K_{av} (av stands for available; see Ref. 23) is the partition coefficient for the solute between the gel phase and the surrounding liquid phase. As long as steric exclusion effects govern the equilibrium, K_{av} is less than 1 at equilibrium: the substance favors the liquid phase over the gel phase, the more so the larger the molecule.

The gel contains gel matrix and also a certain amount of liquid located so that it cannot be replaced even by low-molecular-weight

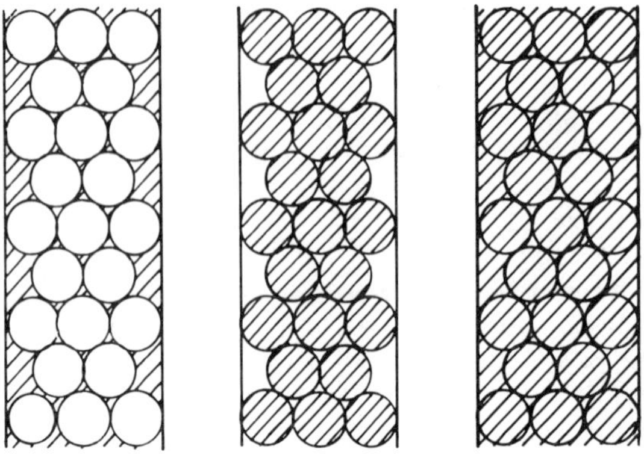

Figure 1. Diagrammatic representation of V_0, V_t, and V_g ($V_g = V_t - V_0$). (From Ref. 15.)

solutes (mostly liquid of solvatization of the matrix). If only steric exclusion effects govern the separation, K_{av} will therefore never reach the value 1, but a value corresponding to the fraction of the whole volume available to low-molecular-weight solutes. For some purposes, particularly to facilitate data reduction, it is convenient to have a parameter that approaches 0 for large molecules and 1 for small molecules. The basic chromatographic equation (1) can be rewritten

$$V_e = V_0 + K_d V_i \tag{2}$$

where V_i is intended to represent the volume of the liquid included in the gel phase. In practice, it is best defined operatively as the parameter that makes $K_d = 1$ in Eq. (2) for low-molecular-weight solutes that do not interact with the matrix. The relationship between the characteristic volumes of a typical SEC column is shown in Fig. 2.

In some work on SEC, V_i is not defined operatively but, instead, as the difference between the volume of the gel phase and the partial volume of the gel matrix in the gel phase. This definition has the drawback that it is difficult to accurately establish the partial specific volume of the gel matrix, and furthermore, it does not guarantee that the parameter K_d takes the value 1 for low-molecular-weight solutes.

In Eqs. (1) and (2), V_0, V_g, and V_i are parameters characteristic for the column, independent of the analyte substance or the flow rate. K_{av} and K_d are parameters characteristic of a given

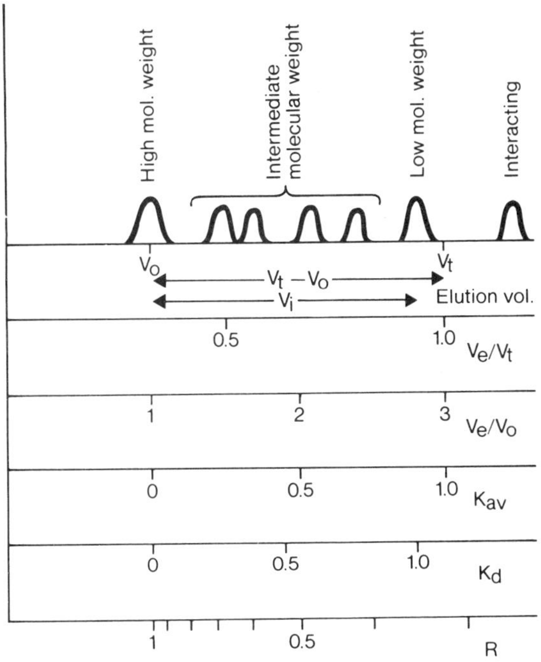

$$K_{av} = \frac{V_e - V_o}{V_t - V_o}$$

$$K_d = \frac{V_e - V_o}{V_t - V_o - V_{gel\ matrix}} = \frac{V_e - V_o}{V_i}$$

R = retention coefficient = V_o/V_e

Figure 2. Relationship between several variables used for normalizing elution behavior. (From the booklet "Gel Filtration Theory and Practice," by Pharmacia Fine Chemicals.)

substance and a given gel type; they are independent of the column dimensions, packing conditions, flow rate, and particle size. The elution volume should therefore be independent of flow rate and the particle size of the packing material. In some investigations, such independence was taken as proof that the separation was based only on steric exclusion; such conclusions are false, however, since Eqs. (1) and (2) are based only on the basic equation for partition chromatography. Only the limitation of K_d to the range from 0 to 1 and K_{av} to the range from 0 to V_i/V_g are related to the steric exclusion mechanism. Other investigators have found variation of V_e with flow rates. In these cases, the cause of the variation was probably their definition of the elution volume: this has been defined as "the maximum of the elution curve," or even, "the front of the elution curve." These can be expected to vary with flow rate and with particle size since they would be influenced by the zone width and shape. If a more adequate measure of the elution volume is used, it can be expected not to vary with flow rate [24, 25]. It is not a prerequisite for Eqs. (1) and (2) to be valid that local equilibrium of the partitioning of the analyte substance is attained [26]. Grubisic and Benoit [27] found a good correlation between partition coefficients determined statically and by chromatography.

II. SEC SELECTIVITY CURVE

A. Mathematical Models

Many models have been set up [28] to explain the dependence of K_{av} (or K_d) quantitatively as a function of molecular properties. The simplest models visualize the gel as a system of pores with diameters distributed according to some regularity and derive from this the proportion of the volume of the gel that can be reached by molecules of a given size. The simplest such model was presented by Porath [29]. In this model, the gel is visualized as a system of conically shaped pores. The model has the advantage of being easy to understand (although it is unrealistic chemically), but the equation derived is too complicated to be practically useful. To make the relationship conform to experimental data, a layer of hydration water was assumed to be so strongly bound as not to be exchangeable with the substance. The presence of such water can in fact be confirmed by washing gel material that has been swollen in water with dioxane: a certain amount of the water cannot be removed but does not contribute to the dielectric properties of the gel [30]. Squire [31] visualized the gel as consisting of equal amounts of cones, crevices, and cylinders. This, as would be expected, yields a large and complicated equation that is not very well suited for data reduction. van Kreveld and van der Hoed [32] describe the gel matrix as consisting of randomly distributed identical spheres, and the solutes are represented by random coils. Ackers [33] and Carmichael [34], to avoid a too rigid mechanical representation of the gel matrix, suggested that the microregions within the gel are randomly distributed with respect to the size of the molecules they can accomodate. Kubin [35] described the gel structure as pores of "irregular cross section."

A model that gives a picture of the gel structure that correlates better with what is known about the structure of gels was suggested by Laurent and Killander [23]. In this model, the molecules are represented by spheres of a given radius and the gel network is represented by infinitely long, straight rods that are randomly distributed in space. The derivation is based on the treatment by Ogston [36] of the fraction of space available to a sphere with radius r_s if straight cylindrical rods with radius r_x are distributed randomly with an average density of L units of rod length per unit volume of space. Ogston used this model to calculate the influence of one macromolecule on the activity of another in solution (see also Hellsing, Ref. 37). Laurent and Killander, using the formula for molecules of infinite length, derived the relationship

$$K_{av} = e^{[-\pi L(r_s + r_x)^2]} \tag{3}$$

This relationship was found to fit well with experimental data for Sephadex gels (L and r_x were not used as free parameters but calculated from the percentage of dry matter in the gel and from the known dimensions of dextran chains). The model can be adapted for other gel types, particularly for gels consisting of irregular aggregates of polymer chains (see page 366), by assuming that the gel matrix polymer chains aggregate into bundles of a given average size.

Most molecules are not spherical. As Ogston's original formula is strictly valid only for spherical molecules, an extension to molecules of more irregular shape could be interesting. In a previous publication I suggested the relation [15]

$$K_{av} = e^{(-LA/4)} \tag{4}$$

where A is the surface area of the molecule and L the average density of rods per unit volume of space. A derivation is given in Sec. V pages 367-371. Similar models in which the interaction between randomly distributed rods and planes were considered have been presented by Giddings [38], Giddings et al. [41], Ogston [40], and Rodbard and Chrambach.

Carmichael [42,43] recast the stochastic theory of chromatography developed by Giddings [44] to fit SEC, making specific assumptions about the rate constants λ_1 and λ_2 and thereby obtained good correlation with experimental data. Mechanisms other than partitioning between stationary and moving phases have been suggested, although none has found widespread acceptance. Thus, Ackers [45] and Ligny [46] proposed that the restricted diffusion rate in and out of the gel may be responsible for the separation. The correlation between the rate of diffusion and SEC behavior was found to be good, but the mechanism whereby the diffusion rate gives chromatographic separation does not seem very convincing. Dimarzio and Guttman [47] and Mori et al. [48] suggested that the separation in SEC is a type of flow separation as liquid and solutes flow through "pores" in the gel. The resistance to flow in the gels concerned is, however, so great that this effect must be negligibly small.

B. Retention Factors Other than Steric Exclusion

From a physicochemical standpoint, factors other than steric exclusion should influence the partitioning of the solutes in SEC. The first such effect is the "osmotic pressure" in the gel (the swelling pressure; see, for example, Refs. 49-51). It is difficult to accurately measure the swelling pressure, but pressures of several hundred kilopascals, which rough calculations indicate to be present in chromatographic gels, are sufficeint to influence the partitioning. It is, however, not realistic to assume that the swelling pressure alone is responsible for the partitioning, as assumed by Ginzburg and Cohen [52] and by Polson and Katz [53].

In aqueous solution, electrical interaction may profoundly influence the distribution of the solutes in SEC [54-56]. If the number of charged gorups is large, the gel must be regarded as an ion exchanger and the methodology of separation modified accordingly. If the eluent has a low ionic strength, even small amounts of charged groups may create considerable differences in the Donnan potential between the interior and exterior of the gel, which will influence the partitioning of charged solutes [57]. The same is true for charged macromolecules: a large charged macromolecule will not penetrate into the gel. The counterions do so until a Donnan potential is reached inside the gel that counterbalances the tendency of the counterions to pass in [58]. In this way an originally neutral gel is converted to an "ion exchanger" and the partitioning of other substances is influenced. Although this effect may in principle be utilized to give chromatographic separations (see, for instance, Ref. 59), all electrical effects are normally considered undesirable. They can be counteracted by increasing the ionic strength of the eluent. In SEC with aqueous solutions, buffer solutions with sufficiently high concentrations of electrolytes to counteract electrical effects are therefore used. If salts are undesired in the final product, one may use volatile electrolytes which can be removed by lypophilization.

A further effect that may influence the separation in SEC is the adsorption of solute onto the matrix. Such adsorption may be nonspecific or specific [60-65]. Nonspecific adsorption due to polarity [54,66-74] or to hydrogen bonding, may be counteracted by a suitable choice of eluent. The ideal eluent for SEC is a good solvent for both the solute and the substance forming the gel matrix [75]. In aqueous solution, hydrogen-bond-breaking agents, chaotropic ions, or surfactants may be added to counteract nonelectrostatic interaction in the few cases in which it is noticeable.

Solute-matrix interaction may actually be used to improve the separation [54,76-81]. Philosophically, such a separation should no longer be termed SEC, but because the chromatographic system is the same, it is usually still included under this heading.

As long as the separation is achieved by steric exclusion only, the elution volumes should ideally remain independent of the temperature. This can be realized both from the mechanistic model and from theromodynamic considerations: thermodynamically, the steric exclusion in SEC is an entropy effect, and equilibria that are governed exclusively by entropy differences are temperature invariant. If a temperature variation has been found, this indicates that other factors in addition to steric exclusion are influencing the separation [80-82]. This statement is not strictly true: it is conceivable that the degree of swelling or the microstructure of the gel is influenced by changes in the temperature [83,84], which in turn change the elution volume by solutes separated by steric exclusion. The three-dimensional structure of the substances to be separated may also be influenced by temperature. If this causes a change in the outer dimensions of the gel substance, elution volume will be influenced.

In practice it has been found that elution volumes in SEC show very little variation with temperature, as expected from theory. This temperature invariance of the elution volume helps to make SEC a very reproducible chromatographic process, thereby contributing to its usefulness. The constant elution volume at different temperature does not, however, imply that the chromatograms remain identical: the diffusion rate, and thereby the zone broadening, varies with temperature.

C. Practical Representation of the Molecular Weight-Elution Volume Relationship and Data Reduction

In principle, the mathematical models discussed in Sec. II.A can be used as a basis for constructing diagrams and mathematical relationships between molecular weights and elution volumes. The models are, however, too simple for this purpose: the gel structure must be assumed to be heterogeneous, and the relative distribution of the matrix density in the gels is not known. Also, the relative importance of the various factors that fundamentally influence size separation (steric exclusion and swelling pressure) is not known. In addition, the relationships are not in a suitable form for use in data reduction and graphic presentation of the relationship. Therefore, ways of representing the relationship are used that conform well with experimental data and are easy to handle mathematically. An excellent discussion is given by Rodbard [85,86].

As the parameter representing the chromatographic behavior of a given substance, the value of K_{av} or K_d is suitable. It is independent of the dimensions of the column and of the detailed experiment setup. K_{av} or K_d may be plotted directly as a function of the molecular weight (Fig. 3). The diagram is easier to use if the molecular weight is given on a logarithmic scale (Fig. 4). This way of representing the relationship gives a sigmoidal curve, which can be crudely divided into three phases: a nearly horizontal part at low molecular weight, with $K_d = 1$; a nearly horizontal part at higher molecular weight, with $K_d = 0$; and a sloping part (the fractionation range) in between them. Determann and Michel [87] used the parameters of the linear equation approximating the sloping part of

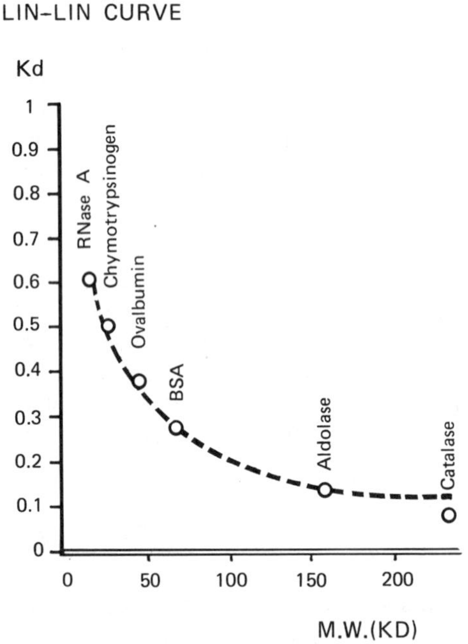

Figure 3. Example of selectivity curve plotted in lin-lin scale. Gel material: Sephacryl S-200 Superfine. Substances used for the calibration: ribonuclease A, MW = 13,700; chymotrypsinogen A, MW = 25,000; ovalbumin, MW = 43,000; bovine serum albumin, MW = 67,000; aldolase, MW = 158,000; catalase, MW = 232,000. The curve was derived from the measuring points by regression analysis, as described in the text. The diagrams in Figures 3-5 are part of the printout from a Sharp PC-1500 hand-held computer using a program for evaluation of molecular weights from SEC data. The abbreviation kD = kilodalton. The data on which the diagrams are based were obtained from Pharmacia Fine Chemicals AB, as described in the source to Table 1 (page 368).

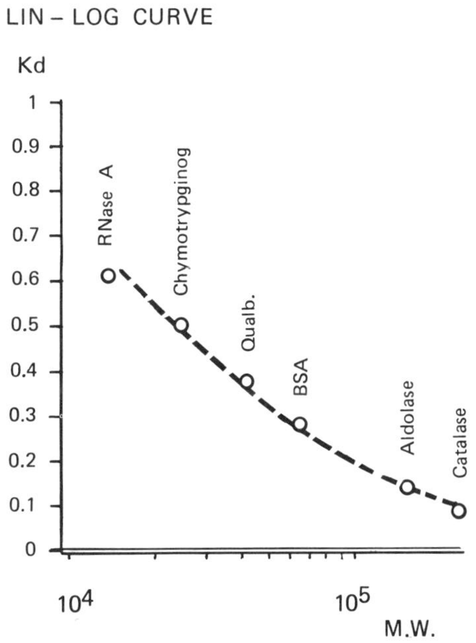

Figure 4. Example of selectivity curve plotted in lin-log scale. Data as in Figure 3.

the curve to characterize the gel materials. Grover and Kapoor [88] improved this by using a third-degree polynomial equation. The rather imprecise upper limit of the fractionation range is called the exclusion limit. The exclusion limit and the slope of the curve in the middle of the fractionation range give a fairly good characterization of gel materials.

It has been customary to approximate the K_d versus log molecular weight curve (called the calibration curve in polymer chemistry and selectivity curve in biochemistry) by three straight lines: a sloping line within the fractionation range, and horizontal lines above and below. This must necessarily be a very crude approximation, and it is not easily handled mathematically.

Sigmoidal functions are often well represented either by a logit or by a probit type of curve. The logit function is defined as

$$\text{logit}\ (x) = \ln\left(\frac{x}{(1 - x)}\right) \tag{5}$$

The function is defined in the range $0 < x < 1$. Logit (0.5) = 0, and the function is symmetrical around this point and goes toward positive infinity as x approaches 1 and toward negative infinity as x approaches 0.

The probit function is based on the error function that, defines normal distributions. The definition of the error function is

$$\mathrm{erf}(x) = \frac{1}{\sqrt{2\pi}} \int_{-\infty}^{x} e^{-t^2/2}\, dt \tag{6}$$

The probit function is obtained if 5 is added to the inverse function to the error function:

$$\mathrm{probit}\ (x) = \mathrm{erf}^{-1}(x) + 5 \tag{7}$$

The function is defined for $0 < x < 1$. Probit (0.5) = 5, and the function is symmetrical around this point and goes toward positive infinity as x approaches 1 and toward negative infinity as x approaches 0.

The selectivity curve is represented as a linear relationship in logit or probit terms:

$$\mathrm{logit}\ (K_d) = a + b \log\ (\mathrm{MW}) \tag{8}$$

$$\mathrm{probit}\ (K_d) = a + b \log\ (\mathrm{MW}) \tag{9}$$

where the parameters a and b are derived by regression analysis. The logit function is easier to calculate, particularly in computer programs. Figure 5 shows a selectivity curve in a logit-log diagram. Notice that the logit or probit functions are used mainly for data reduction, and the corresponding diagrams are preferably plotted, not as straight lines in logit or probit diagrams, but rather as the corresponding sigmoidal curves in lin-log diagrams. For further discussion, see Rodbard [86].

These two ways of representing the selectivity curve have the advantage of permitting the relationship to be easily handled in a simple computer program. The curve can be derived from measuring points by weighted linear regression analysis. Some caution must be used with weighting profiles derived directly from the transformation. Such profiles give heavy weight to points close to the middle of the curve. If the "true" curve is not strictly linear, a limited deviation of the slope in the middle part of the curve may express itself by great deviation of the curve at the ends. Although such nonlinearity should in principle be corrected by the use of nonlinear relationships with further parameters, these parameters require more

LOGIT-LOG CURVE

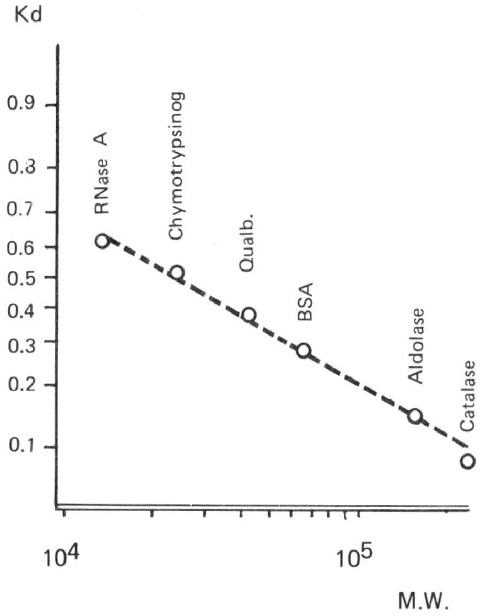

Figure 5. Example of selectivity curve plotted in logit-log scale. Data as in Figure 3.

measuring points than are normally available. A simple solution of this problem is to correct the weighting profile by adding a small constant error, corresponding to the deviation of the true curve from the best straight line. Although this error is, strictly speaking, not random, the curvature of the true curve is not known, and this treatment of the data gives a satisfactory result. A similar problem, related to the treatment of radioimmunoassay data, that also give sigmoidal curves, was discussed by Fischer and Rodbard [89].

The treatment of the selectivity curve by a logit or probit transformation has the further advantage of letting the statistical error of the molecular weight (or molecular size) estimates be easily calculated. Unfortunately, the errors derived in this way by strict statistical methods are considerably larger than the errors of around plus or minus 10% given in the literature (except for Rodbard, no one has indicated how their confidence limits were obtained, if they stand for 1 or 2 standard deviations). If branched macromolecules are studied, a correction must be used for the degree of branching. This may be based on the viscosity [90].

In the preceding discussion, the different geometrical structures of different classes of substances have not been considered. Obviously behavior in an SEC system will be different for molecules with compact or extended structures. As a measure of the extension of the molecular structure, Grubisic et al. [91] used viscosity data, arriving at a universal calibration variable $[\eta]\cdot(MW)$ where $[\eta]$ is the intrinsic viscosity. The use of this universal calibration variable instead of the moelcular weight alone made the SEC selectivity curves for different classes of macromolecules coincide [92-95].

III. RESOLUTION AND SEPARATION IN SEC

The general principles governing resolution in chromatography have been outlined in Chap. 2 and by Giddings [96] (see also Hendrickson, Ref. 97). The comments here will therefore concentrate on those aspects that refer directly to steric exclusion chromatography.

The mechanisms contributing to zone broadening in SEC, as in all types of chromatography in which no gradients are used, are

1. The limited rate at which equilibrium between the stationary and the moving phase is established
2. "Eddy diffusion," or the zone broadening due to differences in length of different stream paths through a bed filled with irregularly packed particles
3. Longitudinal diffusion in the column

The rate of equilibration between the stationary phase and the moving phase depends only on the rate of diffusion within the gel particles; in true SEC, no adsorption-desorption takes place and no hindrance at the "phase" borders can be expected. Therefore, the rate at which equilibrium is established is a direct function of the particle size. In the first attempts to calculate the rate constants for equilibration between stationary and moving phases in SEC, the assumption was made that diffusion in the gel was as rapid as that in free solution. This assumption is not valid, since the gel matrix network forms an obstacle to diffusion. It has been found that the electrophoretic migration rate in a gel, relative to that in free solution, is proportional to the K_{av} value of the same molecule in SEC with the same gel [98,99]. It is reasonable to assume that this is also true for the diffusion rate, although no such measurements have been done.

The eddy diffusion effect in SEC is the same as in all types of chromatography, depending on the particle size, particle size distribution, and regularity of the packing [100]. The situation is somewhat complicated in SEC because most packing materials have soft or elastic particles. The "goodness" of the packing depends very much on the packing procedure. This cannot be derived theoretically but must be found practically for different types of packing materials. In a well-packed column, eddy diffusion effects seem to have a subordinate influence, but many columns used in practice are so badly packed that eddy diffusion is the major factor influencing the resolution. Only long practical experience and critical evaluation of one's own columns will make it possible to pack very good SEC columns.

The longitudinal diffusion in the column, as in most types of liquid chromatography, is of subordinate importance at normal flow rates.

These mechanisms allow transport equations or other models for the chromatographic process (mainly different "theoretical plate" equations) to be set up [101-104]. These theoretical models confirm practical experience with the technique, indicating that the following variables govern the separation.

Flow Rate

At extremely low flow rates, longitudinal diffusion dominates among the causes of zone broadening. Under these conditions, zone broadening decreases as the flow rate increases. At normal flow rates, the restricted rate of equilibrium dominates and eddy diffusion should ideally be independent of flow rate. Zone broadening therefore has a minimum; below this minimum, the zones are broadened mainly by longitudinal diffusion, and above the minimum, they are broadened by the limited rate of equilibration between the phases. The flow rate corresponding to this minimum is so low that it is uninteresting in practical work. A higher flow rate therefore gives a faster experiment at the expense of the separation [105].

Particle Size

Smaller particles give faster equilibration and less eddy diffusion, and therefore resolution in SEC is improved dramatically if the particle size is reduced. Practical problems limit the extent to which the particle size can be reduced. The resistance to flow goes up as the particle size is reduced. Therefore, a higher driving pressure must be used to maintain the flow rate. The particles of many packing materials are soft or elastic, and a higher driving pressure causes the particles to become compressed and the flow to stop completely.

Temperature

A higher temperature will give faster diffusion. As the equilibration rate between the stationary and moving phase, which depends on the diffusion rate, is the major factor influencing the resolution; an increase in temperature generally influences the separation favorably [106]. The stability of the solutes must, however, be considered.

It is always recommended that the temperature be kept constant: changes in temperature with time or between different parts of the column must be avoided. The sample and eluent must have the same temperature as that at which the column is kept. Columns standing in direct sunlight or kept in a cold room, where the temperature goes up and down, seldom work well. The temperature should be kept constant at all times, not only while the column is being used; but at all times since major changes in temperature destroy the quality of the column packing.

Work at high temperatures is difficult, not only because labile substances may be destroyed but also because gas dissolved in the liquid forms bubbles. For work above room temperature, the eluent must be carefully deaerated (for instance, by boiling it before use and keeping it above column temperature until use).

Column Length

The distance between peaks generally increases in proportion to the column length, and the zone width increases in proportion to the square root of the column length. Therefore, under ideal conditions, resolution in SEC improves in proportion to the square root of the column length. Practical problems limit the length of the columns, particularly those filled with soft gel materials. The solution is the use of several shorter columns attached in series (or possibly "recycling," by pumping the effluent from a column back into the same column [107,108]). Another problem is that a longer column gives a greater dilution of the sample, but this can be counteracted by application of a larger or more concentrated sample or the use of columns of smaller diameter.

Sample Volume and Concentration

An infinitely small sample will theoretically give a zone with a shape that depends, among other factors, on column packing and other factors. In SEC, in which no overlaoding effects or nonlinearity of the partition isotherms need be considered, the shape of the zones obtained from samples with volumes that are not negligibly small can be calculated by integration of the eluted zones of the "infinitely small" partial volumes making up the total sample volume. The result is that the eluted zone is widened by approximately the volume of the sample. In preparative work this is an important relationship. Chromatographic separation can be improved at a given sample load by choosing a longer column to increase the separation and reduce the column volume required for a given sample load. This facilitates operation and reduces cost. The upper limit of length of the column to be used is defined by practical considerations and the fact that elution takes longer the longer the column. Another advantage of increasing the volume of the sample relative to the column volume in this way is that the dilution of the sample is reduced.

In preparative work the concentration of the sample should be as high as possible, to increase the capacity and the concentration in the fractions. In SEC the partition isotherms are linear or practically linear and large amounts of substance do not generally give inferior separations. The limit is set by the solubility of the substances being separated and by the viscosity of the sample (see below).

Composition of the Mobile Phase

The influence of the composition of the mobile phase on the separation in SEC is different in organic solvents and in aqueous solution. In organic solvents, the main characteristic influencing the separation is the dissolving power of the eluent on the substances in the sample. In a bad solvent, the latter substances tend to become nonspecifically adsorbed to the gel matrix [95,109,110]. Such solvents should therefore be avoided. If mixtures of solvents are used, one of the components may be enriched in the gel phase. If the solvents have different polarities, a straight-phase or a reversed-phase partitioning system is obtained and the substances will no longer be separated according to size [111-116]. Great care must therefore be exercised in the choice of eluent if mixed solvents are used, to avoid such undesired effects.

In aqeuous systems, electrostatic effects must be considered. Electrically charged groups on the gel matrix [117] and macroions in the solution create Donnan potentials between the gel and surrounding solution (see page 354). Such Donnan potentials give the gels an ion-exchange character. The substances to be separated themselves contribute to the potentials, making the partitioning isotherms nonlinear. These electrical potentials will therefore, in addition, to the unwanted change in the separation pattern according to size, give rise to tailing and zone spreading [106,118,119].

To reduce electrostatic effects in aqueous media, the concentration of low-molecular-weight ions in the eluent and sample should be made sufficiently high. This may be achieved by the use of fairly concentrated buffers as eluents, and often neutral salt is added. Under physiological conditions (0.9% NaCl), electrostatic effects are normally considered negligible. Also, the pH value may be used to reduce ionic interaction [64,120,121]. Charged groups on the gel matrix may be removed chemically [54,122,123].

The use of salts to diminish electrostatic effects inevitably results in salts in the fractions collected. Often this does not interfere with the experiments being done, but it must be considered when choosing the sequence of steps in a purification series. Size exclusion chromatography on a small column with a tightly cross-linked gel and elution with distilled water may be used to remove salts after SEC fractionation: the high-molecular-weight solutes will not pass into the gel, and will therefore elute with little zone broadening in spite of the electrostatic effects. Furthermore, the separation of macromoelcules and salts on this type of column is so good in any case that the zone broadening is of little consequence. Electrolytes may also be removed by dialysis. Volatile electrolytes, such as ammonium formate, ethylene diamine acetate, or ammonium carbonate may be removed by freeze-drying.

If strongly charged substances are to be separated, the addition of salts may not be sufficient. No good separation system has been described for strongly substituted sulfated polysaccharides, and problems are also encountered in the separation of nucleic acids. In the latter case, the choice of a proper pH may help, since the charged groups can be neutralized.

The electrostatic effects in aqeuous systems may in principle be used to achieve a separation. The condition is that a molecule other than the substance to be separated be used to maintain the Donnan potential. The separation obtained is similar to that obtained on ion exhcangers rather than that in normal SEC. An example of such a system is one in which the eluent (and the sample) contains a certain amount of high-molecular-weight dextran sulfate. The gels would act as ion exchangers, and at a pH above the isoelectric point of the substances to be separated, these substances preferentially pass into the gel if their size allows. The substances would be eluted by changing the pH, ionic strength, or concentration of dextran sulfate in the eluent. For preparative purposes a limitation is the presence of dextran sulfate in the fractions, but the systems are interesting for analytic purposes. No such experiments have been published.

Some solvents influence the structure of substances. A special case in this respect—denaturing and reducing solvents—is important for the characterization of proteins. Denaturing solvents, containing, for example, urea, guanidine hydrochloride, or sodium dodecyl sulfate, break up the secondary structure of proteins. The polypeptide chains, which are normally compactly ordered into an intricate structure, are released into a random coil. Such denatured proteins take more space than the corresponding native proteins and therefore in SEC elute as if they were larger. Accurate calculations of this effect are impossible; since the denaturing agents also modify the structure of the gel matrix. The use of denaturing agents in SEC is often practiced to estimate molecular weights of proteins, as under these conditions, individual differences in secondary structure between different proteins no longer matter, only the length of the chain [124-133].

The use of denaturing agents in SEC is often combined with reduction of the disulfide bridges that keep the subunits of proteins together. To prevent the re-formation of disulfide bridges between different chains, reduction may be followed by methylation of the thiol groups of the resultant cysteine residues. Subsequent SEC with a denaturing eluent gives information about the molecular weights of the subunits of the protein under study.

Viscosity Effects

The viscosity of the eluent may in itself influence separation, since diffusion is slower at higher viscosities. This effect has not been studied separately, as the change in viscosity is related to a change in temperature or eluent composition, and it is difficult to separate the effect of viscosity from other effects.

The most important viscosity effect is related to the difference in viscosity between sample and eluent. If the concentration of macromolecules in the sample is so high that the viscosity is markedly higher than that of the eluent, an instability of the liquid flow in the chromatographic bed results. The zone soon becomes very irregular, and zone broadening is observed [134]. That this is due to the increase in viscosity in the sample relative to that in the eluent is easily demonstrated: if the viscosity in the sample only is increased, for example by the addition of dextran, zone broadening takes place, but the addition of dextran to both the sample and eluent does not produce zone broadening.

Linearity of Partition Isotherm

One of the advantages of SEC is that the partition isotherm is linear up to very high concentrations; or in other words, the partitioning coefficient of a given solute between the gel phase and the surrounding liquid is constant, independent of the concentration of the solute. This makes SEC a very reproducible chromatographic process, and also results in sharp peaks with little tailing [135-140].

IV. MATERIALS FOR SEC

The matrix of a gel for SEC must be lyophilic (wetted by the liquid), since lyophobic interactions (hydrophobic interaction in aqueous media and polarity and hydrogen bonding in organic media) would otherwise dominate the separation. In aqueous solvents, polysaccharides or synthetic hydrophilic polymers, such as polyacrylamide, form the basis for the matrix. A few such materials, for example agarose, form gels spontaneously, but most of the polymers used are soluble in water and must be cross-linked to form a gel. This is also true in organic solvents, in which polystyrene or similar polymers must be cross-linked to form a rigid gel.

Some gels, such as dextran gels, can be dried and rewetted reversibly. A distinction is commonly made between aerogels, which maintain their open structure when dried, and xerogels, which shrink on drying and reswell when wetted. Many gel materials lose their structure when dried and cannot be reversibly rewetted.

The most commonly used commercial gel materials for aqueous media are as follows

Sephadex which is made by cross-linking dextran with epichlorohydrin in alkaline solution [2]. A hydride is added to prevent oxidation. To make the particles bead shaped the dextran solution is suspended in an organic solvent with a suitable surfactant before epichlorohydrin is added. Xerogels, delivered dry, may be used in some polar organic solvents [141].
Sephacryl which is made by cross-linking allyl dextran with N-N'-methylene bisacrylamide. The gel is delivered is suspension; no method has been described to reversibly dry an rewet the gel. The particles are more rigid that those of Sephadex. Sephacryl may also be used in some organic solvents [130,142, 143].
Sepharose and Bio-Gel A are beaded agarose gels. Agarose gels have an open structure, as the polysaccharide chains form large aggregates [106,123,144-160]. Therefore, the fractionation ranges extend up to very high molecular weights, often into the range of small virus particles. The selectivity curves are flat, indicating a heterogeneous strucutre. These gels are delivered in suspension.
Sepharose CL is a beaded agarose gel cross-linked with 2,3-dibromopropanol to increase the chemical stability. The fractionation properties are not substantially changed by the cross-linkage.
Superose comprises beaded agarose gels with small particles (around 30 μm) and a narrow particle size distribution. The particles are so rigid that they can be used at pressures up to the megapascal range, although this will give such high flow rates that the resolution will be unsatisfactory [161-167].
Bio-Gel P is a bead-polymerized polyacrylamide cross-linked with methylene-bis-acrylamide. Its fractionation properties are similar to those of Sephadex [60,153,168-176] and comes in the form of xerogels delivered dry.
Ultrogel is an agarose gel with acrylamide polymerized into the large pores of the agarose [177]. The agarose gives rigidity to the gel, and the acrylamide defines the fractionation properties. The selectivity curve is not as steep as for Sephadex or Bio-Gel P with the corresponding exclusion limit. Ultrogel is delivered in suspension.

Fractogel comprises vinyl polymer-based gels of small particle size (around 30 μm), with a narrow particle size range. Particles are rigid.

Other materials that have been suggested for SEC in aqueous solution include porous glass [128,178-183], polyglycol acrylates (also used in contact lenses), polymethyl methacrylates [184], and polymethyglutamate, which gives very rigid small particles [185]. Gel materials for organic solvents include the following

Styragel. These cross-linked polystyrene gels are usually delivered in ready-packed columns and used for the determination of molecular weight distributions of organic polymers [3,4].
Merckogel. These include gels based on polyvinyl alcohol, polyethylene glycol dimethacrylate, and polyvinyl acetate. Some of them are claimed to also be of use in aqueous media [186-191].
Sephadex LH. This substituted dextran gel works in water and polar organic solvents [115, 192-209].

Characteristics of the selectivity curves of some gels are summarized in Table 1.

V. APPENDIX

A. Definition of the Problem

In an infinite space, infinitely long and infinitely thin rods (or rays or shots) are randomly distributed. If an irregularly shaped body is placed somewhere in this space, what is the probability that no rod passes through any part of it? Or, put in another way, in what proportion of the space can the body be placed without interfering with the rods? The average density of rods in the space is L units of length per unit volume, and the body has surface area A. There are no holes, intrusions, or other irregularities in the surface of the body (the surface is everywhere positively curved; no rod can pass out of the body and re-enter).

B. Solution

A positive direction is arbitrarily assigned to every rod. Every surface has one outside and one inside. If the rod passes into a surface, it is considered to pass the surface in the positive direction; if it passes out of the surface, it is considered to pass the surface in the negative direction. The density of rods coming from the element of space angle $d\sigma$ is $\lambda\ d\sigma$: a surface perpendicular to the direction of the rods, will be statistically penetrated by $\lambda\ d\sigma$ rodes originating from the element of space angle $d\sigma$ per unit area. (λ will later be related to the total density of rods in the space L.)

Table 1. Fractionation characteristics of some gel materials[a]

Gel type	a^b	b^b	$K_d = 0.5^c$	$K_d = 0.1^c$
Sephadex G-75	15.43	−3.66	16,500	65,000
Sephadex G-75SF	13.18	−3.24	11,500	55,000
Sephadex G-100	11.36	−2.64	20,000	135,000
Sephadex G-100SF	13.54	−3.16	19,000	95,000
Sephadex G-150	12.69	−2.74	42,000	265,000
Sephadex G-150SF	13.25	−2.99	27,000	150,000
Sephadex G-200	10.71	−2.26	55,000	500,000
Sephadex G-200SF	11.31	−2.47	37,000	300,000
Sephacryl S-200SF	9.69	−2.21	24,000	235,000
Sephacryl S-300SF	7.07	−1.47	65,000	2 million
Sepharose CL-6B	5.95	−1.14	160,000	
Sepharose CL-4B	6.11	−1.05	650,000	

[a]There is some interbatch variation, and variation due to different pretreatment of the gel material occurs. The parameters must be determined separately for each column used for molecular weight estimation.

[b]The equation of the selectivity curve was assumed to be

$$\text{logit}\ (K_d) = a + b \log (MW)$$

where the parameters were derived by regression analysis as described in the legend to Fig. 3.

[c]The molecular weights corresponding to $K_d = 0.5$ (midrange) and $K_d = 0.1$ (practical exclusion limit) were calculated from the preceding equation (note a) using the values for the parameters a and b given in the table.

Source: The data in the table have been calculated from calibration data obtained in the laboratories of Pharmacia Fine Chemicals and kindly provided by Mr. Rolf Berglund.

Now consider a small element of the surface dA somewhere in this space. As a construction for the derivation, half a unit sphere is placed over this surface element (see Fig. 6). Now consider the rods having incidence angles (in radians) between α and $\alpha + d\alpha$. The surface element dA has a projection $dA \sin\alpha$ in the direction of the incoming rods. The circle on the unit sphere through which the rods must pass has radius $\cos\alpha$, and the width of the strip is $d\alpha$. The space angle from which the rods come is therefore $2\pi \cos\alpha \, d\alpha$, corresponding to the area of the strip on the unit sphere. The number of rods passing with incidence angle α is thus $2\pi\lambda \cos\alpha \sin\alpha \, dA \, d\alpha$. The sum of rods passing in a positive direction is obtained by integration:

$$dl = \int_0^{\pi/2} 2\pi\lambda \, dA \cos\alpha \sin\alpha \, d\alpha = \pi\lambda \, dA \qquad (10)$$

Now consider a very large number N_0 of the irregularly shaped bodies with surface area A. On all of these bodies we proceed in the same regular fashion over the surface, and register whether a rod passes in a positive direction through the element of surface area just covered. So far, assume we have reached the point at which surface a has been covered, and N bodies still remain without any rod passing into the surface area covered (we do not consider rods passing out, as they must pass in at some other location, and will be registered there). We now continue over the following surface element da on all of the bodies. The number of bodies that will disappear because of rod passes in through this element of area is

$$-dN = N\pi\lambda da \qquad (11)$$

Using the border condition that $a = 0$ gives $N = N_0$, this differential equation gives

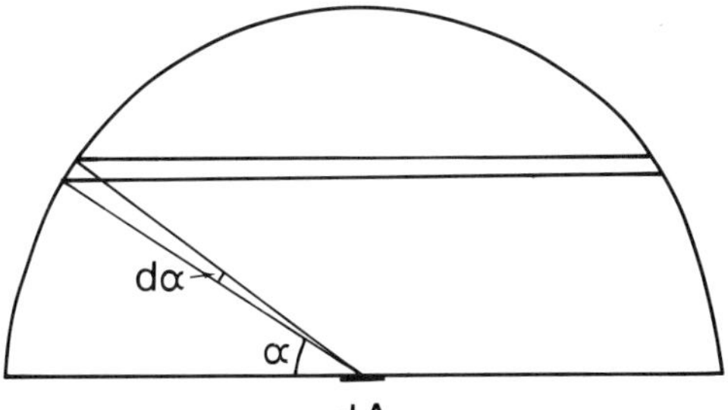

Figure 6. Construction for the derivation of Eq. (10) (see text).

$$\ln \frac{N}{N_0} = -\pi\lambda a \tag{12}$$

When the whole area is covered, a = A,

$$\frac{N}{N_0} = e^{-\pi\lambda A} \tag{13}$$

To relate λ to L, consider an element of space dv, with surface area da and thickness dx. A unit half-sphere is placed over it as shown in Figure 7. Now consider the rods hitting this area with an angle of incidence between α and $\alpha + d\alpha$. The number of rods hitting the volume element is

$$2\pi\lambda dA \cos\alpha \sin\alpha \, d\alpha$$

The length of each rod in the volume element is $dx/\sin\alpha$. Thus, the length of rods from this angle of incidence is

$$2\pi\lambda \cos\alpha \, dv \, d\alpha$$

Integrating with respect to α from 0 to $\pi/2$ yields the total length of rods entering from above:

$$dl_u = 2\pi\lambda \, dv \tag{14}$$

Including the rods entering into the element of volume from below gives

$$dl_{tot} = 4\pi\lambda \, dv \tag{15}$$

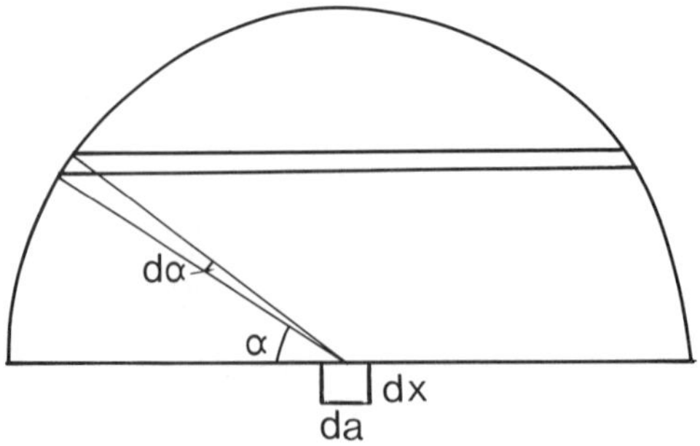

Figure 7. Construction for the derivation of Eq. (16).

$$L = \frac{dl_{tot}}{dv} = 4\pi\lambda \quad (16)$$

Inserting this into Eq. (13) gives

$$\frac{N}{N_0} = e^{-LA/4} \quad (17)$$

which is equivalent to Eq. (4).

Editor's note: The nomenclature for SEC, used in this chapter, is by convention somewhat different from that used in other branches of chromatography. There is a special ASTM standard (D-3016) which only partly agrees with the ASTM standard for liquid chromatography (E-682), which is adopted in this book. To facilitate comparisons with other parts of the book (especially Chap. 2, pages 34-35 and 80-81) the following correspondences are valid:

Chap. 8	Chap. 2
V_0	V_e
V_e	V_R (or V_M)
V_i	$V_0 - V_e$
K_d	k_0

REFERENCES

1. J. Porath, Fractionation of polypeptides and proteins on dextran gels, *Clin. Chim. Acta, 4*:776 (1959).
2. J. Porath and P. Flodin, Gel filtration: A method for desalting and group separation, *Nature, 183*:1657 (1959).
3. J. C. Moore, Gel permeation chromatography, I, A new method for molecular weight distribution of high polymers, *J. Polym. Sci., A2*:835 (1964).
4. J. C. Moore and J. G. Hendrickson, Gel permeation chromatography, II, The nature of the separation, *J. Polym. Sci., C8*:233 (1965).
5. G. K. Ackers, Analytical gel chromatography of proteins, *Adv. Protein Chem., 24*:343 (1970).
6. R. Audebert, Fractionation mechanisms in gel permeation chromatography, *Analusius, 4*:399 (1976).
7. R. Audebert, Nonexclusion effects in g. p. c. A review, *Polymer, 20*:1561 (1979).

8. J. Brewer, Enzyme purificaiton by gel filtration, *Process Biochem.*, *6*:9:39 (1971).
9. J. Coupek, M. Kubin and Z. Deyl, Practice of gel chromatography, in *Liquid Column Chromatography: A Survey of Modern Techniques and Applications* (Z. Deyl, K. Macek and J. Janátc, eds.) Elsevier, Amsterdam, p. 301 (1975).
10. J. M. Curling, The use of Sephadex in the separation, purification and characterization of biological materials, *Exp. Physiol. Biochem.*, *3*:417 (1970).
11. J. M. Curling, Process gel filtration, Part I, From laboratory to process scale, *Am. Lab.*, *8*:5:47 (1976).
12. J. M. Curling and J. M. Cooney, Operation of large scale gel filtration and ion exchange systems, *J. Parenter, Sci. Technol.*, *36*:2:59 (1982).
13. H. Determann and J. E. Brewer, Gel Chromatography, in *Chromatography* (E. Heftmann, ed.) 3rd ed., Reinhold, New York, p. 362 (1975).
14. H. Determann, *Gel Chromatography*, Springer-Verlag, Berlin (German 1967, English 1967 and 1969).
15. L. Fischer, Gel filtration chromatography, in *Laboratory Techniques in Biochemistry and Molecular Biology, Vol. 1, Part 2* (T. S. Work and R. H. Burdon, eds.), Elsevier/North Holland, Amsterdam (1980).
16. P. Flodin, Methodological aspects on gel filtration with special reference to desalting operations, *J. Chromatogr.*, *5*:103 (1961).
17. V. F. Gaylor and H. L. James, Gel permeation chromatography (steric exclusion chromatography), *Anal. Chem.*, *50*:29R (1978).
18. G. L. Hagenauer, Size exclusion chromatography, *Anal. Chem.*, *54*(5):265R (1982).
19. J. Reiland, Gel filtration, *Meth. Enzymol.*, *22*:287 (1971).
20. H. Susskind and W. Becker, Random packing of spheres in nonrigid containers, *Nature*, *212*:1564 (1966).
21. J. Eastwood, E. J. P. Matzen, M. J. Young, and N. Epstein, Random loose porosity of packed beds, *Br. Chem. Eng.*, *14*:1542 (1969).
22. A. R. Dexter and D. W. Tanner, Packing densities of mixtures of spheres with log-normal size distributions, *Nature, Phys. Sci. 238*: 31 (1972).
23. T. C. Laurent and J. Killander, A theory of gel filtration and its experimental verification, *J. Chromatogr.*, *14*:317 (1964).
24. J. Å. Jönsson, Non-ideal effects in linear chromatography, *Chromatographia*, *13*:729 (1980).
25. J. Å. Jönsson, The median of the chromatographic peak as the best measure of retention time, *Chromatographia*, *14*:653 (1981).

26. G. H. Weiss and G. K. Ackers, Magnitude of a finite equilibration effect in analytical gel chromatography, *Biopolymers, 11*: 2125 (1972).
27. Z. Grubisic and H. Benoit, Comparison of partition coefficients determined by static and dynamic methods on porous glass beads, *J. Chromatogr. Sci.*, *9*:262 (1971).
28. A. M. Basedow, K. H. Ebert, H. J. Ederer, and E. Fosshag, Fractionation of polymers by gel filtration chromatography: An experimental and theoretical approach, *J. Chromatogr.*, *192*:259 (1980).
29. J. Porath, Some recently developed fractionation procedures and their application to peptide and protein hormones, *J. Appl. Chem.*, *6*:233 (1963).
30. G. T. Kolde and E. L. Carstensen, Dielectric properties of Sephadex and its water of hydration, *J. Phys. Chem.*, *80*:55 (1976).
31. P. G. Squire, A relationship between the molecular weights of macromolecules and their elution volumes based on a model for Sephadex gel filtration, *Arch. Biochem. Biophys.*, *107*:471 (1964).
32. M. E. van Kreveld and N. van den Hoed, Mechanism of gel permeation chromatography distribution coefficient, *J. Chromatogr.*, *83*:111 (1973).
33. G. K. Ackers, A new calibration procedure for gel filtration columns, *J. Biol. Chem.*, *242*:3237 (1967).
34. J. B. Carmichael, Theory of gel filtration separation of biopolymers assuming a gaussian distribution of pore size, *Biopolymers*, *6*:1497 (1968).
35. M. Kubin, Model of the mechanism of the separation of macromolecules in gel permeation chromatography on a packing with nonhomogeneous pores, *J. Chromatogr.*, *108*:1 (1975).
36. A. G. Ogston, The spaces in a uniform random suspension of fibres, *Trans. Faraday Soc.*, *54*:1754 (1958).
37. K. Hellsing, Gel chromatography in eluants containing polymers, *J. Chromatogr.*, *36*:170 (1968).
38. J. C. Giddings, Resolution and optimization in gel filtration and permeation chromatography, *Anal. Chem.*, *40*:2143 (1968).
39. J. C. Giddings, E. Kucera, C. P. Russell, and N. M. Myers, Statistical theory for the equilibrium distribution of rigid molecules in inert porous networks. Exclusion chromatography, *J. Phys. Chem.*, *72*:4397 (1968).
40. A. G. Ogston, On the interaction of solute molecules with porous networks, *J. Phsy. Chem.*, *74*:668 (1970).
41. D. Rodbard and A. Chrambach, Unified theory for gel electrophoresis and gel filtration, *Proc. Natl. Acad. Sci. USA*, *65*: 970 (1970).

42. J. B. Carmichael, Stochastic model for gel permeation separation of polymers, *J. Polym. Sci., Polym. Phys. Ed., 6*:517 (1968).
43. J. B. Carmichael, Stochastic model for gel permeation chromatography at high flow rates, *J. Phys. Chem., 49*:5161 (1968).
44. J. C. Giddings, *Dynamics of Chromatography*, Part 1. *Principles and Theory*, Marcel Dekker, New York (1965).
45. G. K. Ackers, Molecular exclusion and restricted diffusion processes in molecular sieve chromatography, *Biochemistry, 3*:723 (1964).
46. C. L. de Ligny, A comparative study of the influence of restricted diffusion upon the moving rate of the solute in various forms of chromatorgraphy, *J. Chromatogr., 36*:50 (1968).
47. E. A. Dimarzio and C. M. Guttman, Separation by flow and its application to gel permeation chromatography, *J. Chromatogr., 55*:83 (1971).
48. S. Mori, R. S. Porter, and J. F. Johnson, Column fractionation of polymers, XXXI, Gel permeation chromatography. Mechanism of separation by flow, *Anal. Chem., 46*:1599 (1974).
49. G. E. Boyd and B. A. Soldano, Osmotic free energies of ion exchangers, *Z. Elektrochem., 57*:162 (1973).
50. G. E. Myers and G. E. Boyd, A thermodynamic calculation of cation exchange selectivities, *J. Phys. Chem., 60*:521 (1956).
51. J. Lecourtier, R. Audebert, and C. Quivoron, Theoretical study of chromatographic separations performed with cross-linked organic polymers, *J. Chromatogr., 121*:173 (1976).
52. B. Z. Ginzburg and D. Cohen, Calculation of internal hydrostatic pressure in gels from the distribution coefficients of non-electrolytes between gels and solutions, *Trans. Faraday Soc., 60*:185 (1964).
53. A. Polson and W. Katz, A quantitative theory for gel-exclusion chromatography, *Biochem. J., 112*:387 (1969).
54. D. Eaker and J. Porath, Sorption effects in gel filtration, I, A survey of amino acid behavior on Sephadex G-10, *Separation Sci., 2*:507 (1967).
55. T. Ogata, N. Yoza, and S. Ohashi, The effect of anions on the gel chromatographic behavior of magnesium ions, *J. Chromatogr., 58*:267 (1971).
56. A. Domard, M. Rinaudo, and C. Rochas, Application of gel permeation chromatography to polyelectrolytes: Salt rejection mechanism and molecular-weight distribution, *J. Polym. Sci., Polym. Phys. Ed., 17*:673 (1979).
57. A. N. Glazer and D. Wellner, Adsorption of proteins on "Sephadex," *Nature, 194*:862 (1962).
58. L. W. Nichol, W. H. Sawyer, and D. J. Winzor, Gibbs-Donnan effects in gel chromatography, *Biochem. J., 112*:259 (1969).

59. L. Strid, Separation of peptides according to charge by gel filtration in the presence of charged detergents. *FEBS Lett.*, *33*:192 (1973).
60. A. Pusztai and W. B. Watt, The determination of the molecular size of peptides and proteins by chromatography on Bio-Gel P-100 columns in phenol-acetic acid-water (1:1:1, w/v/v) solvent mixture, *Biochim. Biophys. Acta*, *214*:463 (1970).
61. C. A. Streuli, The interaction of aromatic compounds in alcohol solution with Sephadex LH-20 dextran, *J. Chromatogr.*, *56*:219 (1971).
62. C. A. Streuli, The effect of solvent change on the separation process, *J. Chromatogr.*, *56*:225 (1971).
63. K. W. Williams, Solute-gel interactions in gel filtration, *Lab. Pract.*, *21*:667 (1972).
64. H. M. Ortner, Gel chromatography of rhenium (VII), *J. Chromatogr.*, *107*:335 (1975).
65. H. M. Ortner and H. Dalmonego, Behavior of molybdenum (VI) and molybdenum (V) on dextran and polyacrylamide gels, *J. Chromatogr.*, *107*:341 (1975).
66. B. Gelotte, Studies on gel filtration: Sorption properties of the bed material Sephadex, *J. Chromatogr.*, *3*:330 (1960).
67. H. Determann and I. Walter, Source of aromatic affinity to Sephadex dextran gels, *Nature*, *219*:604 (1968).
68. H. Determann and K. Lampert, Hydrophobic interaction in gel adsorption chromatography, *J. Chromatogr.*, *69*:123 (1972).
69. J.-C. Janson, Adsorption phenomena on Sephadex, *J. Chromatogr.*, *28*:12 (1967).
70. G. I. Glover, P. S. Mariano, and T. J. Wilkinson, The separation of water-soluble organic diastereomers by chromatography on Sephadex, *Separation Sci.*, *10*:795 (1975).
71. G. I. Glover, P. S. Mariano, and S. Cheowtirakul, Separations of diastereomeric organic acids on Sephadex G-10, *Separation Sci.*, *11*:147 (1976).
72. K. Lampert and H. Determann, An aqueous two-phase system of dextran/hydroxypropyldextran as a model in adsorption studies of Sephadex gels, *J. Chromatogr.*, *63*:420 (1971).
73. V. di Gregorio and M. Sinibaldi, Adsorption of inorganic anions on Sephadex gels, *J. Chromatogr.*, *129*:407 (1976).
74. T. Deguchi, A. Hisanaga, and H. Nagai, Chromatographic behavior of inorganic ions on a Sephadex G-15 column, *J. Chromatogr.*, *133*:173 (1977).
75. J. V. Dawkins and M. Hemming, Gel permeation chromatography with cross-linked polystyrene gels and poor and theta solvents for polystyrene. 2. Separation mechanism, *Makromol. Chem.*, *176*:1795 (1975).
76. A. J. W. Brook and S. Housley, The interaction of phenols with Sephadex gels, *J. Chromatogr.*, *41*:200 (1969).

77. A. J. W. Brook and S. Housley, The interaction of organic acids and bases with Sephadex gels, *J. Chromatogr.*, *42*:112 (1969).
78. A. J. W. Brook, The separation of biochemical acids and bases on columns of Sephadex G-10, *J. Chromatogr.*, *47*:100 (1970).
79. A. J. W. Brook and K. C. Munday, The interaction of phenols, anilines and benzoic acids with Sephadex gels, *J. Chromatogr.*, *47*:1 (1970).
80. A. J. W. Brook and K. C. Munday, The effect of temperature on the interaction of phenols with Sephadex gels, *J. Chromatogr.*, *51*:307 (1970).
81. D. Saunders and R. L. Pecsok, Calculation of distribution coefficients in inorganic gel chromatography, *Anal. Chem.*, *40*:44 (1968).
82. B. Öbrink, T. C. Laurent, and R. Rigler, Studies on the temperature dependence of chromatography on a dextran gel (Sephadex G-200), *J. Chromatogr.*, *31*:48 (1967).
83. T. R. C. Boyde, Swelling and contraction of polyacrylamide gel slabs in aqueous solution, *J. Chromatogr.*, *124*:219 (1976).
84. K. Lampert and H. Determann, Effect of temperature on the porosity of dextran gels, *J. Chromatogr.*, *56*:140 (1971).
85. D. Rodbard, Estimation of molecular weight by gel filtration and gel electrophoresis. I. Mathematical principles, *Methods Protein Sep.*, *2*:145 (1976).
86. D. Rodbard, Estimation of molecular weight by gel filtration and gel electrophoresis, II, Statistical and computational considerations, *Methods Protein Sep.*, *2*:181 (1976).
87. H. Determann and W. Michel, The correlation between molecular weight and elution behavior in gel chromatography of proteins, *J. Chromatogr.*, *25*:303 (1966).
88. A. K. Grover and M. Kapoor, Improved treatment of data for estimation of molecular weights by Sephadex gel filtration, *Anal. Biochem.*, *51*:163 (1973).
89. L. Fischer and D. Rodbard, Logit-log radioimmunoassay data reduction: Weighted vs. unweighted, *Clin. Chem.*, *29*:391 (1983).
90. S. Nakano and Y. Goto, Method for correcting molecular weight from gel permeation chromatography, II, Long-chain branching correction of low-density polyethylenes, *J. Appl. Polym. Sci.*, *20*:3313 (1976).
91. Z. Grubisic, P. Rempp, and H. Benoit, A universal calibration for gel permeation chromatography, *J. Polym. Sci., Polym. Lett.*, *5*:753 (1967).
92. L. Z. Vilenchic and B. G. Belenky, Comments on the theoretical basis of gel permeation chromatography of polymers, *J. Chromatogr.*, *56*:13 (1971).

93. J. V. Dawkins, Calibration procedures in gel permeation chromatography, *Br. Polym. J.*, *4*:87 (1972).
94. J. V. Dawkins and M. Hemming, Universal calibration for cyclohexane at 35°C, *Makromol. Chem.*, *176*:1777 (1975).
95. W. Heitz, Mechanism of gel permeation chromatography, *Fresenius' Z. Anal. Chem.*, *277*:323 (1975).
96. J. C. Giddings, Maximum number of components resolvable by gel filtration and other elution chromatographic methods, *Anal. Chem.*, *39*:1027 (1967).
97. J. G. Hendrickson, Basic gel permeation chromatography studies, III, Mathematical analysis of peak spreading series, *J. Polym. Sci. A:2*, *6*:1903 (1968).
98. C. J. O. R. Morris, Gel filtration and gel electrophoresis, *Protides Biol. Fluids Proc. Colloq.*, *14*:543 (1966).
99. C. J. O. R. Morris and P. Morris, Molecular-sieve chromatography and electrophoresis in polyacrylamide gels, *Biochem. J.*, *124*:517 (1971).
100. F. W. Billmeyer, Jr., G. W. Johnson, and R. N. Kelley, Evaluating dispersion in gel-permeation chromatography, I, Theoretical analysis. *J. Chromatogr.*, *34*:316 (1968).
101. M. V. Tracey and D. J. Winzor, Interpretation of zonal asymmetry in gel filtration, *Arch. Biochem. Biophys.*, *117*:184 (1966).
102. T. Nakagawa and H. Jizomoto, Simulation applied to the gel filtration of surfactants. II. Mixture of two non-ionic surfactants, *Kolloid-Z. Z. Polym.*, *234*:1124 (1969).
103. T. Nakagawa and H. Jizomoto, Simulation applied to the gel filtration of surfactants, IV, Mixture of two ionic surfactants. *Kolloid-Z. Z. Polym.*, *239*:606 (1970).
104. A. C. Quano and J. A. Barker, Computer simulation of linear gel permeation chromatography, *Separation Sci.*, *8*:673 (1973).
105. L. Hagel, Comparison of some soft gels for the molecular weight distribution analysis of dextrans at enhanced flow rates, *J. Chromatogr.*, *160*:59 (1978).
106. F. N. Hayes and V. E. Mitchell, Gel filtration chromatography of polydeoxynucleotides using agarose columns, *J. Chromatogr.*, *39*:139 (1969).
107. J. Porath and H. Bennich, Recycling chromatography, *Arch. Biochem. Biophys. (Suppl.)*, *1*:152 (1962).
108. K. I. Bombaugh and R. F. Levangie, High resolution preparative liquid chromatography using recycle, *J. Chromatogr. Sci.*, *8*:560 (1970).
109. W. Heitz, On the mechanism of gel permeation chromatography, *Z. Anal. Chem.*, *277*:323 (1975).
110. A. Campos, V. Soria, and J. E. Figueruelo, Polymer retention mechanism in GPC on active gels. 1. Polystyrene in pure and mixed eluents, *Makromol. Chem.*, *180*:1961 (1979).

111. M. A. Wells and J. C. Dittmer, The use of Sephadex for the removal of nonlipid contaminants from lipid extracts, *Biochemistry, 2*:1259 (1963).
112. L. D. Zeleznick, The use of Sephadex G-25 in partition column chromatography, *J. Chromatogr., 14*:139 (1964).
113. A. N. Siakotos and G. Rouser, Analytical separation of nonlipid water soluble substances and gangliosides from other lipids by dextran gel column chromatography, *J. Amer. Oil Chem. Soc., 42*:913 (1965).
114. B. Bush and T. E. L. Jones, The behavior of some Sephadex gels in dioxane-water mixtures, *J. Chromatogr., 49*:448 (1970).
115. J. G. Bell, Separation of oil-soluble vitamins by partition chromatography on Sephadex LH-20, *Chem. Ind.* (London), *7*:201 (1971).
116. N. V. B. Marsden, Chromatographic determination of the internal solvent composition, solvent regain and distribution coefficients in gels in mixed solvent systems, I, Theory, *J. Chromatogr., 105*:1 (1975).
117. R. L. Pecsok and D. Saunders, On the mechanism of gel chromatography of inorganic salts, *Separation Sci., 3*:325 (1968).
118. H. M. Ortner and O. Pacher, The amount of carboxyl groups in tightly cross-linked gels and its influence on the cation selectivity, *J. Chromatogr., 71*:55 (1972).
119. N. V. B. Marsden, Anionic dominance in ionic interactions with aqueous dextran gel systems, *Naturwissenschaften, 60*:257 (1973).
120. H. J. Cruft, The fractionation of histones on Sephadex G-75, *Biochim. Biophys. Acta, 54*:611 (1961).
121. K. Ujimoto, T. Yoshimura, I. Ando, and H. Kurihara, pH dependence of the distribution coefficients of monomeric oxo anions of phosphorous in gel chromatography with tightly cross-linked gels, *J. Chromatogr., 174*:123 (1979).
122. H.-C. Chen, L. C. Graig, and E. Stoner, On the removal of residual carboxylic acid groups from cellulosic membranes and Sephadex, *Biochemistry, 11*:3559 (1972).
123. T. Låås, Agar derivatives for chromatography, electrophoresis and gel-bound enzymes, II, Charge-free agar, *J. Chromatogr., 66*:347 (1972).
124. P. R. Carnegie, Estimation of molecular size of peptides by gel filtration, *Biochem. J., 95*:9P (1965).
125. P. R. Carnegie, A peptide mapping technique for the estimation of molecular size, *Nature, 206*:1128 (1965).
126. C. F. A. Bryce and R. R. Crichton, Gel filtration of proteins and peptides in the presence of 6 M guanidine hydrochloride, *J. Chromatogr., 63*:267 (1971).

127. K. G. Mann and W. W. Fish, Protein polypeptide chain molecular weights by gel chromatography in guanidinium chloride, *Methods Enzymol.*, *26*:28 (1972).
128. W. Haller, A single equation relating molecular weight, pore size, and elution coefficient in the controlled pore glass chromatography of protein-sodium dodecyl sulfate complexes, *J. Chromatogr.*, *85*:129 (1973).
129. A. A. Ansari and R. G. Mage, Molecular weight estimation of proteins using Sepharose CL-6B in guanidine hydrochloride, *J. Chromatogr.*, *140*:98 (1977).
130. M. Belew, J. Porath, J. Fohlman, and J.-C. Janson, Adsorption phenomena on Sephacryl S-200 Superfine, *J. Chromatogr.*, *147*:205 (1978).
131. J. F. Collawn, D. J. Cox, L. M. Hamlin, and W. W. Fish, Thin layer gel chromatography of proteins in mild and denaturing detergents, *J. Chromatogr.*, *157*:227 (1978).
132. P. F. Davidson, Proteins in denaturing solvents: Gel exclusion studies, *Science*, *161*:906 (1968).
133. W. W. Fish, K. G. Mann, and C. Tanford, The estimation of polypeptide chain molecular weights by gel filtration in 6 M guanidine hydrochloride, *J. Biol. Chem.*, *244*:4989 (1969).
134. K. H. Altgeld, Gel permeation chromatography with high loads, *Separation Sci.*, *5*:777 (1970).
135. D. J. Winzor and L. W. Nichol, Effects of concentration-dependence in gel filtration, *Biochim. Biophys. Acta*, *104*: 1 (1965).
136. W. W. C. Yau, C. P. Malone, and C. H. L. Suchan, Separation mechanisms in gel permeation chromatography, *Separation Sci.*, *5*:259 (1970).
137. L. W. Nichol, M. Janado, and D. J. Winzor, The origin and consequences of concentration dependence in gel chromatography, *Biochem. J.*, *133*:15 (1973).
138. E. P. Otocka and M. Y. Hellman, Vacancy permeation in gel permeation chromatography. Calibration and dispersion, *J. Polym. Sci., Polym. Lett.*, *12*:439 (1974).
139. J. Janča, Theory of concentration effects in gel permeation chromatography, *Anal. Chem.*, *51*:637 (1979).
140. J. Janča, Concentration effects in gel permeation chromatography, IV, Contribution of viscosity phenomena and marcomolecular expansion, *J. Chromatogr.*, *170*:309 (1979).
141. I. J. Galpin, G. W. Kenner, S. R. Ohlsen, and R. Ramage, Gel filtration of protected peptides on Sephadex G-50 in hexamethylphosphoramide containing 5% water, *J. Chromatogr.*, *106*: 125 (1975).
142. L. A. Haff, Fractionation of water-insoluble protein using Sephacryl S-200 in formamide, *Prep. Biochem.*, *8*:99 (1978).

143. G. Morgan and D. B. Ramsden, Comparison of the gel chromatographic properties of Sephacryl S-200 Superfine and Sephadex G-150. *J. Chromatogr.*, *161*:319 (1978).
144. A. Polson, Fractionation of protein mixtures on columns of granulated agar, *Biochim. Biophys. Acta*, *50*:565 (1961).
145. M. Cech, Some notes on the use of agar gel filtration in the study of plant viruses, *Virology*, *18*:487 (1962).
146. S. Hjertén, Chromatographic separation according to size of macromolecules and cell particles on columns of agarose suspensions, *Arch. Biochem. Biophys.*, *99*:466 (1962).
147. S. Hjertén, A new method for preparation of agarose for gel electrophoresis, *Biochim. Biophys. Acta*, *62*:445 (1962).
148. R. L. Steere, Tobacco mosaic virus: Purifying and sorting associated particles according to length, *Science*, *140*:1089 (1963).
149. S. Bengtsson and L. Philipson, Chromatography of animal viruses on pearl-condensed agar, *Biochim. Biophys. Acta*, *79*: 399 (1964).
150. S. Hjertén, The preparation of agarose spheres for chromatography of moelcules and particles, *Biochim. Biophys. Acta*, *79*:393 (1964).
151. B. Russell, T. H. Mead, and A. Polson, A method of preparing agarose, *Biochim. Biophys. Acta*, *86*:169 (1964).
152. S. Bengtsson, Separation of macromolecules by gel filtration on pearl-condensed agar or agarose, *Protides Biol. Fluids Proc. Colloq.*, *14*:583 (1966).
153. S. Hjerten, Molecular and particle sieving (chromatographic and electrophoretic) on agarose and polyacrylamide gels, *Protides Biol. Fluids Proc. Colloq.*, *14*:553 (1966).
154. T. C. Laurent, Determination of the structure of agarose gels by gel chromatography, *Biochim. Biophys. Acta*, *136*:199 (1967).
155. G. A. Locascio, H. A. Tigier, and A. M. Batlle, Estimation of molecular weights of proteins by agarose gel filtration, *J. Chromatogr.*, *40*:453 (1969).
156. M. Duckworth and W. Yaphe, The structure of agar, *Carbohyd. Res.*, *16*:189 (1971).
157. J. Porath, J.-C. Janson, and T. Låås, Agar derivatives for chromatography, electrophoresis and gel-bound enzymes, I, Desulphated and reduced cross-linked agar and agarose in spherical bead form, *J. Chromatogr.*, *60*:167 (1971).
158. S. Arnott, A. Fulmer, W. E. Scott, I. C. M. Dea, R. Moorhouse and D. A. Rees, Agarose double helix and its function in agarose gel structure, *J. Mol. Biol.*, *90*:269 (1974).
159. A. Amsterdam, Z. Er-el, and S. Shaltiel, Ultrastructure of beaded agarose, *Arch. Biochem. Biophys.*, *171*:673 (1975).

160. T. C. J. Gribnau, C. Stumm, and G. I. Tesser, Microscopic observations on commercial Sepharose: Deviations from normal bead structure, *FEBS Lett.*, *57*:301 (1975).
161. T. Andersson, and L. Hagel, Some properties and applications of Superose, *Anal. Biochem.*, *141*:461 (1984).
162. L. Hagel and T. Andersson, Characteristics of a new agarose medium for high-performance gel filtration chromatography, *J. Chromatogr.*, *285*:295 (1984).
163. T. Andersson, M. Carlsson, L. Hagel, P.-Å. Pernemalm, and J.-C. Janson, Agarose-based media for high-resolution gel filtration of biopolymers, *J. Chromatogr.*, *326*:33 (1985).
164. Y. C. Ha and P. J. Barter, Rapid separation of plasma lipoproteins by gel permeation chromatography on agarose gel Superose 6B, *J. Chromatogr.*, *341*:154 (1985).
165. E. H. Cooper, R. Turner, E. A. Johns, and R. A. Crockson, Identification and measurement of paraprotein polymers by high performance gel filtration chromatography, *Biomed. Pharmacother.*, *39*:78 (1985).
166. B.-L. Johansson and C. Ellström, Column lifetime of a new agarose medium for high-performance gel filtration chromatography at basic pH, *J. Chromatogr.* *330*:360 (1985).
167. B.-L. Johansson and L. Åhsberg, Column lifetime of Superose 6 at 37°C and basic pH, *J. Chromatogr.*, *351*:136 (1986).
168. D. J. Lea and A. H. Sehon, Preparation of synthetic gels for chromatography of macromoelcules, *Can. J. Chem.*, *40*:159 (1962).
169. S. Hjertén and R. Mosbach, "Molecular sieve" chromatography of proteins on columns of cross-linked polyacrylamide, *Anal. Biochem.*, *3*:109 (1962).
170. S. Hjertén, "Molecular sieve" chromatography on polyacrylamide gels, prepared according to a simplified method, *Arch. Biochem. Biophys., (Suppl.)*, *1*:147 (1962).
171. K. Sun and A. H. Sehon, The use of polyacrylamide gels for chromatography of proteins, *Can. J. Chem.*, *43*:969 (1965).
172. G. Trenel, M. John, and H. Dellweg, Gel chromatographic separation of oligosaccharides at elevated temperature, *FEBS Lett.*, *2*:74 (1968).
173. H. Dellweg, G. Trenel, M. John, and C. C. Emeis, Determination of oligosaccharides in worth and beer by gel chromatography in an automatic analysis system, *Monatsschr. Brauerei*, *22*(7):177 (1969).
174. M. John, G. Trenel, and H. Dellweg, Quantitative chromatography of homologous glucose oligomers and other saccharides using polyacrylamide gel, *J. Chromatogr.*, *42*:476 (1969).
175. M. John and H. Dellweg, Gel chromatographic separation of oligosaccharides, *Sep. Purif. Methods*, *2*:231 (1973).

176. J. Gressel and A. W. Robards, Polyacrylamide gel structure resolved? *J. Chromatogr.*, *114*:455 (1975).
177. E. Boschetti, R. Tixier, and J. Uriel, Permeation chromatography on acrylamide-agarose gel, *Bull. Soc. Chim. Fr.*, *11*(1): 2295 (1974).
178. W. Haller, Chromatography on glass of controlled pore size, *Nature*, *206*:693 (1965).
179. W. Haller, Correlation between chromatographic and diffusional behavior of substances in beds of pore controlled glass. Contribution to the mechanism of steric chromatography. *J. Chromatogr.*, *32*:676 (1968).
180. C. W. Hiatt, A. Shelokov, E. J. Rosenthal, and J. M. Galimore, Treatment of controlled pore glass with poly(ethylene oxide) to prevent adsorption of rabies virus, *J. Chromatogr.*, *56*:362 (1971).
181. G. L. Hawk, J. A. Cameron, and L. B. Dufault, Chromatography of biological materials on polyethylene glycol treated controlled-pore glass, *Prep. Biochem.*, *2*:193 (1972).
182. M. J. Telepchak, New uses for molecular-size exclusion chromatography, *J. Chromatogr.*, *83*:125 (1973).
183. C. Persiani, P. Cukor, and K. French, Aqueous GPC of water soluble polymers by high pressure liquid chromatography using glyceryl GPC columns, *J. Chromatogr. Sci.*, *14*:417 (1967).
184. H. Kopecka and P. Schneider, Internal diffusion in porous poly(methyl methacrylate) column packings, *J. Chromatogr.*, *174*:13 (1979).
185. C. Hiryama and H. Ihara, Novel gel permeation chromatography packing composed of cross-linked and porous spherical particles of poly-gamma-methyl-L-glutamate, *J. Chromatogr.*, *347*:357 (1985).
186. W. Heitz, Synthesis and properties of gel chromatography materials, *J. Chromatogr.*, *53*:37 (1970).
187. H. H. Oelert, The resolving power by gel chromatography of vinyl acetate gel for hydrocarbons and mineral oils, *J. Chromatogr.*, *53*:241 (1970).
188. D. Randau and H. Bayer, The use of vinyl acetate gels, *J. Chromatogr.*, *53*:63 (1970).
189. D. Randau, H. Bayer, and W. Schnell, Chromatographic applications of the polyethylene glycol dimethylacrylate gel Merckogel PGM 2000 under normal and high pressure, *J. Chromatogr.*, *57*:77 (1971).
190. K. Saitoh, M. Satoh, and N. Suzuki, Distribution coefficients of acetylacetone and tris(acetylacetonato)-chromium(III) in the Merckogel-DR-2000-chloroform system. *J. Chromatogr.*, *92*: 291 (1974).

191. W. Heitz and P. Bier, Gel chromatographic properties of polyvinylalcohol gels, *Makromol. Chem.*, *176*:657 (1975).
192. E. Nyström and J. Sjövall, Separation of lipids on methylated Sephadex, *Anal. Biochem.*, *12*:235 (1965).
193. E. Nyström and J. Sjövall, Methylated Sephadex as support in reversed phase partition chromatography, *J. Chromatogr.*, *17*:574 (1965).
194. V. Mutt, E. Nyström, and J. Sjövall, Use of "lipophilic" Sephadex in peptide synthesis, *J. Chromatogr.*, *24*:205 (1966).
195. E. Nyström and J. Sjövall, Separation of protected peptides on methylated Sephadex, *J. Chromatogr.*, *24*:208 (1966).
196. R. Vihko, Methylated Sephadex and Sephadex LH-20 in steroid separations, *Acta Endocrinol. (Copenhagen) 52(Suppl. 109)*:15 (1966).
197. M. Wilk, J. Rochlitz, and H. Bende, Column chromatography of polycyclic aromatic hydrocarbons on lipophilic Sephadex LH-20, *J. Chromatogr.*, *24*:414 (1966).
198. M. Joustra, B. Söderqvist, and L. Fischer, Gel filtration in organic solvents, *J. Chromatogr.*, *28*:21 (1967).
199. J. Sjövall, E. Nyström, and E. Haahti, Liquid chromatography on lipophilic Sephadex: Column and detection techniques, *Adv. Chromatogr.*, *6*:119 (1968).
200. E. Nyström and J. Sjövall, Chromatography of protected peptides on lipophilic Sephadex, *Protides Biol. Fluids. Proc. Colloq.*, *14*:601 (1968).
201. J. Ellingboe, E. Nyström, and J. Sjövall, A versatile lipophilic Sephadex derivative for "reversed phase" chromatography, *Biochim. Biophys. Acta*, *152*:803 (1968).
202. C. J. W. Brooks and R. A. B. Keates, Gel filtration in lipophilic solvents using hydroxyalkoxypropyl derivatives of Sephadex, *J. Chromatogr.*, *44*:509 (1969).
203. J. Ellingboe, E. Nyström, and J. Sjövall, Liquid-gel chromatography on lipophilic-hydrophobic Sephadex derivatives. *J. Lipid Res.*, *11*:266 (1970).
204. M. Calderon and W. J. Baumann, Fractionation of neutral lipids on a lipophilic dextran gel, *Biochim. Biophys. Acta*, *210*:7 (1970).
205. K. Beijer and E. Nyström, Reversed-phase chromatography of fatty acids on hydrophobic Sephadex, *Anal. Biochem.*, *48*:1 (1972).
206. R. A. Anderson, C. J. W. Brooks, and B. A. Knights, Hydroxyalicyclic derivatives of Sephadex LH-20 for lipophilic gel chromatography, *J. Chromatogr.*, *83*:39 (1973).
207. E. Nyström and J. Sjövall, Chromatography on lipophilic Sephadex, *Methods Enzymol.*, *35(B)*:378 (1975).

208. D. Apter, O. Jänne, P. Korvonen, and R. Vihko, Simultaneous determination of five sex hormones in human serum by radioimmunoassay after chromatography on Lipidex 5000, *Clin. Chem.*, *22*:32 (1976).

209. I. J. Galpin, A. G. Jackson, G. W. Kenner, et al., Improved method of gel filtration of protected peptides using Sephadex LH-60, *J. Chromatogr.*, *147*:424 (1978).

Index

D

E

F

G

O

P

R

W

Z